INTRODUCTION TO
INDUSTRIAL
ENGINEERING

SECOND EDITION

Industrial Innovation Series

Series Editor

Adedeji B. Badiru

Air Force Institute of Technology (AFIT) – Dayton, Ohio

PUBLISHED TITLES

PUBLISHED TITLES

Project Management for the Oil and Gas Industry: A World System Approach, *Adedeji B. Badiru & Samuel O. Osisanya*

Quality Management in Construction Projects, *Abdul Razzak Rumane*

Quality Tools for Managing Construction Projects, *Abdul Razzak Rumane*

Social Responsibility: Failure Mode Effects and Analysis, *Holly Alison Duckworth & Rosemond Ann Moore*

Statistical Techniques for Project Control, *Adedeji B. Badiru & Tina Agustiady*

STEP Project Management: Guide for Science, Technology, and Engineering Projects, *Adedeji B. Badiru*

Sustainability: Utilizing Lean Six Sigma Techniques, *Tina Agustiady & Adedeji B. Badiru*

Systems Thinking: Coping with 21st Century Problems, *John Turner Boardman & Brian J. Sauser*

Techonomics: The Theory of Industrial Evolution, *H. Lee Martin*

Total Productive Maintenance: Strategies and Implementation Guide, *Tina Agustiady & Elizabeth A. Cudney*

Total Project Control: A Practitioner's Guide to Managing Projects as Investments, Second Edition, *Stephen A. Devaux*

Triple C Model of Project Management: Communication, Cooperation, Coordination, *Adedeji B. Badiru*

FORTHCOMING TITLES

3D Printing Handbook: Product Development for the Defense Industry, *Adedeji B. Badiru & Vhance V. Valencia*

Company Success in Manufacturing Organizations: A Holistic Systems Approach, *Ana M. Ferreras & Lesia L. Crumpton-Young*

Design for Profitability: Guidelines to Cost Effectively Management the Development Process of Complex Products, *Salah Ahmed Mohamed Elmoselhy*

Essentials of Engineering Leadership and Innovation, *Pamela McCauley-Bush & Lesia L. Crumpton-Young*

Handbook of Construction Management: Scope, Schedule, and Cost Control, *Abdul Razzak Rumane*

Handbook of Measurements: Benchmarks for Systems Accuracy and Precision, *Adedeji B. Badiru & LeeAnn Racz*

Introduction to Industrial Engineering, Second Edition, *Avraham Shtub & Yuval Cohen*

Manufacturing and Enterprise: An Integrated Systems Approach, *Adedeji B. Badiru, Oye Ibidapo-Obe & Babatunde J. Ayeni*

Project Management for Research: Tools and Techniques for Science and Technology, *Adedeji B. Badiru, Vhance V. Valencia & Christina Rusnock*

Project Management Simplified: A Step-by-Step Process, *Barbara Karten*

A Six Sigma Approach to Sustainability: Continual Improvement for Social Responsibility, *Holly Allison Duckworth & Andrea Hoffmeier Zimmerman*

Work Design: A Systematic Approach, *Adedeji B. Badiru*

INTRODUCTION TO
INDUSTRIAL
ENGINEERING

SECOND EDITION

Avraham Shtub • Yuval Cohen

CRC Press
Taylor & Francis Group
Boca Raton   London   New York

CRC Press is an imprint of the
Taylor & Francis Group, an **informa** business

This book was previously published in Hebrew by The Open University of Israel.

CRC Press
Taylor & Francis Group
6000 Broken Sound Parkway NW, Suite 300
Boca Raton, FL 33487-2742

First issued in paperback 2017

© 2016 by Taylor & Francis Group, LLC
CRC Press is an imprint of Taylor & Francis Group, an Informa business

No claim to original U.S. Government works

ISBN-13: 978-1-4987-0601-8 (hbk)
ISBN-13: 978-1-138-74785-2 (pbk)

Visit the Taylor & Francis Web site at
http://www.taylorandfrancis.com

and the CRC Press Web site at
http://www.crcpress.com

To my wife Ailona Shtub—Avi Shtub

To my family—Yuval Cohen

Contents

Preface

This book presents the major tasks performed by industrial engineers, and the tools that support these tasks. The focus is on the organizational processes for which these tasks are needed, and the terminology used to describe the tasks, tools, and processes. The tools discussed here are basic tools that do not require in-depth knowledge of mathematics, statistics, psychology, or sociology. The book also examines the role of the industrial engineer in the production and service sectors. The intention is to help new students understand current pathways for professional development, and help them decide in which area to specialize during the advanced stages of their studies.

This book delineates the broad scope of areas in which industrial engineers are engaged, including areas that became part of industrial engineering (IE) in recent decades such as information systems, supply chain management, and service engineering. These fields are becoming an important part of the IE profession, alongside the traditional areas of IE such as operations management, project management, quality management, work measurement, and operations research. Industrial engineers require a strong understanding and good knowledge in all of these fields in order to perform their tasks.

This book contains the following chapters.

Chapter 1. Introduction

Here we discuss the nature of the IE profession and provide answers to basic questions such as

- What is engineering?
- What is IE?
- What is the IE profession?
- How do you acquire this profession?

Other points covered in this chapter are

- The system concept and its implementation in manufacturing and service.
- Tools needed by industrial engineers in order to perform their jobs.
- Frequently used methods of teaching in this field.

To give students a historical perspective, we show the development of the profession from its early days until recent years. Today, the profession must take into consideration the intense competition in industry due to globalization. Elements of competition include

- Cost reduction
- Shortening delivery times—time-based competition
- Quality improvement
- Achieving maximum flexibility

These elements of competition are the essence of the challenge facing industrial engineers. They are charged with designing systems and organizations that not only survive in the global competitive environment but also succeed.

Chapter 2. Organizations and Organizational Structures

This chapter deals with the organization of people and resources in order to achieve organizational goals. The chapter begins by explaining the need for a well-designed organization of human resources. Classical organizational structures are presented as

- Functional organization
- Project organization
- Matrix organization

Relative advantages and disadvantages of each of these organizational types are discussed, with an emphasis on communication, responsibility, and authority as tools for achieving a competitive advantage.

The discussion leads to the question of which organizational structure is best for today's competitive environment and the conclusion is that the organization must be (1) modern, (2) process based, and (3) supported by an appropriate information system. These three conditions are essential for success.

In addition to the organization of human resources, other resources such as production resources must also be efficiently organized. Production resources are mainly machines and equipment such as material handling equipment. The chapter reviews different layouts used to organize these resources:

- Flow shop

- Job shop
- Cells of group technology-based layout

Advantages and disadvantages of these layouts are discussed, alongside a survey of the fixed location layout where people, material, and equipment are transported to the place of work. This layout is quite common in nonrepetitive environments or projects such as home or ship building.

The relationship between the organizational structure and equipment layout leads to a discussion on processes in production and services and how these processes should be organized.

Chapter 3. Project Management

This chapter discusses organizations that perform projects (i.e., nonrepetitive undertakings). The discussion opens with a mapping of a project's stakeholders and understanding their needs and expectations from the project. Needs and expectations are translated into a conceptual design, using special decision-making tools to choose between technological and operational alternatives. Analysis of the cost/benefit/risk and time is discussed and the appropriate analytical tools are presented. A review of the project life cycle serves as a guideline for displaying methods for scheduling, budgeting, management, and control of projects, with emphasis on the relatively simple methods used in the industry.

The discussion of project management leads to discussion on information and its use—especially, turning data into information that supports decision making.

Chapter 4. Information and Its Uses

This chapter extends the discussion on data and information. It examines data collection, storage, retrieval, and processing, and using appropriate models to create the information necessary to support decision making.

The discussion emphasizes the following topics:

- Quality of information
- Data collection methods and how to use raw data to create useful information
- How to forecast future data

The chapter aims to develop a basic understanding of the nature of information systems, decision support systems, and database systems. We show the relationship between the knowledge base and models used to analyze data and to support decision-making processes.

Chapter 5. Marketing Considerations

This chapter is the first in a series of two chapters focusing on the interface between the industrial engineer and other professionals within the organization. This chapter deals with the customers, while the following chapter deals with suppliers and subcontractors. We present the tool that links production to marketing—the Master Production Schedule—and discuss the relationship between inventory and delivery times. The chapter introduces the classic dilemma between having high levels of inventory (for which a price must be paid) and the resulting shorter delivery times and lower inventory levels causing longer lead times. We discuss some policies including

- Make to stock
- Make to order
- Assemble to order
- Design/engineer to order

Chapter 6. Purchasing and Inventory Management

The industrial engineer must understand the organization's relationship with suppliers and subcontractors. Procurement is important in the competitive world, and this chapter discusses some key points of this topic:

- What to buy from suppliers and subcontractors and what to make in house—the make or buy problem.
- If the decision is to buy, how to find suitable suppliers to form a list of candidates.
- How is a supplier chosen from the list of suitable suppliers?
- How to manage the relationship with the supplier over time.

When it comes to purchasing materials, inventory management issues are also important such as

- How often to order?
- What quantity to order?
- What are the costs associated with inventories?
- What are the advantages in maintaining inventories?

Resolving these issues is not simple, and there is a need for decision support tools. This chapter presents the basic models and the assumptions underlying each model.

Some purchasing decisions are repetitive, and some are not. How these decisions are made and how to take advantage of procurement and inventory to achieve competitive advantage are the main subjects of this chapter.

Chapter 7. Scheduling

This chapter focuses on scheduling the organization's operations. Scheduling an organization's operations is dependent both on marketing and the interface with customers (Chapter 5) and on procurement and the interface with suppliers (Chapter 6).

Scheduling issues exist in both manufacturing and service systems. Competition drives many scheduling goals and constraints. After setting scheduling goals and constraints, the industrial engineer has to select the right scheduling method.

Our discussion starts with scheduling of the job shop. Next, we discuss the scheduling of the flow shop, and finally we present a general discussion about scheduling, using a concept of the Toyota production system (TPS): Just In Time (JIT), and we also discuss the Theory of Constraints (TOC) that focuses on scheduling bottlenecks.

We explain the logic of simple scheduling methods and provide examples highlighting the effectiveness, advantages, and disadvantages of these methods.

Chapter 8. Material Requirements Planning

This chapter introduces the basic computerized approach for managing production and procurement of material. Material requirements planning (MRP) was developed in the 1970s when the price of computers dropped enough for commercial organizations to be able to buy them. The method is based on simple logic and common processing of data from multiple files including

- The bill of material (BOM) file
- Inventory files
- The master production schedule (MPS)

Processing these files enables detailed planning and coordination between procurement and manufacturing activities. MRP logic comprises several components such as

- Gross to net
- Time phasing
- Lot sizing

MRP systems are the basis for planning and management of material in many organizations. It is important that industrial engineers understand, early on in their studies, the logical principles underlying these systems.

The discussion in this chapter reveals the weakness of the first generation MRP systems, which did not include mechanisms for planning production capacity. Solutions to this problem were developed later in the form of rough-cut capacity planning and capacity requirement planning (CRP).

Chapter 9. Enterprise Resource Planning

This chapter presents a framework of the information systems that manage the entire enterprise: enterprise resource planning (ERP) systems.

ERP systems are advanced organizational information systems, and many organizations have implemented them.

Industrial engineers play a major role in implementing these systems. Their tasks include, among others:

- Setting the system requirements
- Defining organizational processes
- Choosing the suitable system for the organization
- Recruiting participants
- Implementing the ERP system

We explain the principles of ERP systems and discuss their selection and implementation.

Chapter 10. The Human Factor and Its Treatment

This chapter presents principles from psychology, sociology, and ergonomics that industrial engineers use in their work.

Industrial engineers must learn how to integrate the human factor as a key component in their analysis and design. Questions concerning motivation, team-building, leadership, and organizational learning are discussed in this chapter, focusing on integrating the human factor in the different systems. In addition, industrial engineers must have knowledge of ergonomics in order to fit the environment and the task to the human operator. The chapter introduces basic issues in ergonomics and human–machine interface that are often used by industrial engineers.

Chapter 11. Supply Chain Management

In this chapter, we explain the need for the management of supply chains. Each organization is a link in the chain receiving goods and/or services from its suppliers and supplying goods and/or services to its customers. The end customers can choose between products produced by alternative supply chains. Thus, a competitive supply chain must act as a coordinated team to reduce costs and provide high quality products in a competitive time and with flexibility.

Characteristics of supply chains and the bullwhip effect are discussed and explained as the basis for understanding the advantage of managing the chain in an integrated manner. In particular, we emphasize the importance of information systems and their integration along the supply chain.

Primary considerations in the design of supply chains are discussed, including the selection of participants in the chain, and the design of interfaces between participants. Finally, the chapter discusses the necessary information for supply chain design, implementation, monitoring, and control of the chain's operations.

Chapter 12. Service Engineering

This chapter opens with a review of service processes, and the characteristics of service systems and service systems design. Next, we discuss arrival processes and randomness inherent to these processes. The calculation of queue lengths and waiting times in simple systems is discussed and the relationship between the level of service and the workforce is explained, including the tradeoff between waiting time cost and service capacity cost.

The chapter presents simulation as a major tool in analyzing service systems, which have inherent uncertainty in their processes. Simulation is an invaluable tool in analyzing the effect of service capacity (e.g., number of servers) on waiting times, queue length, and service level.

We also introduce basic considerations regarding reliability, maintenance, and issue of warranty.

Finally, the chapter ends with a discussion of the important subject of customer feedback that often needs to be actively explored using various techniques including questionnaires.

Authors

Professor Avraham Shtub holds the Stephen and Sharon Seiden chair in project management at the faculty of Industrial Engineering and Management of the Technion—Israel Institute of Technology. He earned a BSc in electrical engineering from the Technion—Israel Institute of Technology (1974), an MBA from Tel Aviv University (1978), and a PhD in management science and industrial engineering from the University of Washington (1982).

Professor Shtub is a certified Project Management Professional (PMP) and a member of the Project Management Institute (PMI-USA). He is the recipient of the Institute of Industrial Engineering 1995 "Book of the Year Award" for his book *Project Management: Engineering, Technology, and Implementation* (coauthored with Jonathan Bard and Shlomo Globerson), Prentice-Hall, 1994. He is the recipient of the Production Operations Management Society's Wick Skinner Teaching Innovation Achievements Award for his book *Enterprise Resource Planning (ERP): The Dynamics of Operations Management.* His books on project management have been published in English, Hebrew, Greek, and Chinese.

He is the recipient of the 2008 Project Management Institute Professional Development Product of the Year Award for the training simulator *Project Team Builder—PTB.*

Professor Shtub was a department editor for *IIE Transactions* and was on the editorial boards of the *Project Management Journal, The International Journal of Project Management, IIE Transactions,* and the *International Journal of Production Research.* He was a faculty member of the Department of Industrial Engineering at Tel Aviv University from 1984 to 1998 where he also served as a chairman of the department (1993–1996). He joined the Technion in 1998 and was the associate dean and head of the MBA program.

He has been a consultant to industry in the areas of project management, training by simulators, and the design of production operation systems. He has been invited to speak at special seminars on project management and operations in Europe, the Far East, North America, South America, and Australia.

Professor Shtub visited and taught at Vanderbilt University, The University of Pennsylvania, Korean Institute of Technology, Bilkent University in Turkey, Otego University in New Zealand, Yale University, Universidad Politécnica de Valencia, and the University of Bergamo in Italy.

Yuval Cohen is a senior faculty member at the Industrial Engineering Department of the Tel-Aviv Afeka College of Engineering. Previously, he was the head of the Industrial Engineering Program at the Open University of Israel. His areas of expertise are planning and operation of assembly lines, design and management of production and logistic systems, project management, supply chain management, and business decision making. He has published many papers in these areas. Dr. Cohen served several years as a senior operations planner at FedEx Ground (USA) and received several awards for his contributions to hub and terminal network planning. Dr. Cohen earned his PhD from the University of Pittsburgh (USA), his MSc from the Technion—Israel Institute of Technology, and BSc from Ben-Gurion University. Dr. Cohen is a fellow of the Institute of Industrial Engineers (IIE), and a full member of the Institute for Operations Research and Management Sciences (INFORMS).

1

Introduction

Educational Goals

This chapter presents the profession of industrial engineering (IE), the broad scope of areas in which industrial engineers are engaged in manufacturing and services, the market in which they operate, and the roles that they play in the economy.

Understanding the historical background is an important component of the training in IE. The following historical review highlights significant events and people that have contributed to the development of the profession.

We explain the need for integrated processes, supported by modern information systems, with an emphasis on the competitive market today.

1.1 Definitions and Examples Related to Industrial Engineering

1.1.1 Engineering

The Merriam-Webster dictionary defines engineering as the design and creation of large structures such as roads and bridges or new products or systems by using scientific methods.

According to the Merriam-Webster dictionary, "to design" is to plan and make decisions about something that is being built or created—to create the plans, drawings, etc., that show how something will be made.

Engineering design is a process of translation of requirements, specifications, and needs into a language understood by the people responsible for making the new product, service, facility, or system.

For example, the civil engineer translates the requirements for transporting a volume of traffic over a water barrier into the design of a bridge, including the geometry of the bridge, quantities of materials required to construct the bridge, the processing of materials, the layout and assembly of the parts, and finally the testing of the bridge during its construction and after its completion. The design is in a language understood by construction workers, purchasing agents, suppliers, subcontractors, quality control experts, and so on.

1.1.2 Industrial Engineering

According to the Merriam-Webster dictionary, IE deals with the design, improvement, and installation of integrated systems (as of people, materials, and energy) in the industry.

Modern IE is concerned with the design, management, and control of operational processes. For that purpose, IE combines classical knowledge in physics, mathematics, computing, and statistics with tools for incorporating the human factor, ergonomics, sociology, and psychology.

For example, in the design of a new business branch, industrial engineers plan the work packages and their allocation to operators. Industrial engineers also design the work positions using their knowledge of ergonomics, facility layout planning, and efficient work planning.

1.1.3 Industrial Engineers

Industrial engineers design organizational processes and perform projects and ongoing activities that may involve facilities, products, and systems. The facilities, products, systems, and processes are used to supply products or services. Industrial engineers often focus on processes that take into account the human factor. For designing organizational processes, industrial engineers collaborate with managers and their subordinates, and sometimes with peers, such as engineers (e.g., mechanical engineers, electrical engineers, software engineers, etc.).

An industrial engineer is also involved in determining how to best utilize the resources of the organization. Resources such as workers, raw materials, capital, information, buildings, equipment, energy, and technological knowledge are used by industrial engineers to perform their tasks. Industrial engineers translate the goals of the organization, the constraints imposed, and the uncertainty, to find the best solutions for the organization. They do so by utilizing the rules of physics, mathematics, and statistics along with human factor related knowledge, such as ergonomics and psychology, and the rules of law, morality, and ethics.

A relatively new area of industrial engineering is the design and implementation of information systems that supports processes. In the past, industrial engineers integrated material requirement planning (MRP) systems into industrial organizations. Later on, they played a vital role in the incorporation of enterprise resource planning (ERP) systems in many organizations. Today, many industrial engineers work with ERP systems on a daily basis. With the advent of supply chain management, some industrial engineers entered the challenging world of interorganization data sharing.

1.1.4 Production/Service Systems

Some organizations are engaged in the production or supply of products. Other organizations provide services and some do both. These organizations use systems to perform their operations. We define a system as a collection of resources such as people, computers, information, machinery, and facilities working to achieve a common goal. The role of this system is to transform "inputs" such as raw material, energy, and demand information into "outputs," which are products, information, and customer service. The output may include damaged items, which should be avoided or minimized as much as possible.

1.1.5 What Do Industrial Engineers Do?

Industrial engineers are involved in designing organizational processes, performing projects and ongoing activities, and planning their operations. This includes a large variety of activities. A partial example includes, among others, the following: Design of production and service systems, design and implementation of processes, production management, design of supply chains, planning and managing supply-chains operations, project management, economics analysis, quality control, and design and operations of information systems.

1.1.6 Tools Used by the Industrial Engineer

To succeed in his or her job, the industrial engineer needs understanding, skills, tools, and techniques in a variety of fields. We will review the skills necessary and demonstrate how these skills are used in the case of a company in the automotive industry.

1.1.6.1 Understanding "Engineering Language": Drawings, Specifications, etc.

For example, in the production of cars, engineering design passes the detailed drawings of the car to production engineering. An industrial engineer in a production engineering department will translate the design into a bill of material with information regarding the car assemblies, subassemblies, and the parts from which it is composed, determine which machines will be used in order to produce parts, design the supply chain that supplies car parts not manufactured by the factory, etc.

1.1.6.2 Understanding the Physical Processes, Knowledge of the Basic Laws of Physics

Physical processes affect machine operation, maintenance, quality, and efficiency, as well as human performance and the associated ergonomics.

Dexterity requires a compact comfortable work environment. It also requires lightweight tools and parts that are easy to handle and manipulate.

The manufacturing process of a product is a process that combines a sequence of operations, performed by machines and humans, aimed at transforming raw materials into a finished product. The industrial engineer must understand the process and the physical principles involved in the process such as metal cutting, forging, etc.

For example, in car assembly lines such as those used by Toyota or General Motors, quality problems are caused by physical processes. Industrial engineers should be able to analyze these problems and find effective solutions. Physical factors also affect the load and human posture of the assembly operator. The industrial engineer should be able to identify problems along the assembly line, and use ergonomic principles to design solutions for the problematic workstations.

1.1.6.3 Knowledge of Economics and Financial Management

The industrial engineer's main role in many organizations is contribution to the bottom line profits. Therefore, operational decision making involving intensive financial considerations is a main specialty of an industrial engineer. Investment decisions, and the effects of interest rate and taxation, are common problems that an industrial engineer should be able to tackle. Such problems appear in a car assembly line when buying a new machine or improved conveyor system. In service industries, these problems may be connected to the number of waiting positions (such as car spaces in a parking lot, number of chairs in a dentist waiting room, etc.).

Industrial engineers have to decide what to produce inhouse and what to buy. For purchased items, they must decide from which country and which supplier to buy in order to minimize cost and maximize quality. The industrial engineer must make a decision that combines economic considerations, like the cost of production in each country and the cost of transportation of the various parts between the countries, as well as aspects of taxation and customs that should be considered in the decision process.

1.1.6.4 Understanding Mathematical and Statistical Models

For example, in determining the size of production batches on a machine with long set-up time (the time to switch from one operation to another) compared to the time needed to process a single item, it might be better to produce in large batches to reduce the number of set-ups and, thus, the total machine time required. However, large batches create large and expensive inventories.

To solve this problem, mathematical models trade off the desire to minimize set-up times and consequently produce large batches and the desire to minimize inventories by producing small batches "just in time." When demand is random, that is, uncertain, the problem is harder to solve and the industrial engineer must use statistical tools to determine the size of each batch.

1.1.6.5 *Knowledge of Human Resources Management*

Since human resources are the center of any organization, and since industrial engineers are the designers of organizational processes, they need knowledge of human resource management. Some examples of the areas that they need to understand are task design, workplace design (ergonomics), workforce scheduling, determining incentives and remunerations, etc.

For example, the decision to implement an incentive-based payment system and how to apply it is a traditional task of industrial engineers. Understanding the structure and the pros and cons of such systems is important, as different employees performing different tasks in different departments of the same organization may react differently to the same incentive-based payment system. Therefore, proper selection and implementation of such systems is crucial.

1.1.6.6 *Knowledge of Computerized Information Systems*

Organizations handle and generate vast amounts of data in the course of their operations. Daily data updates are commonplace in areas relating to inventories, orders, quality, production, deliveries, workforce, maintenance, sales, payments, etc. No organization can efficiently operate without this data stored and optimally accessed and organized by a computerized information system.

Efficient scheduling of resources such as machines and employees and their integration with the timing of material supply requires advanced information systems that can process huge amounts of data, transforming it into information that supports management decisions.

For example, the final assembly of cars, which consists of thousands of parts manufactured by thousands of operations and supplied by hundreds of suppliers, needs to be scheduled. To schedule such operations efficiently, an enterprise information system is necessary. Industrial engineers define the requirements for these enterprise information systems, and help install, adapt, modify, and integrate such systems for the organization.

1.2 Models

Problem solving and decision making are important parts of the job of the industrial engineer. Textbooks and courses are often organized according to the types of problems. Industrial engineering textbooks have chapters dealing with problems related to inventory, production scheduling, service

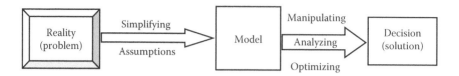

FIGURE 1.1
Models and their use.

system design, procurement, and the like. One of the primary tools presented to students in all their classes and textbooks, which will become a part of their professional toolkit, is problem modeling.

A model is a simplified presentation of reality. Many real problems are very complex due to their size, the number of different factors to consider, and the dynamic and stochastic nature of interactions between many of these factors. By using simplifying assumptions, one develops a model of the problem. A good model of a real problem must not only be simple enough to understand and analyze but also be sufficiently a representative of the real problem so that, its solution can be implemented successfully on the real problem.

Many models are mathematical ones. Most mathematical programming models define an objective function and a set of constraints. Such models are designed to find the values of the "decision variables" that meet the constraints, while minimizing or maximizing the objective function.

Conceptual models are also common. An organizational chart, for example, is a model that conceptually describes the relationships among members in the organization.

When the level of uncertainty is high, statistical or stochastic models that represent the uncertainty of the real problem are used. Such tools include regression analysis and stochastic dynamic programming.

When decision makers analyze a model, they are trying to find a good solution to the problem that the model represents. This solution is appropriate for solving the original problem if it is not too sensitive to the simplifying assumptions underlying the model. Therefore, a sensitivity analysis of the solution obtained must be performed to assess its suitability for solving the original problem. The relationship between the original problem, the model, and the solution is shown in Figure 1.1.

1.2.1 Use of Models

Models are frequently used for routine repetitive decisions. A computer can handle some of these decisions automatically. Inventory management in a supermarket is a typical example of routine decisions: Orders for new shipments are required when the existing inventory level drops below a certain level. The value of this so-called "reorder level" or "order point" is calculated by fitting a model. Computer software can be set up to continuously monitor and update the level of inventory available based on transactions recorded at the checkout counters and order automatically from suppliers when the inventory level drops below the reorder level.

Commercial business software packages are designed to support routine decisions. For example, software systems based on the logic of MRP automatically issue production and purchasing orders of certain items. These systems also provide data about past events, inventory, etc. Historical data are essential for the development of policies and are used to support the cost estimates of labor and materials, needed for bidding and marketing.

Models can be used to solve nonroutine or ad hoc problems as well. A map is an example of such a model. A tourist can find his or her way in a city he or she never visited before using a map that represents real streets and available attractions. The map has only two dimensions and in reality there are three dimensions, it is much smaller than the real city, and it is static while in the real city people and vehicles are moving. Despite all the simplifying assumptions, following the route from an origin to a destination on the map can help the tourist find his or her way in the real city. In much the same way, industrial engineers use various charts that model various aspects of processes (e.g., flow, quality, etc.).

While a map and a chart are static models, industrial engineers often use dynamic simulation modeling to check different scenarios and support a decision-making process seeking the best scenario. This is discussed next.

1.2.2 Dynamic Aspect: Simulation and Dynamics Systems

Simple models such as a map are static in nature, that is, these models present a snapshot at a given moment of the organization and its environment. In reality, time plays a very important role in decision making. Values of key parameters change over time. These include various factors in the organization and its environment, which are dynamic. Information is collected over time. Competitors enter the market and develop new products at certain points along the timeline.

Lecturers and practitioners in industrial engineering strive to integrate the dynamic aspect into their models. In the 1960s, the system dynamics approach was developed as a new tool for analyzing the dynamic nature of systems and processes (Forrester, 1961, 1968). Advanced simulation modeling tools are based on the system dynamics approach.

Forrester modeled systems and processes using two types of entities: Levels and rates. Rates generate changes in the levels, and levels are used as state variables so that the value of the levels at a given time determines the state of the system.

Consider the following example of system dynamics: The fuel system in a car is analyzed to develop the best strategy for refueling the car. The objective is to minimize fuel cost, and the constraints are the capacity of the fuel tank and the location of fueling stations. The rate of the car's fuel consumption is determined by factors such as speed, load, road conditions, etc.

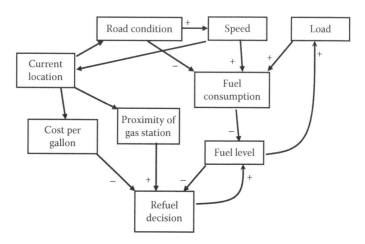

FIGURE 1.2
Levels and rates flow diagram (+ is positive correlation, – is negative).

Most drivers (decision makers) base their decision on when to refuel on factors such as the current fuel level, the distance to the next fueling station, and the cost per gallon/liter of fuel at each station. Figure 1.2 shows the relationship between levels and rates in this example.

Another simulation model is discrete event simulation (DES) developed at IBM in 1961 by Geoffrey Gordon and initially called general purpose simulation system (GPSS). Since the development of DES, many simulation languages have been developed and evolved. For example, SIMULA evolved into BETA and its concepts are used in SimPy (Python based DES library). SLAM (simulation language for alternative modeling) evolved into SLAM-II, which developed into SIMAN, and finally into ARENA simulation software. Since the turn of the twenty-first century, many other DES software packages have emerged: Promodel, Simul8, SimEvent, Plant Simulation, and SimCAD Pro—to name a few. Modern simulation languages are user-friendly, powerful, and flexible. Simulation is considered an important tool that is used by industrial engineers for analyzing complex systems, most of which include random processes.

1.2.3 Simulation Models and Decision Making

When using simulation as a tool for decision support, the simulation model presents three aspects of the real world:

1. The flow of objects such as materials, equipment, and people
2. The flow of information
3. The decision-making process

In the inventory system simulation example, the flow of material deals with materials entering the system, their storage, and exit from the system. Information related to the flow of materials is collected during the simulation run. The user can see the exact time when a unit of material is created, and when it moves within the system or leaves it. The data collected during the simulation serve as a basis for understanding and analyzing the inventory system.

The logic of decision making is part of the simulation model. Such logic can be based, for example, on a simple model that recommends issuing an order for new materials each time the inventory level drops below the designated order point. In this case, the order level is the decision variable, and the simulation compares system performance for various reorder levels. The parameters of the model's input and the logic by which it operates are built into the simulation model. The advantage of this approach is that one can try out a large number of different decision rules by running the simulation. One more advantage is the possibility of running large simulations offline, that is, when the decision makers are not present. For example, the model can run at night, and in the morning the users get the results for analysis.

The disadvantage of using simulation for decision making is that in real life many decisions are based on intuition and experience, and it is very difficult (sometimes impossible) to program a computer to model intuition and experience. Furthermore, in many organizations, decisions are made by a group of employees from different fields. Group decision making is a very complex process, and there is not enough knowledge to model it with a reasonable degree of precision.

1.3 Teaching Industrial Engineering

IE is taught through lectures, books, and projects; mostly using two approaches:

1. *The case study approach:* based on the analysis of events/specific situations. Some case studies describe real problems and some describe imaginary situations. Each event presents a particular problem and students discuss the problem, analyze it, and find an appropriate solution.

2. *The modeling approach:* based on the development of models designed to handle the most important aspects of a problem. Some models are based on mathematical programming tools (e.g., linear programming), statistical tools (e.g., regression analysis), search techniques (e.g., genetic algorithms), and simulation. These tools are used to get insight into the model and the problem it represents.

For example, consider the relationship between the price of a product and the demand for this product. Assuming that there is no randomness, a mathematical expression will find the price that will maximize revenue. The mathematical expression is a simple model that shows the relationship between the price and the resulting demand. Using calculus, it is possible to find the optimal price that will maximize revenues. If randomness is considered, a mathematical expression may enable finding the optimal expected (mean) revenue.

Some industrial engineering courses combine the two approaches and present case studies that can be analyzed using the models learned during the course.

1.3.1 Industrial Engineering Curricula

Curricula in IE programs in academic institutions, though not uniform, are based on similar principles.

Freshman (first year) curriculum usually covers all core subjects including mathematics, physics, and information systems, along with the course "Introduction to Industrial Engineering," which focuses on the role the industrial engineer fills in organizations, the training he or she needs, and the tools that he or she uses.

Sophomore (second year) curriculum focuses on advanced courses in mathematics including courses in operations research, probability and statistics, as well as basic courses in psychology, sociology, and process-product design.

Junior (third year) curriculum moves to cover advanced courses on production and service systems.

Senior (fourth year) curriculum is usually reserved for elective courses. Students select subjects that interest them. In many schools, fourth year students perform a final project. The final project is based on a real IE problem, which gives students an opportunity to get experience in IE, implement the knowledge they learned in the previous three years, and learn about new tools and techniques and new situations.

1.4 Historical Overview

Humanity has been systematically planning major undertakings for thousands of years. The Egyptian pyramids, the two Temples in Jerusalem, ancient Roman roads and structures, the Great Wall in China, and Watt Angkor in Cambodia are examples of huge building projects carried out thousands of years ago. These undertakings required the organization of many people and proper planning. They were probably planned using methods that are very different from the methods used today by industrial engineers. For example, the production of most products in the past was carried out in workshops, by expert craftsmen. These craftsmen acquired their profession by serving for many years as apprentices, and learning how to make the whole product, without any help from other people. Today, factories divide the production process into stages or steps, and most workers work on a small part of the whole production process, in production or assembly lines. Furthermore, computers, automated machines, and robots are parts of many production processes in the modern factory. The transition from the small shops of individual craftsmen to the modern automated factory with multiple workers (dividing the work among them and using automated machines and robots) was a long process spanning over more than 100 years.

1.4.1 Industrial Revolution: Eighteenth Century

A turning point in the development of industrial engineering was the Industrial Revolution that started in England in the eighteenth century. Two main factors led to the transformation. One factor was the introduction of mechanical energy, which replaced human energy or energy from other sources such as animals, water, or wind. The invention of the steam engine in 1764 made this possible. The second factor was the transfer of production from the small workshops of craftsmen into factories where production was done by many workers, applying the principle of division of labor.

In 1776, Adam Smith published a book analyzing the economic benefits of the division of labor. His theory is that breaking up the work required to produce a product into a series of small simple tasks performed by a number of production workers increases efficiency. Each worker needs to know only a small part of the whole process required to manufacture the product, unlike earlier methods where craftsmen had to learn the whole process required to produce and assemble a product from start to finish. The application of division of labor resulted in the need to organize and manage the teams of production workers.

In 1798, the American inventor Eli Whitney developed the idea of standard product parts. Whitney developed a system for producing muskets for the U.S. government. In his system, all the pieces of the same type produced by any worker (e.g., the barrel of the musket) were exactly the same. The finished product could be assembled from standard parts randomly selected from the stock and no special adjustments for each gun were needed. Division of labor and Eli Whitney's standardization of parts are the basis of the modern mass production system: identical parts are manufactured in batches, and a random selection of a set or kit of parts can be assembled into the final product, without needing adjustments.

1.4.2 New Developments in the Early Twentieth Century

In 1911, Frederick Taylor (1856–1915) published his "scientific management theory." His goal was to improve productivity by making employees more efficient. Taylor claimed that management should define the desired output, provide training for employees, and monitor them in order to generate the desired output. In addition, Taylor argued that to manage a production system, quantitative measurements of working time, material, and resources are needed in order to minimize waste and to build an efficient production system.

Taylor brought awareness of the importance of proper production planning to the world of manufacturing and beyond. His planning relies on accurate measurements and deals with employee placement, training, and supervision. Taylor's philosophy highlighted the importance of work standards, which are the point of departure for performance planning and evaluation.

At about the same time, Frank and Lillian Gilbert developed a method for predicting the time it will take to perform a given task. This method and philosophy aligned well with Taylor's scientific management theory and was accepted with enthusiasm by Taylor's followers. However, its application was too detailed and tedious. Ten years later, a more efficient technique called Motion & Time Measurement (MTM) was developed by Westinghouse Corporation and spread throughout the business world.

In 1913, Henry Ford developed the assembly line. His idea was to bring division of labor and work standardization to perfection. The worker repeats short cycles of identical work and the product is conveyed to the worker on the line. Before Ford's car production assembly line began operation, the average time for the assembly of a car was about 12.5 h. Eight months later, when the assembly line was completed, and each worker had a specific role in the line, the average time for the production of a car was reduced to 93 min.

Ford's assembly line applied the principles of scientific management— standard product design, division of labor, mass production of standard parts, and reduction of production costs. Ford's factory integrated these ideas, which led to peak efficiency at the time.

In 1914, Henry Gantt developed a chart for scheduling process activities. The chart used the timeline as the horizontal axis and the vertical axis was dedicated to machines or operators (each one having its own horizontal timeline). In this way, the activities done at any point in time could be easily illustrated. The Gantt graphical model still serves as a popular tool for scheduling workstations, machines, and projects.

In 1917, F. W. Harris developed an optimization model for determining batch sizes for management and control of inventories. Proper management of inventory has a great impact on the profits of enterprises, and hence the importance of this development.

Between 1927 and 1930, Elton Mayo at the Hawthorne plant of Western Electric studied psychological and sociological factors impacting the efficiency of a group of workers. These studies were based on experiments aimed at examining the effects of environmental changes such as lighting and noise on employees' performance. The results were surprising. For example, in some cases, even though the physical conditions of the workers deteriorated, performance of the workers did not change, and sometimes even improved. The conclusion was that the fact that employees were aware that they were part of an experiment raised the level of their motivation and made them work more efficiently. This finding was the trigger leading managers and researchers to realize the importance of psychological and behavioral factors, and their direct impact on employee motivation and work efficiency. Studies of this kind have led to the management focus on the human factor, with an emphasis on creating conditions that encourage the employee to contribute to the organization. These were the early steps in the development of the discipline of management of human resources.

During World War II, armies were faced with complex logistical problems, such as the transfer of aircraft, ships, supplies, and troops between different parts of the world. At the same time, the operation of newly developed complex weapons (such as the new radar system) posed a planning, modeling, and optimizing challenge. To cope with these challenges, operations research methodologies were developed. Operations research is based on the construction of mathematical and statistical models of complex problems. At that period, the model was typically a combination of objective function and constraints, and mathematical analysis was used to find an optimal, implementable solution for the model, and hence for the problem.

After the war, experts in operations research returned to academia and industry and applied the new knowledge in public and private sectors. The wide applicability of operations research and its fast evolution has made it into a separate field of research and study. It has developed rapidly and continues to do so today.

Work in operations research requires complex calculations and nowadays is based on extensive computer use due to the need to deal with complex, large mathematical models. Operations research models are used to support complex decision making.

A typical example of operations research models is the case of fashion goods supply chain management by international companies. Sewing factories located in different countries sell their products around the world. The decision of which products to produce in each factory and how to ship the products from factories to the wholesale and retail stores is a very complex one that involves considerations such as the cost of transportation, the production capacity of the factories, the distances between the factories and the shops, the cost of labor and raw material in each country, the demand for different products in every country, etc. Operations research models support decision making in such complex problems.

After World War II, a significant expansion of the public and private service sectors motivated the development of tools and techniques for planning and managing service operations. Today, more than two-thirds of the workforce in the United States is employed in the service sector. Services are characterized by uncertainty related to customer arrivals and requirements. Operations research models were developed for analyzing queues and dealing with demand uncertainty, for these kinds of problems.

Techniques for planning and managing projects were developed during the 1950s, including the program evaluation and review technique (PERT). This tool is designed for project scheduling method and uses a statistical model to estimate the likelihood (or probability) that the project will end on a certain date. It was developed by the US Navy to support the Polaris nuclear submarine construction project.

At about the same time, the critical path method (CPM) was developed by Union Carbide Corporation. CPM focuses on a project's critical activities—activities that management should focus on, as any delay in a critical activity will cause a delay in the entire project.

In the 1970s and early 1980s, with the development of relatively inexpensive computers, industrial engineers started to use computers and software to solve complicated, large production and logistics problems. The development of MRP software that manages material in production facilities helped industrial engineers to quickly adapt production schedules and procurement planning to the dynamic needs of the market.

In parallel to what was going on in the West, interesting developments also took place in Japan after World War II. Following the war, Japanese industry focused on the quality of its products, adopting the approach of experts such as Edward Deming and Joseph Juran. The total quality management (TQM) approach was developed here and later adopted by Western industry as well.

The high quality of Japanese products eventually led to their success in the competitive world markets. Their cameras and cars were perceived to have higher quality than their competitors did, and, thus, were preferred by many consumers.

As part of their efforts, the Japanese stopped using traditional quality targets: the ratio (percentage) of defective items to nondefective items in a production batch. In many Western companies, a defect ratio of 1% was considered acceptable. The Japanese, seeing defects as a waste, were able to reach a ratio of few defective parts per million (PPM). This achievement was the result of using TQM aimed at continuous quality improvement in all processes within the organization, from the design process to the production process through the supply process.

The TQM approach has become a central part of the way many organizations throughout the world operate. In the United States, for example, a national effort to improve quality led to the Malcolm Baldrige National Quality Improvement Act of 1987, signed into law on August 20, 1987. The Malcolm Baldrige National Quality Award recognizes U.S. organizations in the business, health care, education, and nonprofit sectors for performance excellence.

In the 1990s, TQM became so important to international trade that it evolved and became the international ISO-9000 family of standards. This standard includes a certification part, which is managed globally. Since its initiation, ISO-9000 has been updated several times and is still a prominent standard used worldwide.

An important development that also emerged from Japan a decade after World War II is the Toyota production system (TPS). TPS is based on the idea of maximizing value to customers while minimizing waste of any kind. To achieve this goal, techniques for inventory management and production scheduling were developed under the inclusive title of Just in Time—JIT. By the end of the 1970s, JIT and TPS were adopted and implemented successfully throughout Japan. Soon after, companies all over the world embraced them.

At the same time, there was significant development in the use of computers in design, planning, manufacturing, and production management. Robotic systems, systems for computer aided design (CAD), and computer aided manufacturing (CAM) along with computerized flexible manufacturing systems (FMS) were introduced. These systems have evolved and thrived and are still prominent in the realm of manufacturing and production.

In recent years, the business world has become "flat." Globalization has emphasized the need for rapid response to constant change. The need to compete in the global market led to the development of business process reengineering (BPR), an effort to improve processes by developing and implementing solutions that change organizational management practices significantly (Cha et al., 2003). Organizations began using ERP at the turn of the last century. Such systems, based on an integrated management approach, provide an integrated solution for many organizational functions: production, purchasing, finance, marketing, human resources, plant maintenance, service and project management. The advantage of all the aforementioned systems is that they create synchronization and coordination between the various parts of the organization, with the aim of improving the level of quality and competitiveness.

Following the direction of TPS, the effort is now on creating value for the customer while minimizing waste. A simple definition of the term value is "what the customer is willing to pay for." Anything that does not generate customer value is a waste. The purpose of the industrial engineer is to bring the waste to a minimum and to build manufacturing and service systems that produce maximum value and minimal waste.

In the last years of the twentieth century, the development of high-speed electronic communication networks, the Internet, facilitating direct contact between the supplier and the customer from anywhere over the globe, has made E-business boom and released an avalanche of information to be mined in aid of organizational growth and benefit.

In addition, integration of the organization with its business environment, that is, its partners—customers, suppliers, service recipients, and service providers, also became a necessity. Out of this need, information management systems were developed. Examples include

MES: Manufacturing Execution Systems

ERP: Enterprise Resource Planning

CRM: Customer Relations Management Systems

SRM: Supplier Relations Management

One especially important type of approach, supply chain management (SCM), was developed to help handle all the partners and stages of business practice. SCM deals with the acquisition of material from raw material by the manufacturer through to the customer's final product. They make use of hardware and software, and electronic communication to enhance productivity, reduce response time, and increase quality while minimizing cost. These systems are continuously developing because of the global competition.

The large amount of data (big data) being generated and the need to analyze it, has led to the rapid development of the field of data mining (DM) as well as the transformation of operations research (OR). Together, these developments have produced a new area of research, business analytics, which combines OR and the processing of large data sets.

1.4.3 Historical Timeline

1911: Frederick Taylor publishes the "scientific management theory" and establishes the principles of the study of time and labor.

Frank and Lillian Gilbert develop the study of movement and lay the foundation for industrial psychology.

1913: Henry Ford develops the moving assembly line.

1914: Henry Gantt develops a chart for scheduling activities with technological links between them.

1917: F. W. Harris develops models to determine batch size for the management of inventory.

1930s: Application of statistical methods in quality control:

1927–1933—Elton Mayo, at the Hawthorne plant of Western Electric, examines the impact of environmental changes on employee performance.

1940s: The development of Operations Research: Mathematical models based on simplifying assumptions, models that allow analysis of complex problems in quantitative terms.

1947: George Denzig develops the simplex method for solving systems that can be described using linear models.

1950s and 1960s: The development of digital simulation; development of PERT and CPM methods for project management.

1970s: Development of a variety of software packages to manage routine tasks; development of MRP.

Latter part of the twentieth century:

Adoption of Japanese manufacturing methods such as JIT, TQC.

Development of CAD/CAM, FMS, CIM, and robots.

Development and standardization of process development (ISO-9000), and streamline systems and processes such as BPR.

Development of E-business and supply chain management methodologies.

1.5 Impact of Globalization on the Industrial Engineering Profession

Globalization and the development of the Internet have created new challenges and pushed many organizations facing competition from all over the world into a state of constant struggle. This rapidly changing environment, where product life cycles are short and global competition is fierce, forced many organizations to search for ways to increase competitiveness in order to survive. Competition can be expressed and appear in one or more of four dimensions: Cost, quality, time, and flexibility. Industrial engineering provides the building blocks, tools, and techniques that organizations are using to improve their performance in these four dimensions. These dimensions are described in detail in the following sections.

1.5.1 Cost

Cost has a major effect on the price of a product/service and reducing the cost enables the seller to reduce the price. In many markets, the price of the product or service has a decisive influence on the customer's decision to choose this product or service. To compete successfully in such markets, the organizations' production costs must be low in order to sell at a low price and still generate a profit. When the market cannot distinguish between the quality of one organization's product (or service) and that of another organization, price becomes a major factor in the customer's decision. Cost-based competition is tough, and only the most efficient organizations survive, when the goods or services of different organizations are perceived to be of the same quality.

For example, in the PC industry Lenovo, HP, COMPAQ, and Dell as well as many small companies that offer similar products without having a famous brand name are all competing against each other. Clients who want a leading brand's product are willing to pay more. Many customers are price sensitive and buy computers from less known companies that charge less, assuming that the components of these computers and their performances are not much different from the brand name computers. These customers are searching for the lowest cost computer with sufficient performance for their requirements.

1.5.2 Quality

Quality pertains to both the offered product/service and the operational processes by which the product/service is attained. There is a difference between the product/service quality and the quality of the operational process through which the product/service is produced. Service/product quality should fit the intended market. The focus is on customer needs, expectations, and fitness for use of the service/product. In a quality driven market, quality products and services can demand higher prices. The quality of the manufacturing process, however, is always critical, because poor quality creates waste of labor and material in the form of defective products. Quality-based competition is thoroughly discussed in Goetsch and Davis (2012) and Edwards Deming (1982).

The Japanese automotive industry is a good example of an industry that gained market advantage due to the high quality standards of its products. Historically, people who bought Japanese cars experienced fewer breakdowns compared to those who owned American-made cars. For many years, Japanese cars were ranked the highest in publications such as *Consumer Report*. Market reaction, in the United States, was a shift from buying American cars to buying Japanese cars.

Standards such as ISO-9000 were developed to ensure compliance with process quality threshold requirements. High operational process quality is related to organization's proficiency, maturity, competitiveness, and reliability. Many companies require their long-term suppliers and supply contracts that span significant periods of time to meet standards that the suppliers might not have otherwise maintained.

The aviation industry is a good example of the importance of process quality. While quality of components and products is important due to safety concerns, checking the finished parts of an aircraft does not ensure the required quality is attained. For instance, paint may cover manufacturing defects, badly welded spots may pass the test, but break after a few flights, an assembly that looks fine may conceal loose coupling of components, etc. Therefore, in the aircraft industry, quality is monitored continuously throughout the manufacturing processes from raw materials to the finished product. Products are purchased only from factories that can prove that their production processes meet the high quality requirements of the aircraft manufacturers.

1.5.3 Time

Time plays a key role when there is competition. It is significant in three major places: (1) Customer waiting time, (2) supply lead-time, and (3) production time. Customers do not like to wait, and the faster they are served, the more competitive is the business. In many markets, supply lead-time (the time it takes to supply the product) is a key to competition. Sometimes customers are willing to pay a higher price for a product with a shorter lead-time. Production time is indicative of production costs. Production time could be shortened in many cases, using effective scheduling techniques.

McDonald's is an example of a thriving business that focuses on short customer waiting times. The importance of lead-time is obvious in retail sales before the holiday season in the United States. Retailers demand on time delivery, as goods supplied after the holiday season are much more difficult to sell, and sometimes even become impossible to sell.

An organization that implements the TPS/JIT technique, discussed earlier, in its production plants (e.g., Toyota) is one that is focusing on production time. For such organization, the final assembly process of their products depends on the timely delivery of raw materials, parts, and subassemblies, by subcontractors and suppliers. These organizations, in order to reduce costs, purchase raw materials parts and components exactly when they are needed in the production process. No inventories are kept to guard against delays in supply. Therefore, the on-time delivery of raw materials parts and components is crucial; any significant delay can stop an entire assembly plant.

1.5.4 Flexibility

Flexibility is defined as the ability of an organization to adapt itself in a short time to the changing demands of its customers and to the changing environment. Flexibility can be measured, for example, by how long it takes to change a car assembly line or adding the ability to assemble a new car model. A more extreme example is being able (or unable) to convert a production line producing one product to manufacture a new generation of products. Flexibility is important when market needs or the environment change frequently and the organization has to adapt itself constantly to these changes.

The German car manufacturer Mercedes Benz is an example of flexibility. It is able to offer the individual customer the opportunity to have a car built according to his or her specifications, which are limited to the variety of models the company produces. For each model, the customer can choose different options for color, upholstery, stereo system, and other extras, and order a car based on his or her needs and expectations. To enable mass production of car models with different characteristics, the production line and the procurement process are designed for flexibility and quick transition from the production of one model to another model with some different specifications.

It is important to note that each organization has to analyze the markets in which it competes, and to decide how it should compete in each market. The right balance between the four dimensions of competitiveness is market- and organization-specific.

1.6 Industrial Engineering and Systems

Industrial engineers design and manage production and service systems. The boundary of the system and the boundary of subsystems is an important issue and two extreme approaches exist with respect to this issue. The first approach is to view a production or a service system as an open system; the second approach views it as a closed system. Closed systems have clear boundaries with their surroundings and for each organizational unit within the system. In a closed system, manufacturing has no direct relationship with customers. Manufacturing gets requirements from marketing. In a similar way, manufacturing has no direct relationship with suppliers. The purchasing department serves as the contact between manufacturing and suppliers. In closed systems, each organizational unit defines the way it performs its tasks and the appropriate performance measures. This approach may lead to local solutions that are inadequate for the whole system. Consider, for example, a closed system in which production's local goal is to maximize the utilization of production resources such as machines and labor. The result may be overproduction, or production that exceeds demand and generates expensive inventories. Thus, the local goal of high utilization of resources creates a system-wide problem—inventory accumulation.

In open systems, different organizational units work as a team to achieve common goals throughout the organization, and to find solutions that are good for the whole system.

Closed systems were very common because information on the entire production or service system was very hard to collect, store, retrieve, and analyze when the information was kept on paper. Development of inexpensive fast computer systems solved the problem, and today an organization's entire information library can be made available to all parties involved. In this case, the industrial engineer has access to all the information related to all the processes taking place in the organization. This is the foundation of supply chain management where organizations along the supply chain share information to improve the performance of the overall system, while different organizational units share information and set up common goals and performance measures.

1.7 Industrial Engineering and Process Design

Industrial engineers design and implement processes in organizations. Although there is a large variety of processes, it is common to use Hammer and Champy (1993) five "basic processes," which is generally found in most organizations:

1. *The development process:* The process starts with an idea for a new product or service and ends with the design of the new product or service and a working prototype.
2. *Preparation of infrastructure:* The process starts with a working prototype of a new product and ends with the successful completion and testing of the production facility for the product.
3. *Sales:* The process starts with market research and ends with an order from a customer.
4. *Delivery:* The process starts with an order from a customer and ends with a delivery and receipt of payment from the customer who received the requested products.
5. *Service:* The process starts with a customer's request for service and ends when the service is provided to his or her satisfaction.

Methods have been developed to support the planning of processes aimed at maximizing the value received by the customer by mapping the value chain. Industrial engineers plan the processes in the organization to achieve organizational goals and customer satisfaction. This role requires a thorough understanding of the organization and its environment, and, accordingly, the industrial engineer must cooperate and collaborate with other professionals in the organization, and people from other units such as

- *Marketing:* This unit is responsible for contact with customers and processing of customer orders.
- *Purchasing:* This unit handles relationships with the external sources involving supply of products and services.
- *Engineering:* This unit is responsible for product design and the design of production-service systems.
- *Finance:* This unit is responsible for the organization's budget and management of cash flow including relationships with banks, payments to suppliers, payments received from customers, etc.
- *Production:* This unit is responsible for the proper operation of the production system.

1.8 Need for Integrated and Dynamic Processes

To address the competition in the markets, processes should be designed taking into account time, cost, flexibility, and quality. For example, the process of new product development should focus on the changing needs of customers while minimizing development time and the cost of development, but at the same time minimizing the cost of manufacturing and servicing the new product. The development team should understand and satisfy the variety of customer needs in different markets by introducing flexibility in the product design. This process must be based on new product development teams that include marketing professionals who know customer needs and expectations, engineers and designers who know how to translate customer needs into specifications, plans, and assembly instructions, etc., and experts in the operations and service of the new product focusing on developing a quality product at a reasonable price to give the customers the best value for their money for the whole product life cycle.

A similar approach is needed in all processes within the organization. It enables to react quickly to changing markets, reduce delivery times, lower costs, improve quality and flexibility, and improve cash flows.

1.9 Summary

In summary, the role of the industrial engineer in the twenty-first century is to ensure the organization's success in dynamic and competitive markets. The emphasis is on the design of processes using information systems to cope with cost, quality, time, and flexibility requirements.

This book will focus on the main processes in the organization, and the role of industrial engineers in planning these processes using simple models and information systems that support these models.

References

Cha, J., Jardim-Gonclaves, R., and Steiger-Garcao, A. (Eds.). 2003. *Concurrent Engineering.* Boca Raton, FL: Taylor & Francis.

Edwards Deming, W. 1982. *Out of the Crisis.* Cambridge, MA: MIT Press.

Forrester, J.W. 1961. *Industrial Dynamics.* Cambridge, MA: MIT Press.

Forrester, J.W. 1968. *Principles of Systems.* Cambridge, MA: MIT Press.

Goetsch, D.L. and Davis, S. 2012. *Quality Management for Organizational Excellence: Introduction to Total Quality.* 7th edn. Englewood Cliffs, NJ: Prentice Hall.

Hammer, M. and Champy, J. 1993. *Reengineering the Corporation, a Manifesto for Business Revolution.* New York, NY: Harper Business.

2

Modeling the Organizational Structure and the Facility Layout

---

Educational Goals

This chapter discusses the ways people and equipment are organized in the modern world so that they work together to accomplish assigned tasks. The discussion starts with an examination of organizational needs, and continues with a description of organizational structure models, concentrating on the way these models depict the relationships between members of the formal organization. The focus then shifts to the individuals—the members of the organization. This discussion includes the following subjects:

- Human physiology
- Human psychology
- Rewarding people
- Motivating people
- Learning and training

Following this, the equipment used by modern organizations to perform production and service processes is examined. In many organizations, equipment is selected and laid out by industrial engineers. Designing production and service facilities and locating equipment and machinery in the facility is known as layout planning. Finally, the chapter reviews the relationship between machines and equipment layout and the organizational structure.

2.1 Introduction

In Chapter 1, we introduced five basic processes that exist in many organizations:

1. *The development process:* This process starts with an idea for a new product or service and ends with the design of the new product or service and a working prototype.

2. *Preparation of infrastructure:* This process starts with a working prototype of a new product and ends with the successful completion and testing of the product's production facility.

3. *Sales:* This process starts with market research and ends with an order from a customer.

4. *Supply:* This process starts with an order from a customer and ends with delivery and receipt of payment from the customer who received the products he or she ordered.

5. *Service:* This process starts with a customer request for service and ends when service is provided to the satisfaction of the customer.

These processes are usually carried out by teams (although in some organizations a single person performs the whole process), using different types of resources, for example, materials, machinery, and equipment. The formal relationships among the team members are presented by a model called the organizational structure model. The organizational structure model answers questions such as who reports to whom, who has the responsibility to perform a specific task, and who has the authority to use the organizational resources. The building block of the organizational structure model is a human being. To understand this building block and use it properly, we must have some understanding of human physiology as well as some understanding of human psychology. Understanding is the key to selecting the right organizational structure, and addressing the issues of authority, responsibility, compensation, motivation, learning, training, and, most importantly, the proper design of the workplace. Industrial engineers address all these issues as discussed in this chapter and in Chapter 10.

In many organizations, the selection of the most appropriate equipment to perform the different parts of a process as well as the arrangement of equipment in the facility (known as the facility layout) is a task performed by industrial engineers, and therefore, we will discuss it as well.

In doing so, we will see how the organizational structure model and the layout model are related.

2.1.1 What Is an Organization?

Roughly speaking, an organization is the unification of a group of people for carrying out processes or activities to achieve certain purposes, typically on a continuing basis. More formal and detailed definitions follow.

Louis (1982) suggested that the term organization could have different meanings depending on the context to which the definition refers. Each meaning describes some aspects and together they answer the question of what an organization is.

2.1.2 Organizations as a Human Creation

Organizations are created by people, in order to fulfill specific needs. For example, the supply of products or services or providing protection to the community can serve as the impetus for setting up an organization. As a human creation, organizations are influenced by the culture and beliefs of the society in which they were created. The organizational structure is determined by the purpose for which it was established. Over time, the organizational structure should be adapted to problems facing the organization, and changes occurring in its environment. In a dynamic environment, the requirement for the organization to respond and adapt to environmental changes can become a matter of life or death.

2.1.3 Organizations as Production Systems

Organizations are systems where inputs such as material and information are combined to create outputs such as products and services. The purpose of the organization in this regard is conversion of inputs into outputs. To survive, the organization must be able to make this conversion effectively without hurting its ability to perform this conversion in the future. The flexibility of the organization is measured by its ability to increase or decrease the rate of conversion as necessary, and by the ability to change the range of products and services it offers, according to customer needs and requirements. In order to achieve flexibility, the organization may encompass additional functions aside from those needed to perform the conversion (e.g., management functions).

2.1.4 Organizations as Economic Entities

Organizations are economic entities, aimed at providing economic goals. A common goal is to maximize profit in the short and long terms. In cases where revenue is a result of external demand, maximizing profit translates into minimizing the total cost of the resources used by the organization (in order to minimize expenses) so that these costs are smaller than the total revenues. If expenses are higher than revenues, the organization will be considered profitable only if the value of its assets increases accordingly. In a dynamic environment, an organization should check and control its efficiency continually, and constantly measure and manage the cost of the resources that it is using.

2.1.5 Organizations as Social Systems

Organizations are a framework in which people are grouped together to work and achieve common goals, financial or otherwise. Members develop ways to work together to resolve problems and deal with conflicts. These patterns of interaction between different members of the organization are the basis of the organizational structure consisting of roles and interactions between members of the organization. A person performing a role in the organization has an impact on the conversion process. The same function may be performed in different ways by different people. The nature of the interactions between people depends on the people doing the jobs.

Organizations provide a social arena in which their members may achieve their personal goals. Operating within this arena also facilitates members' self-realization and attainment of compensation by being recognized within the organization and outside it and conferred social status.

2.1.6 Organizations as Goal-Oriented Systems

An organization is a collection of individuals, each of whom has needs, expectations, and personal goals in addition to the goals defined by the organization. It can be expected that individuals will try to achieve some of their goals and not always give priority to the organization's goals. In a well-designed organizational structure, the organization facilitates the achievement of individual goals through the achievement of the organization's goals. Organizations should strive to satisfy its members' goals so that they have a personal stake in the organization's continued existence. This aspect of the organization has led to the development of an approach that focuses on quality of life at work.

2.1.7 Organizations as Open Systems

This definition includes all of the above definitions. Organizations are subject to forces and requirements dictated by their environment. To survive over time, organizations have to match and adapt to internal changes arising from being a social system. The aspiration to achieve desired outputs, which usually occur under uncertainty—caused by dynamic and changing environments—means that organizations should strive to be in a constant state of internal and external adjustment.

Successful organizations are systems with the ability to adapt to new situations. Organizations that lack the ability to adapt themselves to changes fast enough, will find it difficult to succeed, or even survive in today's dynamic business environment.

2.2 Development of Organizations

Organizations have existed from early in the history of human race. The first organizations—families or groups of hunters—evolved into farming communities, tribes, kingdoms, and empires. The need to perform tasks that a single person could not perform alone and to share or barter scarce resources are just some factors that encouraged the creation of the first organizations.

Modern society's rapidly evolving technology motivates professionalization in very narrow fields of knowledge. These factors constitute more recent impetus for creating organizations. Today, most products and services are based on the integration of hardware, data, software, and human knowhow. A single person cannot possess all the knowledge and expertise required to manufacture complete products or to deliver complete services. Organizations are established when experts in different areas must work together in order to succeed in today's markets.

Organizations designed to perform major projects are not new. Examples of ancient organizational undertakings are the construction of the pyramids in Egypt, the Great Wall of China, and the Temple in Jerusalem. In these cases, the coordinated work of many people was required to complete the project. Principles of division of labor and specialization are fundamental principles in many organizations. Adam Smith, in his book "The Wealth of Nations" (1776), described the production of pins using these principles. The process he described is the result of a well-designed division of labor: Each team member does a small part of the work required for the production of pins. Repeating the same operation again and again and developing greater proficiency in carrying out his or her assignment (in fact, this phenomenon of learning through repetition is modeled by a model known as the learning curve, which is discussed later). The increased efficiency of each team member means improved efficiency of all members and raises the chances that the project will be a success.

The benefits from applying the principles of division of labor and specialization are realized only if good coordination exists between the various people and organizational components engaged in the process. Coordination is a result of process and organizational design—which is generally carried out by industrial engineers. It is relatively easy to coordinate tasks in a repetitive environment (e.g., mass production) where each team member performs exactly the same steps repeatedly. If the enterprise provides a wide range of products and services, and in particular if new products and services are developed, the operation is not repetitive—more effort is needed and coordination is more difficult. Coordination is most difficult when the required goods or services are unique and there is almost no repetition. In this case, the coordination occurs in the project environment, and special methods for planning and controlling are required as explained in Chapter 3.

There is a difference between formal organizations and nonformal organizations. Most formal organizations have a clear delineation of responsibilities, authority, and reporting requirements, while nonformal organizations are based on common interests, shared beliefs, social values, emotions, traditions, and the like. The following discussion focuses on formal organizations, as many industrial engineers are employed in this type of organization. Within formal organizations, however, there are often nonformal organizations. Thus, for example, in a commercial company, which is a formal organization, some members may get together every day to pray. In such a case, they form a nonformal organization within the formal organization.

2.3 Examples of Organizational Structures

There are many types of formal organizations. A relatively small number of models describe the basic theoretical structures of these organizations. Organizations implement many variations of these basic structures in business and industry. These implementations are never exact copies of the models described here, because models are a simplified presentation of reality. In addition, as stated before, organizations are dynamic and continually have to cope with changing environment by constantly adapting their organizational structure to the environment in which they operate.

The organizational structure is related to both the quantity and variety of products and services offered by the organization as depicted in Figure 2.1. A project organization is designed to produce a variety of unique products in very small quantities, while the functional organization is designed to

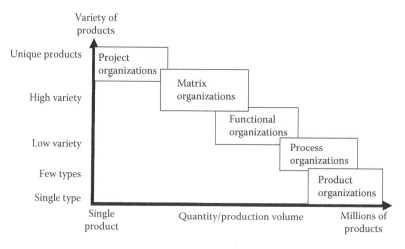

FIGURE 2.1
Typical organizational structures versus variety and quantity of products/services.

cope with a variety of products in medium quantities and product-oriented organizations are designed to produce and supply large quantities of the same product or a family of similar products. We shall now discuss in more detail the organizational structures presented in Figure 2.1.

2.3.1 Project-Oriented Organizational Structure

The building block of a project-oriented organizational structure is a team of people working on the same mission or project. This building block can comprise a team of experts from various fields working to achieve the common goals of a project. A task force is a typical example. When an entire organization is divided into task forces, it is a project-oriented organization. Each project usually has a start time and a due date. Thus, the structure of the project-oriented organization is temporary by definition. Organizations adopting a project-oriented structure bring together people with the required skills and knowledge for the project. They group them into a team that is managed by a project manager who is responsible for the entire project.

The project-oriented organizational structure is described in Figure 2.2.

The main benefits of a project-oriented organizational structure are

- All team members focus on the same mission.
- Stakeholders for whom the project is performed have a point of contact—the project manager—who has the power to manage project resources and influence the success of the project.
- Composition of the project team is optimized for the project mission. Thus, a research and development (R&D) project will be assigned to a team that consists mainly of scientists and engineers, and the design team of a large construction project will include mainly architects and construction engineers.
- The project organization does not suffer from the problem of boundaries separating organizational units (silos) as is frequently the situation in the functional organization; because all team members serve one purpose and are supervised by the same project manager.

The main problems typical of the project-oriented organization are

- There is no pooling of similar resources, and therefore, the benefit of using experts tends to be lower than the benefit achieved when similar resources are pooled together.
- The experts in a given area are not in the same organizational unit, and thus, they do not have an opportunity to learn from each other and to share knowledge with their peers.

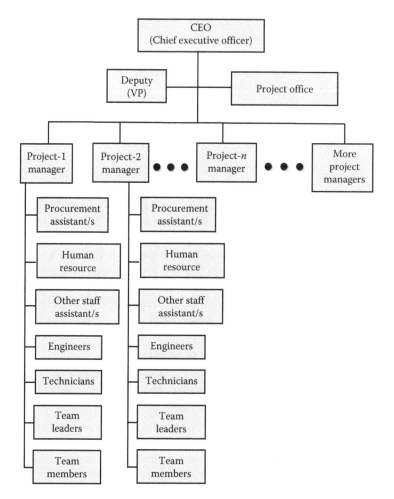

FIGURE 2.2
An example of a typical project-oriented organization.

- The organizational structure is not stable, as projects have limited duration.
- The communication between teams working on different projects is limited.

2.3.2 Matrix Organization

A matrix organizational structure is mainly used when there are many projects with many similarities, requiring several areas of expertise. In such environments, there is an advantage in maintaining several professional units that provide service to various projects. The hybrid organizational model—the matrix structure—which was developed in response to this need, combines the organization's permanent professional units (functional structure) and the project-oriented structure. This arrangement maximizes the use of existing professional expertise while leveraging the advantages of the project-oriented structure.

The organizational structure of a typical matrix organization is described in Figure 2.3.

In Figure 2.3, the vertical dashed lines of the matrix belong to the functional units, whereas the horizontal dotted lines belong to the particular project. The squares with names in the junctions of the horizontal and vertical lines are work packages performed by that person. In this way, the project manager and the permanent professional unit manager simultaneously manage each person. The project manager lays out the needs and the professional manager sees to their being met in a professional way that best serves the organization. Project teams are established when necessary by "assigning" (seconding) experts from the functional units to a project team.

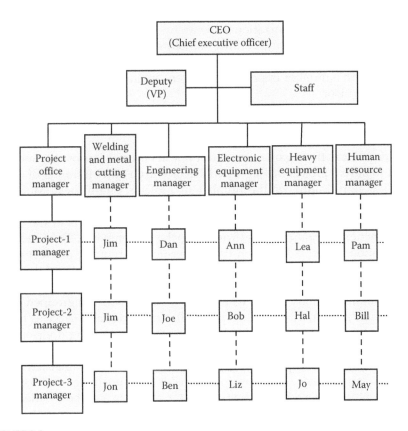

FIGURE 2.3
An example of a typical matrix organization.

These experts are fully engaged in the project, but retain their ties with the functional structure by reporting to the functional manager. The customer has a well-defined point of contact—the project manager.

A matrix structure can exist in many different forms: for projects of any size, complexity, or duration. If the project is small, the project coordinator may devote only part of his time to the project and continue to operate within the functional unit as required. If large and complex projects are performed, a core team, which includes a number of experts headed by a project manager, is assigned to the project and they will work exclusively on the project. The project team receives support from the functional part of the organization by resources assigned to the project only part of the time, as needed. Thus, we can see that a matrix structure is very versatile.

The main disadvantage of this structure is the situation where project members report to two or more managers ("bosses"): the functional manager, and one or more project managers. This "dual boss" arrangement can create communication problems. Namely, the classic principle of management theory of "unity of command"—according to which an employee reports to only one supervisor, who in turn reports to only one supervisor, and so on, up the organizational hierarchy—is broken. This breach may result in disagreements and conflict of interest. The problem worsens when an employee is assigned part-time to a number of projects and is, therefore, working with a number of project managers, in addition to his or her functional supervisor.

2.3.3 Functional Organization

At some point in the organization's life cycle, it becomes impossible and inefficient to have a project manager dedicated to each order (number of orders increases significantly or the number of projects diminishes). Moreover, when different products/services have the same repetitive or similar requirements, it becomes more efficient to standardize parts of the operation. Functional organizations thrive in such environments. The functional organization model is based on the clustering or grouping of members of the organization according to their expertise and the functions they perform. Each group or cluster is an organizational unit and its members perform similar functions: Those who deal with customers are grouped in the marketing division; those engage in purchasing goods and services for the organization are grouped in the procurement or purchasing division; and engineers belong to the engineering division.

Due to the limited span of control (the number of subordinates a single manager can manage and coordinate effectively), functional divisions are divided into smaller groups. Engineers, for example, can be divided into groups engage in developing new products, groups dealing with ongoing production, and yet other groups dealing with quality. Another way to cluster engineers can be by their specialization: A group of electrical and electronics engineers, a group of mechanical engineers, and a group of industrial engineers.

The optimal number of subordinates varies depending on the nature of the organization, the tasks, the environment and people in the organization and, of course, the managers' abilities. The span of control can take different values. If, for example, we assume that it is 10, this means that a single supervisor can manage effectively about 10 subordinates. Hence, organizations with a large number of employees need a hierarchy or a pyramid-like structure. Figure 2.4 depicts a typical functional organization chart.

Advantages of the functional structure are as follows:

- Pooling together similar resources. In many cases, organizations utilize their resources more efficiently by pooling people who have the same expertise and perform similar tasks. As noted above, this pooling constitutes the basic unit of a functional structure and such units are characterized by strong professional capabilities. Moreover, management of people with the same expertise who are likely to have to tackle similar work problems is easier, as is management of similar machines. When similar resources are pooled together, members of the unit, who may be overloaded at a particular time, can get assistance from other group members who have a lighter load at that time.

- Flow of information within each group in the functional organization is usually good because people in each group have a common background; they use a common terminology and share common interests. Good information flow is an important factor affecting the efficiency of teamwork and the creation of a learning organization.

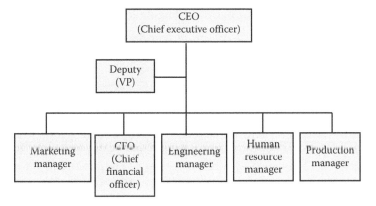

FIGURE 2.4
An example of a functional organization structure.

- The creation of groups of people with similar backgrounds and expertise who are managed by a person with a similar background and expertise facilitates better and easier coordination within the group and thus it is likely that the span of control will be larger.
- The organization is stable, given that its functions do not change rapidly even if the technology, the market, or the products do change.
- The career path is easier to forecast and people joining the organization can plan their future career based on their functional expertise, and based on the clear definitions of responsibility and stability.

Disadvantages of the functional structure are as follows:

- Communication between different groups (different functions) in the organization may be difficult because members of different groups do not speak the same professional language, do not share the same background, and may have different goals.
- Cooperation between employees in different groups in the organization may be poor because of the different interests, different goals of individual employees and their groups, and communication barriers between groups focusing on different functions.
- Clients with special needs—a special service, a special product, or special information—may have problems if their needs can be satisfied only by members of different functions in the organization. Since there may be difficulties in communication between different organizational units (detailed above), the client may have difficulty in obtaining what he or she needs.

2.3.4 Process-Based and Concurrent Engineering Organizations

Large quantities of identical or similar products or services justify focus on operational processes. The process-based organization succeeds when product/service variety exists to a small degree, and the quantity of each type is very large. Management in this case is divided according to the main processes.

Maintaining the best possible organizational structure is an important and difficult task. A structure that was very effective for a long period may not work properly when the environment changes. For example, the delivery process that was successful for a functional organization when it was a monopoly and served a stable market may fail suddenly if the market is opened up to competition. When competition increases, it creates pressure to increase the range of products and improve service delivery time to customers. An example from the distant past is the telephone companies such as Bell or AT&T that lost their monopoly and had to compete in the open market. An example from the more recent past is the development of cell phone technology that forced the traditional phone companies to change the old ways of management.

Increased competition may also lead to pressure for shorter lead times. A functional organization may have difficulties meeting this demand.

Consider the process of developing new products. For many years, the functional organization included special organizational units whose role was to develop new products. When marketing people recognized the need for a new product, the new product development department or R&D department focused on the design issues, producing the required drawings and documents for the new product, defining its physical and functional attributes. Based on product definition, manufacturing engineering developed processes for manufacturing, assembly, and testing of the product, and logistics found suppliers for raw materials and components, developed packaging and transport processes, etc. This process is very slow and suffers from communication problems between functional units. The process does not provide enough "value" to customers (customer value can be defined as what the customer is willing to pay for when buying the product or service).

The traditional process is slow because it is built on a set of actions performed in sequence by different organizational units. Each organizational unit is a link in this chain of activities. It gets its input from the unit preceding it in the process, and produces output, which is the input of the succeeding organizational unit in the process. Since each unit operates independently, the duration of the development cycle is the sum of the duration of the processes performed in sequence, and therefore, it creates relatively long development times.

Another reason for the length of this development process is uncertainty. Some activities may have to be repeated, as early development efforts fail and this may cause cycles and repetitions in the process. A typical reason for such failures is communication problems among the various organizational units, especially in the transition from development to production. Product prototypes, built during development and that may have successfully passed tests, may be very expensive to manufacture or to maintain, and therefore, the design must be changed. This problem stems from the serial structure of the process. The solution is to integrate all the knowledge required for development, design, production, and assembly to reduce or eliminate development cycles and to shorten the duration of the process. Decisions regarding product features, made early on in the product life cycle (product definition phase and the phases of product development) must take into account the ability to manufacture, serve, and maintain the product. Although the cost of new product design processes may be relatively small, most future costs associated with its manufacture, maintenance, and operation are determined at this stage. Since new product development experts are not always well versed in methods to reducing production costs, operation, and maintenance of products, their decisions are based on a local optimum—which often means a product with a wide range of operating options, but very expensive to manufacture, operate, and maintain.

In today's competitive markets, companies have abandoned this slow route and moved, by necessity, to a rapid process of developing new products of high value to the customer. Many organizations are implementing a new approach to new product development dubbed concurrent engineering (CE), which is process based. Experts in various areas—development, production, operation, and maintenance of products—work together as a team (in many cases, also in cooperation with the client) in the product design process to develop high-value products for the customer. The focus is on delivering to the client what he or she wants, at the lowest cost of manufacturing, maintenance, and service.

CE, though process based, differs from the project structure, the matrix structure, and functional structure because it has a different *raison d'être*: to develop a new product. Unlike the functional structure, a CE team includes experts from various fields, is assembled for a particular project, and disbands on completion of the project. The team follows the product throughout its life cycle, including manufacturing, operation, and maintenance.

All five basic processes defined in Chapter 1 can adopt an approach similar to that used by CE for developing new products.

1. *The development process:* The process starts with an idea for a new product or service and ends with the design of the new product or service and a working prototype.
2. *Preparation of infrastructure:* The process starts with a working prototype of a new product and ends with the successful completion and testing of the production facility for the product.
3. *Sales:* The process starts with market research and ends with an order from a customer.
4. *Delivery:* The process starts with an order from a customer and ends with a delivery and receipt of payment from the customer who received the requested products.
5. *Service:* The process starts with a customer's request for service and ends when the service is provided to his or her satisfaction.

Process-based organization is the key to success of the development process in today's competitive market. The objective of CE is to develop high-quality products with the lowest life cycle cost in the fastest way, while maintaining the flexibility necessary in the constantly changing environment. A similar goal exists for the other four processes and similar organizational models can be used.

2.3.5 Product/Service-Based Organizations

When the organization supplies a commodity such as soft drinks, cheese, or paper (or a service such as Internet access), it specializes in producing this special commodity. The organization is constructed according to the main processes required for producing and distributing the commodity. For example, in the automotive industry, it is customary to have different departments for procurement, marketing, and for production processes such as engines, gears, chassis, white body, paint, etc.

The product-based organization is very similar to the process-based organization, but the latter works with a single product type or close variations of the same product.

Because of the similarity between the two organization types—product and process—no further elaboration will be provided.

The effect of the product oriented organization on processes is as follows:

The product organizational structure is stable and rigid, so it is suitable for organizations that functional units cover most of their activities, and organizations that do not have to adapt to changing conditions. Examples are monopolistic organizations or organizations that operate in a very stable market. Over the years, functional organizational units develop their own ways and means to carry out their roles. Awareness of the importance of "local" objectives develops; and a "tradition" emerges that could result in slow response to changes in the environment, such as new needs of customers, competition, changing markets, and changes in the technology. On the other hand, this stability may be an advantage in a very stable mass production environment, where dedicated facilities that require a large capital investment for the production of large quantities of products over time are used.

Organizational processes often cross the boundaries between different functions. In a functional organization, marketing people, trying to get orders, deal with customers; the purchasing department is responsible for timely delivery of raw materials and components needed for production; and operations experts are responsible for scheduling and monitoring of work performed by the various resources. Communication among these different organizational units may be difficult in a functional organization because of the different objectives, nonuniform performance measures, and different terminology used by the different functions. These difficulties are known as the "silo effect." Each organizational unit acts as if it were located in an isolated silo.

The operations side of an organization typically seeks performance measures via output, productivity, efficiency, and utilization of resources. However, other areas use different performance measures: Revenues from sales per month (marketing), number of orders executed in a certain period (purchasing), etc.

The use of performance measures that encourage high utilization of resources may create a situation where good performance is achieved at a local level, but the overall performance of the organization is poor. This can happen if in order to achieve high productivity, for example, production produces its products ahead of schedule, in an effort to run large production batches and to minimize time spent on changeover or set up of the production system. This local level high productivity would create excess inventories that are expensive to carry, may be perishable, stolen, or simply become dead stock because there is no demand for these products. Similarly, the marketing department may promise unrealistic delivery times to increase sales volume, ignoring the fact that production may have difficulties fulfilling their promises.

A product organizational structure may be effective in a mass production environment that produces a single product or family of products on dedicated facilities with a constant demand over a very long period of time. Today, competition on cost, time, quality, and flexibility has eliminated many such markets and pushed many product-based organizations to change their organizational structure.

Assembly plants for cars in the early twentieth century faced constant demand for the same product over a long period of time. In such plants, each division operated relatively independently. The disadvantage of the product-oriented organizational structure—its lack of flexibility—became clear in the latter part of the century. When these assembly plants were required to deal with changing environment (demand, cost, design) and to compete based on flexibility, they were very hard-pressed to respond quickly to new needs.

Possible structure of the product-oriented organization: The organization will be headed by the Chief Executive Officer (CEO) and under him or her will work division managers. Together, they constitute the company's management. Department heads work under the division managers, etc.

2.3.6 Humans and Organizations

All organizations have something in common: Their success is dependent on humans that perform different roles in the organization (Burton et al., 2011). The industrial engineer must understand the human factor, and learn how to design work environments where employees can take advantage of their abilities for their own benefit and the benefit of the organization.

Industrial engineers should aim at several goals when designing the workplace and the organization:

- To increase the productivity and efficiency with which people's activities are conducted in organizations.

- To maintain and strengthen a number of important values, such as human health and human safety in the organization.

- To increase the motivation of all employees in the organization to achieve the organization's goals.
- To strengthen the ties between the employee and the organization in general, and the ties between the employee and other employees involved in the former's specific role in particular.

2.3.6.1 Human Factor Industrial Engineering Courses

To achieve these goals, the industrial engineer has to study issues related to organizational psychology, organizational sociology, psychology, ergonomics, physiology, and anatomy and to understand the different tools available to him or her in each area. With the appropriate knowledge, the industrial engineer can design a system that motivates employees and encourages learning.

1. *Organizational psychology:* Individuals in the organization interact with others—peers, subordinates, superiors, customers, and suppliers. Relations between each person and other people in the organization are studied within the framework of organizational psychology, the science of human relations in organizations. Organizational psychology courses seek to identify factors that affect the relationships between people in the organization with the aim of improving the performances of the organization and strengthening these relationships. Similarly, factors that interfere with the performance of the organization or damage the relationships between its members should be identified and reduced as much as possible.

2. *Organizational sociology:* Sociology deals with group dynamics of various formal and informal groups and the effects of external trends in expectations and beliefs. Sociology includes relationships between the people in the organization and the organization itself, for example, employee–employer relations. Organizational sociology courses examine the relationships between humans and the organization and the factors that affect these relationships.

3. *Psychology:* Industrial engineering focuses especially on aspects related to the work environment and learning, such as the ability to remember or distinguish between details, perception, and motivation.

4. *Ergonomics:* People work on computers, machines, and other equipment to perform their tasks in the organization. Making the interface between the operator and the machines more productive and efficient is one task of industrial engineers. To enable this interface improvement, industrial engineers take ergonomics courses where they examine the relationship between members of the organization and their environment, especially the equipment they use. For

example, proper design of a chair for the supermarket cashier can improve efficiency and reduce fatigue, in a similar way as the proper design of the interface between a person and his or her computer can significantly reduce the time required to learn a new application and the time required to perform various tasks using the computer.

5. *Physiology:* This field (physiology) and anatomy are tightly related to ergonomics. Industrial engineers need to understand the human body and its relation to the work environment, such as mobility and ability to lift weights.

6. *Anatomy:* Again, to design a healthy work environment, industrial engineers must understand the distribution of the human body's dimensions in different populations, and the ability of the human body to apply force.

In Sections 2.3.6.2 and 2.3.6.3, we briefly discuss the issue of industrial and organizational learning and the issue of motivation. These common human factors have grown into important parts of the industrial engineering curriculum.

2.3.6.2 Learning

Learning is the phenomenon of improving performance through repetition. The principle of division of labor increases repetition, and therefore, increases the learning and enhances performance. When a person repeats the same task over and over again, he or she learns to perform it in an effective, quick, and efficient way. In 1936, a researcher named Theodore Paul Wright examined the relationship between the number of repetitions and the time required for each repetition. He collected data on the time needed to perform an operation as function of the number of repetitions performed, and plotted the duration as a function of the number of repetitions. It turned out that the time required for each repetition drops in a nonlinear fashion with increasing number of repetitions (Anzanello and Fogliatto, 2011). The relationship between the time it takes to perform a task and the number of repetitions is described in the learning curve model.

To explain the model, which takes the form of an equation, we use the following notation:

x—number of repetitions

$t(1)$—duration of the first repetition

$t(x)$—duration of repetition x

b—a constant controlling the slope of the learning curve (typically between 0.1 and 0.5); each task has its specific b value

$T(x)$—cumulative time to produce x units

The learning curve equation is

$$t(x) = t(1)x^{-b}$$

In this equation, the learning coefficient (b) is a constant. In addition, the percentage of improvement in time for a unit numbered $2x$ versus time for a unit numbered x (each time the quantity doubles) is a constant: $\phi = 2^{-b}$.

This constant is called the learning curve slope and presents the execution time improvement when the quantity is multiplied. Over the years, tables representing different learning slopes were developed. Table 2.1 is an example of such a table. The right column of the table represents the number of repetitions x. The second column shows the coefficient x^{-b} of learning for the specific number of repetitions and the selected slope (70%) value.

To demonstrate the use of the model and the table, consider the following example:

The production time of the first unit was 10 h, while the time of the second unit was 7 h. What is the forecast production time of the first five units?

From the data for repetitions 1 and 2, the slope of the learning curve is 0.7, and we get the constant for unit 3, which is 0.568.

That is, the predicted time for the production unit 3 can be calculated by

$$t(x) = t(1)x^{-b}$$

$$t(3) = t(1)0.568 = 5.68$$

Moreover, for the next two units:

$$t(4) = t(1)0.490 = 4.90$$

$$t(5) = t(1)0.437 = 4.37$$

TABLE 2.1

Sample of Learning Coefficient Table for 70% Learning Slope

Number of Repetitions x	Coefficient $x^{-b} = x^{-0.512}$
1	1
2	0.700
3	0.568
4	0.490
5	0.437

TABLE 2.2

An Example of a Learning Coefficient Table for a 70% Learning Slope

Slope (x)	70%		75%		80%	
	$t_{(x)}$	$T_{(x)}$	$t_{(x)}$	$T_{(x)}$	$t_{(x)}$	$T_{(x)}$
1	1.000	1.000	1.000	1.000	1.000	1.000
2	0.700	1.700	0.750	1.750	0.800	1.800
3	0.568	2.268	0.634	2.384	0.702	2.502
4	0.490	2.758	0.563	2.946	0.640	3.142
5	0.437	3.195	0.513	3.459	0.596	3.738
6	0.398	3.593	0.475	3.934	0.562	4.299
7	0.367	3.960	0.446	4.380	0.534	4.834
8	0.343	4.303	0.422	4.802	0.512	5.346
9	0.323	4.626	0.402	5.204	0.493	5.839
10	0.306	4.932	0.385	5.589	0.477	6.315
11	0.291	5.223	0.370	5.958	0.462	6.777
12	0.278	5.501	0.357	6.315	0.449	7.227
13	0.267	5.769	0.345	6.660	0.438	7.665
14	0.257	6.026	0.334	6.994	0.428	8.092
15	0.248	6.274	0.325	7.319	0.418	8.511
16	0.240	6.514	0.316	7.635	0.410	8.920
17	0.233	6.747	0.309	7.944	0.402	9.322
18	0.226	6.973	0.301	8.245	0.394	9.716
19	0.220	7.192	0.295	8.540	0.388	10.104
20	0.214	7.407	0.288	8.828	0.381	10.485
21	0.209	7.615	0.283	9.111	0.375	10.860
22	0.204	7.819	0.277	9.388	0.370	11.230
23	0.199	8.018	0.272	9.660	0.364	11.594
24	0.195	8.213	0.267	9.928	0.359	11.954
25	0.191	8.404	0.263	10.191	0.355	12.309
26	0.187	8.591	0.259	10.449	0.350	12.659

The total time for the production of the first five units—$T(5)$—can be calculated by summing up the durations of the five repetitions:

$$T(5) = 4.37 + 4.90 + 5.68 + 7 + 10 = 31.95 \text{ h.}$$

A simple way to calculate this is to use the cumulative coefficient table to calculate the cumulative time of the five units.

In the cumulative table (Table 2.2), we replace $b = 0.512$ with $\phi = 2^{-b} = 70\%$, the coefficient for five repetitions is 3.195. Since $t(1) = 10$, the predicted time for the production of the first five units is

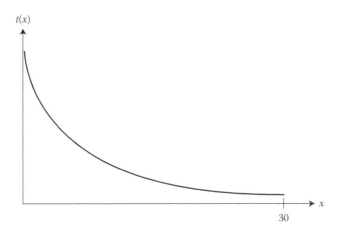

FIGURE 2.5
An example of a learning curve.

$$3.195 * 10 = 31.95 \text{ h.}$$

In Table 2.2, learning coefficients for a range of learning slopes (70%, 75%, 80%) are presented.

For a given slope, the relationship shown in Table 2.2 between the number of repetitions and the duration of the execution of each repetition is depicted in Figure 2.5.

Over the years, industrial engineers collected much data about learning curves, and the learning coefficients corresponding to different types of tasks. They developed models that describe more complex relationships between the number of repetitions and the duration of the performance.

The learning model is used for production planning, to predict the expected output from a given operation performed repeatedly. Based on such predictions and the wage rates of the people performing the operations, the model is used to estimate the cost of labor per unit of product in the future.

2.3.6.3 Motivation

Observations and experiments have shown that the level of employee motivation influences the slope of the learning curve. An early model that dealt with the issue of motivation is Maslow's pyramid of needs (Maslow, 1943), which assumes that there is a hierarchy of needs. Each employee is in a position where he or she wants to meet certain needs, and thus by meeting these needs, the organization can increase employee motivation. The basic model has the shape of a pyramid that consists of several levels (a hierarchy). The needs at the different levels from bottom up are as follows:

1. *Physiological needs:* These are at the bottom of the list and include, for example, food, clothing, and shelter.
2. *Security and safety requirements:* These include personal security, and the ability to anticipate and plan for the future.
3. *Social needs:* These are the needs for a reference group.
4. *Status:* The need for recognition, respect, and status in the reference group.
5. *Self-actualization:* The need for a job where one can express one's capabilities and fulfill oneself and one's ambitions.

This model is shown in Figure 2.6.

The model is based on the observation that different employees are at different levels of needs.

Robbins and Judge (2012) observed that when one need is satisfied, the need above it becomes dominant. In terms of Figure 2.6, the individual moves up the hierarchy. In other words, when a lower need in the pyramid's hierarchy is satisfied (even if not in full), the next level need arises and becomes dominant. Thus, according to this theory, if the organization wants to motivate people, it should understand what level in the hierarchy they are in, and focus on satisfying the needs at that level, and the next higher one.

For example, in times of unemployment and financial difficulties, an income that provides economic security is the most important need for the employee. Accordingly, an important tool to increase motivation is the wage system. This is the case, for example, of a student seeking part-time work during his or her studies whose main interest is financial income. For the same person, after graduation or at the height of his or her career, the most important need may be status and self-satisfaction at work.

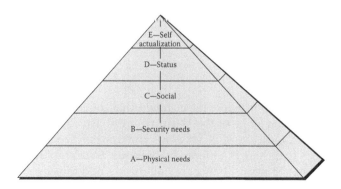

FIGURE 2.6
Maslow's pyramid of needs.

MacGregor (1960) developed another model handling the subject of motivation. This model is based on a completely different set of assumptions and it describes two different theories to explain the motivation of different employees. These theories, known as Theory X and Theory Y, represent two extreme points on a continuum (spectrum) on which different employees may be situated.

- Theory X assumes that workers do not like to work and will avoid work if they can. Therefore, most people have to work under tight control and supervision. The way to get more out of the workers is by standards and threats. The employee prefers to avoid work and responsibility and to do the minimum required to ensure job security because security is the main motive of most people.

- Theory Y assumes that employees want to work, and physical or mental efforts are perceived as a good challenge and a natural work environment. Employees tend to demonstrate initiative and motivation at their job. This commitment is only partly a function of the compensation that the employee receives at work. Under suitable conditions, the employee will not only accept authority but also develop personal responsibility for the efficient execution of his or her job.

Maslow's model can be related to that of MacGregor. We can say that Theory X characterizes the needs at the bottom of the Maslow's pyramid, while Theory Y characterizes the needs at the top of Maslow's pyramid. Both theories are early models of how to motivate employees. More recent models are presented in the book by Robbins and Judge (2012).

Maslow continued to work with Ouchi and developed Theory Z (Ouchi, 1981).

- Theory Z assumes that most managers and employees are interested in advancing the organization. Employees want the security and benefits provided by the company, while the organization wants to provide a warm home for its employees. Managers want to keep employees devoted and hardworking with high morale, and therefore, organizations take care of employees at work and provide adequate working conditions, compensation, and career planning.

To summarize, over the years many models describing the human factor at work have been developed and validated. These models are an important tool-set that helps industrial engineers in the design of organizational structures, role definitions, design of the work environment, and building systems supporting learning and motivation of the members of the organization.

The design of the human environment is discussed in more detail in Chapter 10, which expands on the subject of ergonomics.

2.4 Organizing Workplace Equipment and Machinery

Industrial engineers learn how to organize the physical resources that are used on the production floor or for the provision of services. These resources may include machinery, tools, inventories of raw materials, work in process and finished goods, material handling systems, furniture, office equipment, etc. Some models are similar to those used to organize people by project, function, process, or product. Other optimization models, were also developed to design the layout of physical resources in manufacturing or service organizations. Figure 2.7 shows the layout type related to the product/service variety and the production/service volume.

Organizations that provide services use physical resources such as office furniture and equipment and machines of various types (food processing equipment and refrigerators for restaurants, or x-ray machines, CT scanners, and magnetic resonance imaging [MRI] devices and surgical equipment in hospitals). These physical resources can be arranged or laid out in different

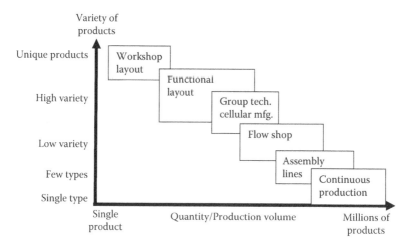

FIGURE 2.7
Layout type relationship to product/service variety and volume.

ways. Preparing the physical organization of the machines, workstations, office equipment, etc. is called layout planning.

2.4.1 Workshop Layout

In a workshop layout, a few machines are spread across the shop floor to enable both buffer accumulation in front of each machine and easy movements between machines. When many projects use the same machines and processes, it makes sense to build a workshop to cater to the various projects or customer orders. A workshop layout is also suitable for small job shop environments. Workshops typically have a few machines of different types that are arranged in a way so that material flow should be minimized. This entails grouping together machines that are used for consecutive processing in various projects. Such a workshop layout is suitable, for example, for a company that produces stage decoration and stage design. Its workshop would have several cloth and wood processing machines as well as welding and lighting posts. Other examples of companies that would set up a workshop are a custom kitchen designing company, a racing car shop, and a company producing wedding decorations and wedding dresses. In workshops, there is no structured flow and the products may be processed in various machines according to process precedence constraint and machine availability (Figure 2.8).

2.4.2 Functional Layout (for Job Shop and Batch Processing)

When production/service volume requires having many machines and at least several machines of the same type, it makes sense to group similar machines under a unified management (Heragu, 2008). For example, the shop floor would be divided into departmental areas (e.g., a department for milling machines, a department for lathes and turning machines, and a department for drilling). This type of layout is called a functional layout. For example, a milling work center in a metalworking factory where various milling machines (small and big, manual, semi-automatic, and fully automatic) are located would have a functional layout. In such a work center, the operators will be machinist experts in operating the various milling machines, and therefore, a pool of experts and related equipment is formed. The functional layout can process a variety of products with different machines that perform the same function. Each product has a specific order of processes required for its production. These processes are performed at various work centers. A functional layout is depicted in Figure 2.9.

The application of a functional layout in a service organization is based on similar models. For example, a hospital's imaging department, which contains all kinds of imaging equipment, can have a functional layout. Patients undergo a series of examinations or procedures, in accordance with

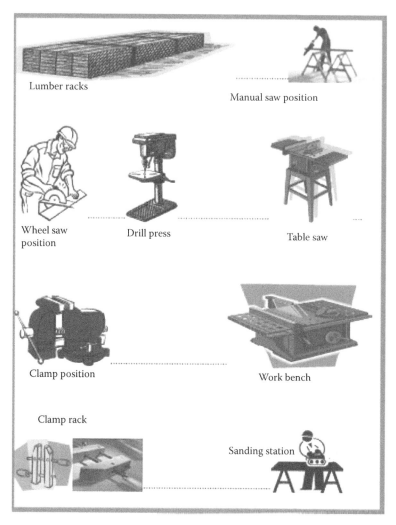

Lumber racks

Manual saw position

Wheel saw position

Drill press

Table saw

Clamp position

Work bench

Clamp rack

Sanding station

FIGURE 2.8
An example of a workshop layout for wood processing.

doctor's orders, performed by experts and equipment found in the imaging department.

Benefits of the functional layout are as follows:

- *Pooling of similar resources:* Similar to the functional organization, there is also a pool of similar resources in one location operating under a common management. As a result, management can better distribute the load among the resources; when one machine breaks down, a similar machine in the same work center can replace it.

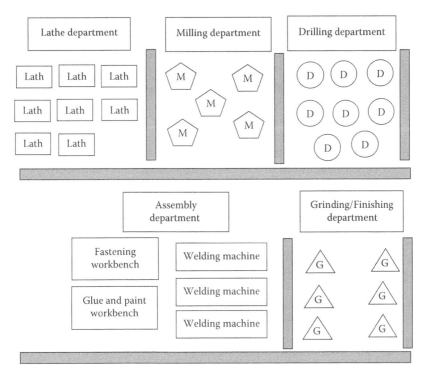

FIGURE 2.9
Functional layout (each shape represents a machine type).

- *Flexibility:* The functional layout can process a wide range of products. Each product will be routed to the proper work center in the right sequence of operations needed for its production (called routing). Product-specific routing to the appropriate machines and setup of each machine for the appropriate task ensure the flexibility of the functional layout.

Disadvantages of the functional layout are as follows:

- *Management is difficult:* It is difficult to coordinate the processing of many different parts routed in different ways through their production processes. Each product type requires special routing and processing—a special sequence of special operations in the work centers. When a large number of different products are produced at the same time, complex scheduling problems are created and the load on different machines tends to be imbalanced. As a result, some resources may be idle, while others will be overloaded, creating bottlenecks, queues will form in front of these bottlenecks, with products waiting to be processed.

- *Protracted setup time (1):* When a wide range of products is produced on general-purpose machines, it is necessary to setup the machines each time when switching from one product type to another. These setups take time and require well-trained and experienced operators.

- *Protracted setup time (2):* Converting machines from the production of one product type to another often takes a relatively long time, so products wait for long periods of time for handling. To save time, large batches of identical products are produced, thus requiring a single setup before each batch. However, the downside is long waiting times; sometimes products spend most of their time waiting and not being processed. Another disadvantage of large batches is inventory buildup as discussed in Chapter 10, which discusses procurement and inventory.

- *System complexity:* Because of the inherent complexity of the system, time spent on transportation of parts between machines, storing and retrieving parts, and waiting in line is very high. This causes long lead times and hurts the ability to compete in the market.

- *Everyone did it so no one is responsible:* In many cases, no one is responsible for the complete product, as products move between work centers.

An example of a functional layout workshop is found in the car-parts manufacturing department producing the body parts. Machines needed to perform operations such as metal cutting, pressing, bending, and drilling are grouped together. Parts are routed between the machines in the order required for their production. In this way, the department has the flexibility to produce parts for different car models. These parts are routed between work center departments in the order of operations required for each specific item.

Functional layout (job shop) with four departments is depicted in Figure 2.9.

2.4.3 Group Technology: Cellular Manufacturing

While a functional layout can serve a large variety of products, the material handling between the functional units is a real barrier to productivity. In a functional layout, every time a part has to be transferred from one process type to another, it has to wait with other products to be delivered to the next department, and then wait to be assigned to a specific machine and enter its waiting list. The excessive waiting is inefficient, and could be drastically improved if the quantities of similar processes justify their clustering into small, dedicated machine groups (cells). This approach, called group technology (GT), was created to solve this problem by enabling seamless flow of parts through the machining processes. Using machine cells creates a sequential flow of products in each cell, which is the most efficient pattern

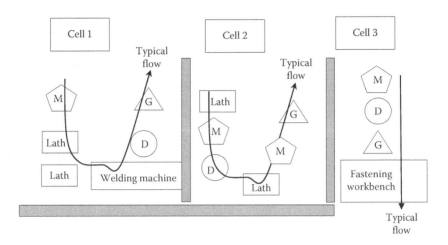

FIGURE 2.10
An example of a cellular layout in group technology (each shape represent a machine type).

of material flow. An example of three cells in GT formation is depicted in Figure 2.10. Each cell is dedicated to a family of products.

A cellular layout design is a hybrid model similar to the matrix organization model. The idea is to combine the workshop model and the flow shop model by dividing the products and parts manufactured by a facility into families of products with similar routing and to group specific machines and equipment used to process each such family into manufacturing cells. The result is a layout where parts and products with similar processing requirements are grouped into families processed by machines laid out in production cells (Heragu, 2008). The use of group technology leads to the flexibility similar to that of a functional system by facilitating the processing of a number of different product families. Within each cell, machines are ordered according to the routing of the family of products that require similar operations in a similar order.

Each cell has dedicated operators led by the cell supervisor, and therefore, each part and each product that are processed in the cell are "owned" by the cell supervisor. Due to the similar routing of the family of products assigned to a cell, management of the operation in each cell is easier and better planning and control can be achieved.

As an example, consider a furniture factory, which produces two types of furniture: Metal furniture and wooden furniture; some furniture is upholstered and some are not.

In the first cell, the metal furniture is processed. All the machines and equipment required for the processing of metal are grouped together including sheet-cutting machines, bending machines, welders, drills, etc.

In the second cell, the wooden furniture is produced, and therefore, all the machines that process wooden parts are grouped here, including saws, polishing machines, drilling machines, and assembly and painting equipment.

Finally, a third cell will specialize in upholstering furniture that can be made of wood or metal. This furniture will be transported to the third cell from the other two cells and will be processed in here because this cell has all the equipment needed for upholstering furniture.

By dividing the machines, equipment, and operators into three cells, the layout consists of three flow shops. Each shop has special machines and equipment along with the specialists and the cell supervisor required for the specific operations that it performs.

By applying the principles of group technology properly, manufacturing cell workers can easily manage the manufacturing process of the products produced in the cell and can specialize in the operations performed in the cell.

2.4.4 Flow Shop

Producing very large quantities of products requires significant specialization. Therefore, production specialization is correlated with the quantity of products to be produced. Thus, organizations that specialize in producing large quantities of products are characterized by a reduced variety of products, and usually their product offering has closely related production processes that enable producing them on a sequential flow line. Flow shop layout is based on the sequence of operations required for the processing of a particular product or a family of products with similar operations performed in the same order. Products are routed and produced on different machines laid out in the order required by their specific routing.

Benefits of the flow shop include the following:

- Movement of products on the production floor is made simple because equipment and machines are arranged according to the routing of the product, and therefore, it is easier to manage the process. This layout is very effective when the organization produces a single product in large quantities and at a high rate over time or when an organization produces a family of products with similar processing requirements and little setup when switching between them and, therefore, a layout that is based on the products routing is justified.

- When the processing times of the product on the different machines is about the same, it is possible to reach a point where the line is balanced—that is, all machines face the same load, and consequently, it is possible to minimize idle time of machines and other means of production.

- Workers in the flow shop are processing similar products in a similar manner repetitively and, therefore, the skill level required is limited and the new workers are riding the learning curve. The experienced workers already reached the flat part of the learning curve, where standard times apply.

• It is possible to assign clear responsibility for each production stage as products are produced by a well-defined set of machines assigned to the flow shop.

Disadvantages of the flow shop include the following:

• By definition, the products manufactured by the flow shop have the same (or similar) routing, and therefore, the range of products manufactured by the same flow shop is limited. New products can be added to and manufactured by the flow shop only if their routing requirements follow the layout of the flow shop.

• By its very nature, the flow shop requires considerable investment in machinery and specialized tools for each product or family of similar products with the same routing. Because the variety of products that can be processed in a flow shop is limited, to cover the required investment, a large number of units have to be produced.

• The flexibility of the flow shop to produce new products is relatively low. Figure 2.11 depicts a typical flow shop layout.

An example of a flow shop is an assembly line used by a car manufacturer. At the beginning of the line, the naked chassis starts to move along the line, along which other parts are assembled. At the end of the line, a complete car is ready. Each such line is dedicated to a family of similar car models that share the same operations in the same order (same routing).

Note: As we will see later, automation can increase the flexibility of flow shops. Automation helps mainly by reducing setup time, as part of the automation setup is done automatically in a very short time. Robots are a typical example of automation in a flow-based layout. Robots are used for assembly, welding, and painting of different car models on the same assembly line.

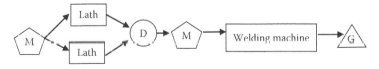

FIGURE 2.11
An example of a flow shop (the arrows represent the flow direction, the shapes represent machine types).

FIGURE 2.12
Examples of automotive assembly line work stations.

2.4.5 Assembly Line

As discussed earlier, assembly lines are a specific example of a flow shop. When mass production volume is required, the product has to be assembled, and product variety reduces to a single product (or several models of the same product), the assembly line becomes the most productive layout. Assembly lines can be single- or double-sided, or may even have a U shape, but their advantage is they have a sequential flow of materials (Figure 2.12).

2.4.6 Continuous Production Layout

Continuous production is a continuous process that usually characterizes chemical processing, plastic plant, or soft drink and other beverage production. In such processes, there is no human intervention at any intermediate stage (Figure 2.13).

FIGURE 2.13
Examples of continuous production processes.

2.5 Organization-Wide Processes

2.5.1 Overcoming the Organizational Structure and Walls for Better Competitiveness

Coordinating the work of different organizational units participating in the same processes in a functional organization is a difficult task. The difficulty stems from the fact that people in the same organizational unit tend to develop their own goals, use the same terminology, and have similar experiences, but such similarities do not exist between people belonging to different organizational units. A process that cuts across the border between different organizational units is in many cases termed "thrown over the wall" from one unit to the next, since each unit, as noted earlier, is a "silo" isolated from other organizational units. To improve the performance of processes in the organization, better coordination between organizational units that participate in the same process is required. Such coordination is achieved in the concurrent engineering new product development process and in the cellular manufacturing layout. Where a group of people from different functions take responsibility for the entire manufacturing process of a part or a product, be it new product development in concurrent engineering or the manufacturing of a product in cellular manufacturing; it takes a team and a team leader to achieve better coordination among all those involved in the process.

For example, consider a customer's order fulfillment and delivery process. To properly design this process, one starts by understanding customer needs. Suppose that a particular organization finds that the most important features to its customers is on-time delivery and minimum product cost. In a functional organization, we assume that a salesperson can get accurate data from the cost accounting system regarding the exact cost of a specific special product order, as well as accurate information from the operations people about the expected delivery time of the customer order. In reality, cost and delivery time change over time and the instantaneous value of both depends on the current load on the production floor and the actual cost of raw materials plus several other parameters.

Furthermore, in a functional organization, different functions may have only partial information. For example, decisions about procurement of raw materials and parts are made by procurement personnel who decide what to order, from whom, in what quantities, and for what due date. The assumption that operations can provide an answer to the question "When will the order be shipped?" is correct only if the procurement people update operations people as to the expected date of delivery of raw materials, and continually update this information, as it is changed by the suppliers. Clearly, there is a need for coordination among different functions, and a need for an information system that stores all relevant information and processes it, so that an accurate answer can be provided to questions such as "When will the order be delivered to the customer?" and "How much does it cost to produce the order?" It is the task of industrial engineers to design the right organizational structure, and to select and operationalize the information system that supports the team involved in each process. Industrial engineers are also responsible for maintaining the information system and the organizational structure and keep changing both in response to the dynamic environment and technology.

The delivery process—starting with understanding customer needs through procurement and delivery of work to subcontractors and production—is subject to uncertainty. Customers can change or cancel orders; forecasts of demand are subject to errors. Availability of production resources depends on a variety of factors: whether machines breakdown or malfunction, whether staff is at work, and whether the delivery times of suppliers is accurate. Quantity and quality are all subject to uncertain or random changes.

Once a well-structured process is defined, clear answers should be available to these questions:

- What to do to ensure that the process is successful?
- When to do it?
- How to do it?
- Who should do it?
- What information should be available, to whom and in what format?
- What performance measures should be used?

Answering these questions is the task of industrial engineers who conduct process design.

The transition to a process-based organization is an opportunity for a thorough examination of equipment and other resources. In some cases, the layout is also examined and changed as part of the organizational change.

The information system that supports execution, monitoring, and control of processes is also examined as part of the change. In the following chapters, we will discuss each aspect of organizational process design including the information system that supports processes and integrates the information throughout the organization—the enterprise resource planning (ERP) system. Industrial engineers involved in setting up ERP systems are engaged in the requirements definition, selection, implementation, and operation of these systems, and therefore, we will discuss the principles of information technology and how to design processes that integrate advanced information systems.

Varieties of resources are used in production and service processes. Choosing the most appropriate resources when planning a new process is part of the industrial engineer's task. To perform this task successfully, the industrial engineer should know the advantages and disadvantages of different types of resources. To select the right means of production or service, alternative resources should be compared. The tools for the analysis include deterministic models (models that do not take into account uncertainty, assuming that we can forecast the future with complete accuracy), stochastic models (where uncertainty is taken into account), simulation tools that simulate the planned system under different conditions, and tools for analyzing economic or financial aspects. Industrial engineering students study these different kinds of tools.

2.5.2 Automation

A major system design issue is determining the best level of automation for a given process. Using modern technology, it is possible to replace manual labor by automatic means such as industrial robots, computer-controlled machines, and computer-controlled production systems composed of machinery and means of conveyance managed by one or more computers (flexible manufacturing system [FMS] and flexible manufacturing cells [FMC]).

Each alternative has its advantages and disadvantages in terms of cost, quality, response time, and flexibility. The industrial engineer learns to compare the alternatives based on technological, operational, and economic considerations.

2.6 Summary

Chapter 2 discussed the organization, looking at several aspects:

- Organizations as human creations that aim to fill a specific need
- Organizations as transforming systems processing raw inputs to provide finished outputs
- Organizations as vehicles aimed at making economic gain
- Organizations as social entities, whose members work together to achieve a common goal
- Organizations as a collection of individuals with private goals
- Organizations as open systems that change the environment

The discussion focused on common models of organizations that provide a clear definition of responsibilities, authority, and reporting requirements.

Determining the structure of an organization is part of the management process. The organizational structure should support effective control over the operations performed by the organization, and efficient use of resources.

Several models of common organizational structures were presented:

- Project-oriented organizations
- Matrix organizations
- Functional organizations
- Process-based organizations
- Product-based organizations

Selecting the right organizational structure that fits the objectives and environment of the organization and enhances its ability to compete in the areas of quality, cost, time, and flexibility is a task frequently performed by industrial engineers.

Organizations are staffed by human beings and, therefore, the operational framework design must take into account the human factor, including organizational psychology, organizational sociology, and ergonomics. Specific issues such as learning (the learning curve model), motivation (Maslow theory, theories X and Y) are also important.

Planning the layout of organizational units, process stages, equipment, and other means of production are other aspects of the industrial engineer's role. The physical resources in an organization can be arranged in several ways, with the goal of efficient use of resources and material flow to provide a competitive advantage. There is some correlation between organizational structure and layout. Both are correlated with production volume and product variety.

The following types of layouts were presented:

- *Workshop layout:* Suitable for multiple projects and a job shop
- *Functional layout:* Suitable for a job shop and batch processing
- *Group technology layout:* Suitable when parts can be grouped into families each of which can be processed by a group of machines
- *Flow shop layout:* Suitable when parts/products can be processed sequentially (with possible bypass) on the line of machines
- *Assembly line layout:* Suitable for large volumes when there are a few different models of the same/similar product

These layouts are ordered in an increasing production volume and decreasing variety of product types.

References

Anzanello, M.J. and Fogliatto, F.S. 2011. Learning curve models and applications: Literature review and research directions. *International Journal of Industrial Ergonomics*, 41(5), 573–583.

Burton, R.M., Børge, O., and DeSanctis, G. 2011. *Organizational Design: A Step-by-Step Approach.* 2nd edn. Cambridge, MA: Cambridge University Press.

Heragu, S. 2008. *Facilities Design.* 3rd edn. Boca Raton, FL: CRC Press.

Louis, E.D. 1982. *Organization Design, Handbook of Industrial Engineering,* Chapter 2.1, New York, NY: Wiley Interscience, pp. 1–29.

Maslow, A.H. 1943. A theory of human motivation. *Psychological Review,* 50(4), 370–396.

McGregor, D. 1960. The human side of enterprise. Chapter 2, paper 14. In: Shafritz J.M., Ott J.S., Jang Y.S. 2015. *Classics of Organization Theory.* 8th edn. Boston, MA: Cengage Learning Publishing, pp. 154–160.

Ouchi, W.G. 1981. *Theory Z.* New York: Avon Books.

Robbins, S.P. and Judge, T.A. 2012. *Organizational Behavior.* 15th edn. Englewood Cliffs, NJ: Prentice-Hall.

Smith, A. 1776. *An Inquiry into the Nature and Causes of the Wealth of Nations.* London: A. Strahan & T. Cadell.

3

Project Management

Educational Goals

This chapter presents the problems that industrial engineers face when dealing with nonrepetitive effort. Examples of situations where this occurs are the development of a new product or service and the building of a new manufacturing or service facility. Project management models, tools, and techniques are the basis of the planning, execution, monitoring, and control of such nonrepetitive efforts.

We focus on the following aspects of project management:

- Definition of a project
- Dealing with uncertainty and risk
- Project life cycle
- Feasibility analysis of alternatives
- Structuring the work content and the organization performing the project
- Scheduling projects
- Assigning and managing resources
- Performing the project work while monitoring and controlling the project
- Computerized project management systems

3.1 Introduction

Industrial engineers fulfill a crucial function in managing nonrepetitive effort to achieve specific goals. Such efforts are organized as projects that, in most cases, are performed under time, budget, and resource constraints and often aim at developing a product, service, or system that must meet technical specifications and requirements. Some of the processes listed in the earlier chapters are managed as projects due to their nonrepetitive nature; the rest may be repetitive or nonrepetitive as explained next:

The five process types listed in Chapters 1 and 2 are as follows:

1. *Development process (nonrepetitive):* It starts with an idea for a new product or service and ends with the design of the new product or service and a working prototype. Since a product or a service can be developed only once, this is a nonrepetitive process and it is managed as a project.

2. *Infrastructure preparation (nonrepetitive):* The preparation process starts with a working prototype of a new product and ends with the successful completion and testing of the product's production facility. Since the infrastructure preparation is done once, this is a nonrepetitive process and it is managed as a project.

3. *Sales (either nonrepetitive or repetitive):* The sales process starts with market research and ends with an order from a customer. In some industries such as the food industry, this is a repetitive process, whereas in other industries, this is a nonrepetitive process as each product is unique and it is designed to satisfy specific needs and expectations of a specific customer (e.g., sales of ships by a shipyard).

4. *Supply (either nonrepetitive or repetitive):* The supply process starts with an order from a customer and ends with delivery and receipt of payment from the customer who received the requested products. In some industries such as the food industry, this is a repetitive process, whereas in other industries, this is a nonrepetitive process as each product is unique and it is designed to satisfy specific needs and expectations of supplies according to the contract with a specific customer (e.g., supply of ships by a shipyard).

5. *Service (either nonrepetitive or repetitive):* The service process starts with a customer request for service and ends when service is provided to the satisfaction of the customer. In some industries such as domestic appliances, this is a repetitive process, whereas in other industries, this is a nonrepetitive process as each customer requires a specific service (e.g., medical services for an area hit by an earthquake or another major disaster).

To summarize this point, it is clear that development processes are projects, given that a product or a service is developed once. Similarly, the infrastructure preparation process is nonrepetitive, and therefore, a project. The other three processes may be repetitive (and in this case are not projects), or nonrepetitive (and in this case they are projects). For example, contracting with a customer to provide a unique product or service in accordance with contract specifications is a nonrepetitive project. However, selling flight tickets to customers is a repetitive ongoing process and, therefore, is not considered a project. Organizations that sell products to customers may do so in a repetitive environment or in a nonrepetitive environment. In a similar way, a service is sometimes an effort to give a unique and one time only experience to the customer and, accordingly, it is nonrepetitive and managed as a project; for example, the management of the visit of the Pope to a foreign country is a project. Likewise, managing the relocation of a family is a project, and managing the Olympics is a project.

Throughout Chapter 3, we will illustrate the principles, models, tools, and techniques for project management using a variety of examples.

Production and service systems can be clustered into the following three groups:

1. Production and service systems supplying a large number of identical products or services to a large number of customers over a long period of time; for example, the mass production of bread in a bakery or milk cartons in the dairy. In such systems, which produce a single product at a high rate over time, or provide uniform service at a high rate over time, the main consideration is maximum efficiency and low cost while satisfying the required quality, often ceding flexibility.

2. Production and service systems supplying a limited variety of products or services in batches. These systems are designed to perform a variety of different production and service operations using pools of resources. Setup of resources is usually required when switching from one operation to another. The setup time and associated setup cost reduce the efficiency and increase cost but the flexibility is higher. The need to make adjustments (setup) in the transition from producing one product to another or providing one type of service to another is a major reason for losing efficiency.

3. Production and service systems that carry out projects. These are nonrepetitive undertakings requiring special resources and knowledge. In Chapters 1 and 2, we saw that project-oriented organizations enjoy maximum flexibility but, generally, resource utilization may be low; therefore, costs tend to be high.

3.1.1 What Is a Project?

A project is a one-time undertaking to achieve a set of objectives (such as cost, time, deliverables, and quality) under constraints. Project work content includes a series of activities carried out in a specific order. The following three types of constraints are common in projects:

1. Time-related constraints such as the required project start and end dates or specific dates (milestones) that specify ahead of time when deliverables must be ready.
2. Budget-related constraints such as the available budget and cash flow constraints.
3. Resource-related constraints such as the availability of personnel, equipment, and/or materials, during specific time periods or throughout the project.

In addition, environmental constraints, legal constraints, and political constraints may be preset, as well as project-specific constraints such as technological constraints.

Organization of work in the form of a project has several advantages. They are listed as follows:

- *Flexibility:* The project manager is able to use the project resources to achieve project objectives while satisfying the project constraints in an optimal manner. This includes changing plans in accordance with dynamic changes in the project environment.

- *A clear point of contact for the customer:* The project manager serves as a point of contact.

- *Dealing with uncertainty:* Managing uncertainty and risk are integral parts of project management. The nonrepetitive nature of projects reduces the availability of past information and the use of such information to improve planning. The knowledge gap created by the lack of information is a source of uncertainty and risk. The project organization is designed to handle such risks by using specific risk management processes as will be discussed later in this chapter.

- *Effective teamwork* based on the common goals and working procedures of the project team members who report to the same project manager—these project procedures include a clear definition of responsibilities and communication channels.

- The project team tends to create a *common language* among the team members that work on the same task to achieve the same goals, despite different backgrounds and education.

Not every nonrepetitive task justifies organization of work in a project form. As we saw in the previous chapters, when many small projects are performed simultaneously along with repetitive work, a matrix organization may be the preferred structure. When projects are smaller, performed in larger quantity, and require similar resources, the functional organization is the preferred structure. Project organization is usually better when the project is complex, important, and large enough to justify the expense of assigning resources full time to the project and developing a special organization for the mission.

Project organizations are typical in the construction and infrastructure industries. Projects such as the construction of a new shopping center, the development of a new oil field, or the construction of a new hospital are typically large and complex enough to justify a project organization.

In addition to the organizational structure, we can classify projects by their initiation process. Within this subset, the first class of projects is when an organization identifies a need and decides on the implementation of an internal project to satisfy that need. The second class is when an organization initiates a project due to a request by another organization that issues a request for proposal (RFP) and chooses the best bidder to execute the outsourced project. A third class is a project initiated and executed by an organization to meet the needs of customers outside the organization. The first and third classes are called internal projects, and the second class is an external project.

Other classifications of projects are as follows:

1. *Research and development (R&D) projects:* Developing a new product or service using new technology.
2. *Execution/construction projects:* Constructing a new facility or modifying a facility, for example, a facility used to manufacture a product or system or to provide a service.
3. *Improvement and maintenance projects:* Upgrading or maintaining the organization's assets, equipment, and facilities; for example, a marketing campaign is a project aimed at improving sales.
4. *Ad hoc projects:* To temporarily improve a situation, for example, a fun day for the employees, a cleaning project, or painting/whitewashing walls.

Projects that fall into these classifications are differentiated by their rationale and level of uncertainty. Uncertainty is present in most projects due to their nonrepetitive nature and the limited ability to collect information needed for planning based on past experience. This uncertainty may impact the ability to estimate the time, resources, and cost required for the implementation of the project. In R&D projects, technological uncertainty may play a major role. Many projects combine elements of research and development with elements of manufacturing and construction.

3.1.2 Uncertainty and Risks

As explained earlier, projects are subject to uncertainty due to their one-time nature and the resulting lack of historical information to support future decisions. Special planning and control methods were developed for project management. These planning and control methods depend on the following alternative assumptions regarding uncertainty:

- *There is negligible uncertainty:* When the level of uncertainty is relatively low, commonly, predictions are made regarding how long each project activity will take, how much it will cost, and how many resource units will be required to perform it. If the actual level of uncertainty is not very low, "the usefulness of plans" based on these assumptions is limited, as their ability to accurately predict the course of the project will be low.

- There is significant uncertainty that can be assessed correctly and taken into account in planning the project, evaluating its cost, and predicting its duration. This is a very difficult task. Not only does the uniqueness of a project make it hard to assess its uncertainties, the frequent strong correlation between uncertainty in time and uncertainty in cost must also be taken into account. The longer the duration of the project, the more some of the costs increase (e.g., the cost of full-time employees, rent of facilities, and leased equipment).

Models for planning and control are selected based on the sources and the level of uncertainty. For example, when uncertainty is low, assuming that the task durations are following the beta distribution enables using the program evaluation and review technique (PERT) model (to be discussed later in this chapter), whereas higher uncertainty may require using simulation.

Sources of uncertainty include the following:

- *Availability of resources:* It is difficult to predict the availability of resources because unexpected changes in their availability may occur. For example, the availability of human resources is very often subject to fluctuation as a consequence of sickness, resignation, or assignment of the needed person to another task. Materials may be in short supply and equipment may break down or be in short supply for some reason. Lack of resources can cause delays and cost overruns in a project. As a result of uncertainty, a need for additional resources or other resources than those planned for may arise, and such resources may not be readily available.

- *Uncertainty in the environment:* This uncertainty is present in all types of projects and refers to uncertainty related to such factors as weather, market condition, changes in laws and regulations, changes in political conditions, changes in the economy, etc.

- In addition to the above sources of uncertainty that are common to all projects (including lack of past information due to the one-time nature of projects), R&D projects are especially susceptible to *technological uncertainty* as well.

- *Technological uncertainty:* This uncertainty is common to development projects based on new technology, and knowledge that does not exist within the organization. The time to develop new technologies and the probability of success are both uncertain.

It is possible to deal with uncertainty and the risk that it generates in projects in different ways:

1. Accept the risk and treat it if it materializes. This is typical of risks that are associated with low damage and have a low probability of occurring. This is reactive risk management. Reactive risk management is an effort to deal with problems caused by risks that have already occurred. For example, project managers may be aware beforehand that there is a chance that a building project may run out of a resource but they will also know that it might be possible to get the needed resource from a local, more expensive supplier who has inventories of such material on hand and can supply the resource immediately from his inventory. In this case, the cost might be the difference between the cost of the local expensive supplier and the cost of the less expensive but late supplier.

2. *Transfer the risk:* Usually, the risk is transferred to insurance companies. This is typical of risks that are associated with large damage but slim probability. For example, insuring construction workers against injuries transfers the risk of an accident to the insurance company.

3. *Share the risk:* This includes not only sharing the ownership on the project, but also its financing. Taking loans from a bank to finance projects effectively shares the risk of the project with the bank. Sharing ownership means spreading the risk across partners.

4. *Reduce the risk:* Proactively reduce the risk in an effort to identify possible risks and protect the project against those risks as part of the project plan. An example is an effort to protect a construction project against the risk of delayed delivery by a supplier of building material. The protection can be in the form of ordering the building material earlier than needed, and keeping it in inventory until it is used. If the supplier ships the order late, having such inventories at the project's end can protect the project from delays. Most proactive risk management activities have costs. In the above example, the cost of carrying inventory and the cost of money spent earlier than needed must be traded off against the cost of delays.

The decision regarding the way to deal with risk is based on statistical analysis of the risks and the cost of managing them.

3.1.3 Project Life Cycle

The need to deal with the constraints affecting projects and the inherent uncertainty in most projects resulted in the development of project management methodologies comprising a variety of tools and techniques. Most of these tools and techniques are based on models. One example is the project life cycle model that presents the project as a series of steps, also called phases, and recommends specific management actions in each phase. There are several such models. Next, we review a basic version of this model, which divides the project duration into four phases performed in sequence.

- *Project initiation:* This is the first phase that starts with the recognition that there is a need or a problem, and a project may be needed. After identification of the need, feasibility studies are conducted to identify possible solutions. If possible solutions are identified, they are analyzed to select the best solution. The analysis is multidimensional. Technology and resources, economics and finance, time, and marketing and sales dimensions of each possible solution are analyzed. The risk associated with each possible solution is also estimated. The analysis focuses on finding the best solution. When a solution is selected based on the above analysis, two documents are prepared: *Specifications* or *specs*, which describes the end product, and *Scope of Work* or *Statement of Work* (*SOW*), which describes the contents of the work to be performed within the project. The SOW specifies the work that needs to be done and the deliverables that the project will produce. At the end of the initiation phase, both the specifications and the SOW are ready.

- *Project planning phase:* The specifications and the SOW are the basis of a plan that will serve as a road map for execution of the project. Planning has several dimensions including time, budget, and resources. The estimated time to perform the tasks that comprise the project work content as specified in the SOW is used as a basis for estimating the time for the entire project. Project planning must keep in mind budget and resource constraints when deciding when the work content will be performed and by what resources. The planning phase also focuses on the organizational structure of the project, on who will participate in the project, and what system of reporting and control will be used throughout the project. The result of this phase is a project plan detailing what should be done (detailed work packages), when it should be done (schedule), using what resources, and how much it is expected to cost (detailed budget).

- *Project execution:* The project plan is the roadmap for execution—implementation of the plans and production of the project deliverables. Due to uncertainty and risk, tracking and monitoring the implementation is important. The goal of tracking and monitoring performance, schedule, and budget is to identify any deviations from the plan, as early as possible, and adjust the program in response to unexpected events. During the execution phase, resources are allocated for performing the project plan (resources execute the work content to produce the desired deliverables adhering to the schedule as much as possible).

- *Project termination:* The end of the project requires releasing of resources, transfer of project deliverables to the customer, training the customer in operation and support of the delivered products and systems, preparation of documentation, and closing of the project accounts. It is very important to learn from any project how to improve the management of similar projects in the future. Lessons learned in the form of best practices and improved tools and techniques are the basis of ongoing improvements, in the nonrepetitive project management world.

A detailed review of the initiation stage in the project life cycle is presented in Section 3.2.

3.2 Project Initiation

3.2.1 Gathering Information

Projects are performed to satisfy the needs and expectations of stakeholders. Stakeholders are individuals and organizations that are related to or interested in the project for any reason. The obvious stakeholders of all projects are the initiator of the project (and his or her organization), the project manager, the customers, the project team, and the project suppliers and subcontractors. While it is important to analyze the need and attitudes of these stakeholders, it is crucial to identify other stakeholders. Most stakeholders are affected by the problem the project is supposed to solve or by the proposed solution—or by both (e.g., neighbors, municipality, customers, suppliers). However, there may be other stakeholders motivated by beliefs and ideology, politics, opportunism, or personal reasons. Gathering information about stakeholders is a crucial step during project initiation. Gathering information is relatively easy when stakeholders state their needs. It is more difficult when they have needs and expectations from the project that are not stated openly. It is important to analyze the stakeholders early on, to understand their needs and expectations as well as their ability to influence the project. Understanding stakeholders' willingness to object or support the project is a key for addressing them successfully. In some projects, the only way to collect information on potential stakeholders is by using a third party. For example, if the stakeholders are the customers, gathering useful information may require market research, performed by a marketing company specializing in market studies. Information is collected to answer questions such as who are the potential customers, what are the needs and expectations of these customers with respect to the problem the project tries to solve, or the product the project is supposed to deliver. Information is also needed with respect to competitors in the market, the suitable technology, the economic implications of implementing the project, and the risks involved.

In some projects in which an organization is trying to find a solution to a specific problem—not within its specialization—a request for information (RFI) is published asking potential suppliers to examine the problem, and to propose ways to solve it. Some projects are outsourced by issuing a request for proposal (RFP). In the RFP, the customer specifies the work content, technical details, time constraints and budget constraints, and all other information needed to enable potential bidders to prepare a detailed proposal for the project. Potential bidders get the RFP and decide whether to submit a proposal.

The information gathered during the initiation phase is used to support the following decisions:

- To do or not to do the proposed project?
- What technological and operational alternatives to use?
- What should the project life cycle phases and its goals be?
- What is the work required during each phase in the life of the project?
- What relevant documents such as standards are applicable to the project?
- What specifications of deliverables (or specs) should be prepared during the initiation phase? The spec document contains a detailed description of the technical requirements of the product, a description of its main features and functions, technical characteristics derived from the main requirements of the deliverables, including required performance and environmental conditions in which the product should operate, and the tests required to ensure compliance with these requirements.
- What is the desired timetable? The answers to this question take the form of a master schedule detailing the dates on which the different phases of the project should be carried out, and the required schedule for major deliverables.
- What is the desired budget? This is detailed in the form of cost breakdown at some level of detail of major cost components such as the cost of human resources, material, and overhead. In addition to the detailed expenses, the desired budget should include the sources of funds as well.

- What are the applicable organization's policies (e.g., FDA approval of supplies in case of medical business, or a security clearance in the case of a defense industry project) and/or restrictions?

The purpose of the information gathered during the initiation stage is to prepare the foundation for detailed planning and execution. The project manager is also selected at this time.

3.2.2 Selection of Alternatives within the Project Scope

In many projects, several technological and operational alternatives exist. An example of technological alternatives is building material. Will wood or cement blocks be used for the building of a new family house? An example of operational alternatives is the choice between using a main contractor who will manage the whole building project or self-managing it using the help of subcontractors. These alternatives should be considered in the initiation phase. It is important to remember that at this stage one alternative is always to abandon the project and not to start it at all (also known as the go/no go decision). There are many reasons for not starting a project, ranging from technological difficulties to competition in the market, lack of financial resources, or lack of economic attractiveness. All these considerations should be taken into account early on during the project initiation phase, based on the data collected.

The analysis starts by transforming the information about stakeholders into a set of criteria and the relative importance or weight of each criterion. For example, in the house-building project, some possible criteria are listed as follows:

- Style and look
- Required maintenance
- Project cost
- Cost of heating and cooling the new house
- Expected life of the house

There are four combinations of alternatives in this example:

1. Using a main contractor and building a wooden structure
2. Using a main contractor and using cement blocks
3. Self-managing the building of a wooden structure
4. Self-managing the building of a structure using cement blocks

A simple model for the selection of alternatives is known as the weights and scores model. Criteria are ranked according to their relative importance

TABLE 3.1

Criteria and Their Weights

Criteria	Relative Weights
Style and look	0.30
Required maintenance	0.20
Cost of the project	0.20
Cost of heating and cooling the new house	0.10
Expected life of the house	0.10

to the stakeholders and each alternative is ranked with respect to each criteria. The weighted sum of the scores is used as a measure of the value or benefit to the stakeholders of each alternative.

Table 3.1 presents the relative weights of the criteria associated with the house-building example.

Next, each alternative gets a score between 0 and 100 with respect to each criterion. The higher the score, the better is the alternative with respect to that criterion:

1. Main contractor building a wooden structure (Table 3.2)
2. Main contractor building a structure using cement blocks (Table 3.3)
3. Self-managing the building of a wooden structure (Table 3.4)
4. Self-managing the building of a structure using cement blocks (Table 3.5)

By multiplying the score of each criterion by the weight of the criterion, a weighted score is calculated for each alternative (Tables 3.6 through 3.9).

The two combinations with the highest value are self-managing the building of a structure using cement blocks and main contractor building a wooden structure. The weighted score of both is 74 out of 100. Naturally, only one can be the final choice. To make this selection, the risk of each alternative is evaluated. Since using a main contractor reduces risk to the stakeholders, given that the contractor shares the risks with them, the main contractor building a wooden structure alternative is selected.

TABLE 3.2

Scoring for Main Contractor Building a Wooden Structure

Criteria	Score
Style and look	90
Required maintenance	70
Cost of the project	90
Cost of heating and cooling the new house	90
Expected life of the house	60

TABLE 3.3

Scoring for Main Contractor Building a Structure Using Cement Blocks

Criteria	Score
Style and look	80
Required maintenance	80
Cost of the project	70
Cost of heating and cooling the new house	80
Expected life of the house	90

TABLE 3.4

Scoring for Self-Managing the Building of a Wooden Structure

Criteria	Score
Style and look	80
Required maintenance	70
Cost of the project	100
Cost of heating and cooling the new house	90
Expected life of the house	50

TABLE 3.5

Scoring for Self-Managing the Building of a Structure Using Cement Blocks

Criteria	Score
Style and look	80
Required maintenance	80
Cost of the project	90
Cost of heating and cooling the new house	80
Expected life of the house	80

TABLE 3.6

Weighted Score for Main Contractor Building a Wooden Structure

Criteria	Weight	Score	Weighted Score
Style and look	0.3	90	27
Required maintenance	0.2	70	14
Cost of the project	0.2	90	18
Cost of heating and cooling the new house	0.1	90	9
Expected life of the house	0.1	60	6
Total score			74

TABLE 3.7

Weighted Score for Main Contractor Building a Structure Using Cement Blocks

Criteria	Weight	Score	Weighted Score
Style and look	0.3	80	24
Required maintenance	0.2	80	16
Cost of the project	0.2	70	14
Cost of heating and cooling the new house	0.1	80	8
Expected life of the house	0.1	90	9
Total score			71

TABLE 3.8

Weighted Score for Self-Managing the Building of a Wooden Structure

Criteria	Weight	Score	Weighted Score
Style and look	0.3	80	24
Required maintenance	0.2	70	14
Cost of the project	0.2	100	20
Cost of heating and cooling the new house	0.1	90	9
Expected life of the house	0.1	50	5
Total score			72

TABLE 3.9

Weighted Score for Self-Managing the Building of a Structure Using Cement Blocks

Criteria	Weight	Score	Weighted Score
Style and look	0.3	80	24
Required maintenance	0.2	80	16
Cost of the project	0.2	90	18
Cost of heating and cooling the new house	0.1	80	8
Expected life of the house	0.1	80	8
Total score			74

3.3 Project Scheduling

The statement of work (SOW) defines the project's work content. To perform this work content, resources are needed. Human resources such as engineers, technicians, and scientists are needed for R&D projects, while heavy equipment such as trucks and tractors are needed for construction projects along with their operators. Due to the nonrepetitive nature of projects, the type and quantity of required resources must be estimated, as well as the time when the resources are needed. Timing is an important consideration in this process. Consider, for example, essential material needed for construction projects, which is ordered from suppliers. It takes time for suppliers to process the order, prepare it, and deliver it. The total of these times is known as lead time. Some materials are available off-the-shelf and then lead time is simply the transportation time. Other material must be prepared according to the engineering design and the lead time might be much longer. It is important to get the required resources when needed as having resources too early when the material or work space is not ready can create idle time of machines and human resources, while getting material too early may necessitate expensive storage of material. Scheduling is an effort to synchronize project activities and resource requirements in order to determine when each activity will be performed and when the resources required to perform it should be ordered and/or delivered.

In order to schedule the project, the SOW is decomposed into the project activities. Assuming that each activity has a known work content and it is possible to estimate the duration of the activity (the time it will take to perform it) based on the resources allocated to it, a project schedule can be developed based on the precedence relations among activities. In most projects, there are precedence relations among activities; for example, in a construction project, it is necessary to design the building and to get permission to build it before construction work can start.

The list of activities, estimated durations, and the precedence relations among activities is the basic information required for project scheduling.

3.3.1 Project-Scheduling Models

There are several project-scheduling models. We shall review some of these models in detail and illustrate the simplest models using a construction project scenario. This project contains two main components that are executed in parallel. The first component is the construction of a water pumping system, and the second component is the construction of a piping system to deliver the pumped water.

The project activities (of the example) are listed as follows:

A: Selecting a contractor and signing a contract to purchase and install the pumping facility

B: Selecting a contractor and signing a contract to purchase and install the pipeline

C: Preparing the pumping facility infrastructure

D: Preparing the pipeline infrastructure

E: Installing the pumping facility

F: Installing the plumbing

G: Operations and testing

TABLE 3.10

Activities' Precedence Relations and Durations for the Example

Activity	Predecessors	Duration (Weeks)
A	–	2
B	–	3
C	B, A	3
D	B, A	5
E	D, C	1
F	D, C	3
G	F, E	1

Activities A and B (selecting a contractor and signing a contract) are performed in parallel. Activities C and D (preparing the pumping facility infrastructure and preparing the pipeline infrastructure) cannot start before Activities A and B are finished. As soon as Activities C and D are finished, Activities E and F (installing the pumping facility and installing the plumbing) can start. Activity G (operations and testing) can start only after Activities E and F are finished.

Table 3.10 lists the predecessors and duration for each activity.

In Table 3.10, the project activities appear in the left-hand column. In the middle column, activities that need to be finished before the activity in the left column can start (predecessors) are shown. The right-hand column shows the estimated number of weeks to complete the corresponding activity (the duration of the activity).

While many project-scheduling models exist, we present only two models and leave the rest for more advanced textbooks. Data in Table 3.10 are used in two simple scheduling models: (1) the Gantt chart developed by Henry Gantt in 1916 and (2) the critical path method (CPM), developed in the late 1950s.

3.3.2 Gantt Chart

The Gantt chart is a diagram showing the planned start time and the planned finish time of each activity based on the estimated duration and precedence constraints. The horizontal axis is the time axis, and the vertical axis shows project activities. Thus, each activity has a full timeline at its corresponding vertical height.

The following assumptions are the basis of the Gantt chart:

- All project activities are known.
- There are no resource or cost constraints.
- The duration of each activity is known and there is no uncertainty.
- All activities should start as soon as soon as possible.

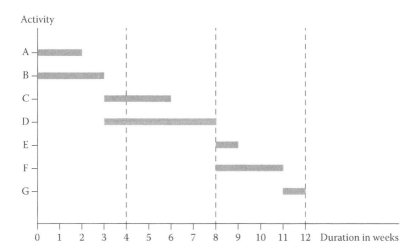

FIGURE 3.1
Gantt chart for the example in Table 3.10.

The Gantt chart shows how long the project will take, taking into account the activities, their duration, and the precedence constraints. Note that sometimes the assumptions underlying the Gantt chart model are not valid. For example, it assumes early implementation or early start of activities (budget-wise this is a drawback because the aspiration is to spend money as late as possible; likewise, it is a waste to install a product component earlier than needed). Also, it assumes that there is no resource constraint while in many projects resource availability is limited.

The Gantt chart in Figure 3.1 is based on the data in Table 3.10. It shows that the project will take 12 weeks.

3.3.3 Critical Path Method and Network Models

The CPM is a scheduling model, which like the Gantt chart, assumes that activity durations and precedence relations are known. This method was developed in the late 1950s for the purpose of scheduling the construction of large chemical plants. Unlike the Gantt model, the CPM does not assume that all activities should start as early as possible. Generally, this method is applied using a network project model, for example, in which each activity is represented by a network node and each precedence relation is represented by an arrow or an arc. Each arrow starts at the preceding activity and points to the succeeding activity.

CPM model assumptions are as follows:

- All project activities are known.
- There are no resource or cost constraints.
- The duration of each activity is known and there is no uncertainty.

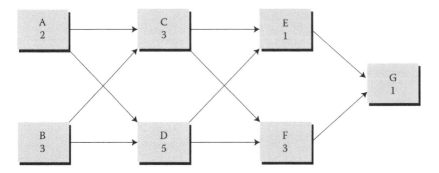

FIGURE 3.2
Precedence diagram for the example of Table 3.10.

The model focuses on the critical path of the project, which is the longest sequence of activities connecting the start of the project to its end. Using this method, the critical activities or activities that cannot be delayed without delaying the project are identified. Noncritical activities can be delayed without delaying the project.

Network CPM analysis is performed in two passes:

1. The forward pass: From the start of the project to its end, this pass calculates for each activity the earliest start time and the earliest finish time, taking into account the precedence relations, which is a very similar analysis to the Gantt chart.

2. The backward pass: From the end of the project to its start, this pass calculates for each activity the latest finish time and the latest start time, taking into account the precedence relations.

The example used to illustrate the Gantt chart is used next to illustrate the CPM method. In Figure 3.2, each node (rectangle) represents an activity. The activity's name and duration are presented inside the corresponding node.

Each arc or arrow represents the precedence relations between its starting node (the preceding activity) and its ending node (the succeeding activity).

3.3.3.1 CPM Analysis

For each activity, we calculate the earliest start time possible, without violating the precedence constraints. This time is marked as the Early Start—ES. The earliest start time of all operations that have no predecessors is assumed to be zero. These activities, which may start at time zero, have a known duration and, their Earliest Finish—EF—is their duration. For each of the activities that have one or more predecessors, the largest EF of their predecessors becomes their early start (ES) and adding their duration to their ES, we get their Early Finish—EF. We continue in iterations of calculating ES and

EF for the successor activities of the succeeding activities. The earliest start time of activities with predecessors is the latest earliest finish time of all its predecessors. The earliest finish time of every activity is equal to the earliest start time of the activity plus its duration.

If we assume that the project starts at time zero and the unit of time is a week of work (i.e., holidays or vacations are not present on the time line—the X axis), the resulting schedule is known as an ARO (after receiving an order) schedule. Using these simple rules, the *ES* and *EF* times of the ARO of each activity can be calculated as part of the forward pass:

$ES\ (A) = 0,\ ES\ (B) = 0,\ ES\ (C) = 3,\ ES\ (D) = 3,\ ES\ (E) = 8,\ ES\ (F) = 8,$
$ES\ (G) = 11$

$EF\ (A) = 2,\ EF\ (B) = 3,\ EF\ (C) = 6,\ EF\ (D) = 8,\ EF\ (E) = 9,\ EF\ (F) = 11,$
$EF\ (G) = 12$

Next, the backward pass from the end of the project to its beginning is performed. This time we first calculate the latest finish time of each activity, which is the latest time an activity can be finished without delaying the project, that is, the project still finishes in the earliest possible time. We denote the Late Finish as *LF* and Late Start as *LS*. The *LS* time of each activity is also calculated in the backward pass: the *LS* is the latest time an activity can start so that the project still finishes in the earliest possible time.

Based on the above assumptions, the *LF* of the last activity is equal to its *EF*.

The late start *LS* of each activity is calculated by subtracting the duration of the activity from its late finish time. (This is the latest time that the activity can start without delaying the project.)

For all activities that have successor activities, the *LF* of each such activity is equal to the earliest *LS* of the succeeding activities.

Using these simple rules, the *LS* and *LF* times of the ARO of each activity can be calculated as part of the forward pass:

$LS\ (A) = 1\ LS\ (B) = 0\ LS\ (C) = 5\ LS\ (D) = 3\ LS\ (E) = 10\ LS\ (F) = 8\ LS\ (G) = 11$
$LF\ (A) = 3\ LF\ (B) = 3\ LF\ (C) = 8\ LF\ (D) = 8\ LF\ (E) = 11\ LF\ (F) = 11\ LF\ (G) = 12$

It is convenient to arrange the data of each activity as shown in Table 3.11.

Using this convention for each activity and calculating the values yields a network model with all the data (Figure 3.3).

TABLE 3.11

The Convention of Depicting
Values of ES, EF, LS, and LF

ES	EF
LS	LF

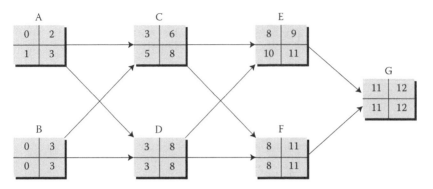

FIGURE 3.3
Activities on nodes (AON) network depicting values of ES, EF, LS, and LF for each node.

3.3.4 Critical Path Network Analysis

The difference between the latest start time and the earliest start time of each activity is called the total slack of the activity. This is the amount of time the activity can be delayed and the project will still finish as early as possible. The total slack of an activity is also equal to the difference between the late finish and the early finish times of that activity. The slack of critical activities is zero and the critical path is defined as the longest path of activities connecting the start of the project network to its end. In our example, the critical path is G, F, D, and B and its length is equal to the project duration of 12 weeks.

3.3.5 Uncertainty and Project Duration

The planned durations of the activities in a project are typically rough estimates with a considerable amount of uncertainty. When the uncertainty is small, the mean or expected times of the activities could be used for the CPM durations. However, when the uncertainty is large, other techniques are necessary for planning and analyzing the schedule. A prominent technique is the PERT. However, PERT has many deficiencies. We, therefore, advocate the use of Monte Carlo simulation for analyzing a project schedule. This method is taught in advanced industrial engineering courses. In this method, the distribution (spread) of each activity's duration is estimated, and the simulation chooses random numbers related to these distributions to calculate a sample of the project duration. Running the project simulation thousands of times creates thousands of project samples. By using these samples, reliable statistics are gathered regarding the population of these project samples, and the uncertainty of projects' duration can be characterized.

3.3.6 Resource Scheduling

In some situations, there are project activities that cannot be executed in parallel even though there are no precedence constraints between these activities. This is a result of constraints such as resource constraints. For example, the company engaged in road construction has two road rollers, but it has begun a project to pave three different roads. Although the three roads are independent and can be paved simultaneously, in practice, only two roads can be paved at the same time due to the limited number of road rollers.

Since resources are limited in many real projects, the next step in our analysis is to introduce resource constraints into scheduling considerations.

Consider the same project example with the additional information that a resource type, "workers," is required to perform each activity. The number of "workers" needed depends on the activity as summarized in Table 3.12.

To analyze this situation, it is convenient to use a graphical model. For each resource, a chart called the resource profile is created. This chart shows the required amount of each resource as a function of time. The resource profile is created by adding, period by period, all the requirements for the resource by all activities scheduled for that period (see Table 3.13 and Figure 3.4). The resource profile can be compared to the number of resource units available. When the number of available resource units is larger than the number of units required for the period, resource idle time is created. When the number of available resource units is smaller than the number of units required for the period, a shortage of resource problem is created and must be solved.

To solve the resource shortage problem, one can choose between several alternatives. Two important alternatives are as follows:

1. Acquisition of additional resources (hiring people, purchasing machines, using subcontractors)
2. Delaying some activities from periods of resource overload to periods when resources are idle

TABLE 3.12

Number of Workers Required for Each Activity

Activity	Predecessors	Duration (Weeks)	Workers
A	–	2	2
B	–	3	1
C	B, A	3	3
D	B, A	5	2
E	D, C	1	1
F	D, C	3	2
G	F, E	1	3

TABLE 3.13

Total Number of Workers per Day Required by the ES
Schedule

Week	Activities Performed	Workers Required
1	A, B	3
2	A, B	3
3	B	1
4	C, D	5
5	C, D	5
6	C, D	5
7	D	2
8	D	2
9	E, F	3
10	F	2
11	F	2
12	G	3

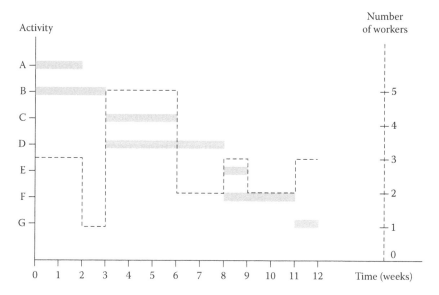

FIGURE 3.4
Early start Gantt chart along with the resource profile.

In most cases, additional resources are not available immediately due to the
"lead time"—the time it takes to acquire additional resources. Furthermore,
hiring and firing as well as purchasing and selling machinery might be quite
expensive. Therefore, it is desirable to keep the resource level as constant as
possible. In many projects, an additional scheduling objective is to create

a situation where demand for resources is uniform throughout the project. Please note that in reality, a completely balanced resource profile is almost impossible to achieve.

We illustrate the construction of the resource profile by using the last example. We start by summing up the number of workers needed to perform all the activities performed in each time period according to their earliest start times (Table 3.13 and Figure 3.4).

Figure 3.4 shows the early start Gantt chart along with the resource profile.

3.4 Implementation Phase—Project Execution Monitoring and Control

3.4.1 Monitoring and Control

The implementation of the project can start after the detailed project plan is ready and each participant knows what activities are assigned to him or her and the schedule for these activities. During the implementation phase, the project manager should monitor the actual progress of the project to ensure that it advances according to plan. Due to unplanned events and randomness, deviations from the project plan are common. In some cases, new information may become available during implementation, and the project plan can be updated accordingly.

In many cases, no new information indicates that the project plan has to be updated, and yet a deviation from the project plan takes place. It is the project manager's task to identify such deviations as early as possible and to take the necessary corrective actions to bring the project back on track.

The process of monitoring actual progress, comparing it to the plan to identify deviations early on, and taking corrective action is an important part of project management. The main goal of this process is to ensure that despite uncertainty, the project achieves its objectives, and does not violate time, budget, and resource constraints. There are numerous sources of uncertainty in projects. For example:

- Absenteeism as a result of illness, accidents, personal issues, errands, etc.
- Unpredictable events such as blackouts, floods, strikes, and accidents
- Knowledge gaps regarding the technology that is used
- Planning based on imperfect estimates of activity duration, resource availability, and costs
- Unexpected breakdowns of machines
- Changing conditions and regulations

Monitoring and control is based on a continuous comparison between the project plans and the actual results. When deviations are detected between planning and implementation, the project manager has to decide whether to take corrective action. Possible corrective actions include changes to the work content and specifications of the project deliveries, time extensions, additional resources, budget additions, replacement of certain resources, etc. Monitoring and control is part of project risk management and requires constant awareness by the project team of events that may affect the project, and of deviations between project plans and actual results.

3.4.2 Testing

Tests are an important part of project control. These tests are done to ensure project deliverables meet the requirements. Some tests are conducted toward the end of the project, but in many projects, tests are performed early on, as part of the implementation phase. The early tests are performed during the implementation phase in order to avoid late detection of major problems toward the end of the project. In some projects, the ability to test the quality of deliverables may be limited by some operations, and therefore, the tests must be performed prior to such operations. Painting is a typical example. It may be impossible to test a surface for defects once it is painted.

3.4.3 Project Ending

This is the last phase of the project and overseeing it is the responsibility of the project manager. The point where the final phase begins is hard to define exactly. A good sign that the project is winding down is the release of resources that are no longer needed and can be assigned to other projects. Another sign is the acceptance of the required deliverables by the customers. As the project begins to end, several things should be done:

1. Orderly release of the resources allocated to the project. The organization should decide which resources—personnel, equipment, and material—will be released and when. Typically, most resources are redeployed, but some resources will no longer be needed by the organization so people will be laid off and machinery sold. Other resources may be allocated to functional units or to other projects and will continue to serve the organization.

2. The project deliverables should be delivered to and accepted by the customers. The deliverables may include a combination of hardware, software, data, and services. Examples of data items are the results of the tests conducted and the literature required for maintenance and operation of products. Examples of services are help provided in the commissioning of the delivered products and warranty services.

3. Ensure that the organization learns from the project. Analyze what happened, what went wrong and why, and what was successful and why. This is important not only if the organization plans to perform similar projects in the future but also as a basis for improving the project management methodology used by the organization. The information obtained from the analysis of the project must be collected, organized, and stored, to enable easy retrieval as a basis for lessons learned. These lessons should be remembered in order to improve the organization's performance when carrying out projects in the future.

4. If the project developed a new product designed to replace an existing product, the logistic support for the new product has to be implemented, including spare parts, maintenance facilities, and documentation. Furthermore, the supply chain for the new product should be in place before the project ends. Preparations needed to remove the existing product from the market, and the introduction of the new product should be made.

3.5 Computerized Systems for Project Management

As in most areas of industrial engineering, today the tools, techniques, and models developed to support project management are computerized or automated. These may be available in specialized software packages. As already discussed, the success of the project depends on many parameters and on large amounts of data. Project management software is designed to collect data and to process it into useful information that can support decisions and enable better planning and control of projects.

A large variety of project management software tools exist, and these tools are used by many organizations. These tools allow organizations to work with large amounts of data and information to support project planning, monitoring, and control. Software packages available on the market such as Microsoft-project and Oracle Primavera enable proper integration between the various elements of project management such as time, cost, and resources, as well as support communication between the project manager, his or her team, and the stakeholders. Most software packages support graphical models, such as a Gantt chart, network diagrams, and resource profiles.

3.6 Summary

Many organizations undertake one-time missions to achieve specific goals. Development of new products and services or establishing new facilities is a typical example. Because of uncertainty associated with the lack of histori-cal information, special tools and techniques are needed for these types of tasks. The tools presented in this chapter are used for initiating, planning, monitoring, and controlling projects. More sophisticated tools for managing risks and integrating requirements with scope, time, and costs are based on advanced math and statistics and are usually taught in advanced courses on project management. Industrial engineers are very often involved in projects, and must master the tools for planning, controlling, and managing projects.

Further Reading

Nicholas, J.M. and Steyn, H. 2012. *Project Management for Business Engineering Business and Technology.* Portsmouth, NH: Butterworth-Heinemann.

Project Management Institute. 2008. *A Guide to the Project Management Body of Knowledge: PMBOK(R) Guide.* 5th edn. PA, USA: Project Management Institute (PMI).

Shtub, A.F., Bard, J.F., and Globerson, S. 2004. *Project Management: Processes, Methodologies, and Economics, 2/E,* Upper Saddle River, NJ, USA: Prentice Hall.

Shtub, A., Cohen, Y., and Keren, B.K. 2008. *Project Management for Industrial Engineering Students.* Raanana, IL: The Open University of Israel.

Shtub, A. 2012. *Project Management Simulation with PTB Project Team Builder.* New York: Springer.

Winston, W. and Ayne, L.W. 2000. *Simulation Modeling Using @ RISK for version 4.* Pacific Grove, CA, USA: Duxbury Press.

4

Information Systems

Educational Goals

This chapter introduces the topic of information systems and discusses their importance in the work of industrial engineers. We present technological aspects such as data collection technologies, databases, and decision-support aspects of information systems as well as application aspects such as forecasting models and their use.

The focus is on the importance of information systems for the tasks performed by industrial engineers.

4.1 Introduction

The development of computer technology along with the parallel development of information systems that are based on computer technology ignited a revolution in the industrial and business worlds. This revolution resulted in changes both in the manufacturing and in the service sectors. Today, many decisions, which in the past were based on "gut feelings," are grounded on hard data and information provided by information systems.

Modern supply chains (materials moving through stages of a production process from raw materials to final goods) could not be managed without information systems revolution. Many other processes, performed as part of the manufacturing and service systems, also produce huge amounts of data. In the past, it was hard to deal manually with all the data collected and the information generated as part of the operation of the organization (Figure 4.1). Today, modern computerized information systems collect data across the chain or from other processes, store it, process it, and share it among the participants in the supply chain/process to support their decisions. The ability of computers to collect relevant data and process it into useful information, for example, inventory status, financial performance measures, and demand projections, supports myriads of decisions. While

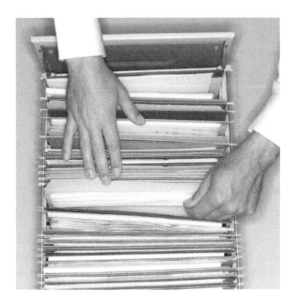

FIGURE 4.1
Manual extraction of data and information.

some decisions may seem inconsequential, others are critical. It is not an exaggeration to say that the success or failure of many businesses depends on computers and information systems.

Information is gathered quickly and accurately at a very low cost through advanced data collection technologies such as bar codes, magnetic stripes, optical character recognition (OCR), and radio frequency identification (RFID) tags. New technologies for image and sound recognition and processing are continually being developed and refined and will further enhance our ability to collect information from new sources in the future.

In the early days of computer technology, in most organizations, each organizational unit had its own information system that operated separately from other units in the organization. This was wasteful and inefficient. It was wasteful because it did not optimally utilize available data. An organization's efficiency becomes compromised when information collected in different organizational units is not integrated because decision makers, at best, are only able to make local optimum (for their organizational units) decisions instead of making global optimum decisions for the whole organization or even better for the whole supply chain. The same is true for organizations trying to solve problems that may be limited to one unit or spread through many units or levels. Without information reflecting the entire organization, the true status of the organization can never be known. Technological developments in database technology, computer networks, computer memory, and the drastic cost reduction of computation power enabled integration of

FIGURE 4.2
Automated extraction of data and information is a key to efficient data collection.

information systems not only within the organization but also between different organizations across the supply chain (Figures 4.2 and 4.3).

In the supply chain, for example, the following three interrelated parts are present:

- *Data:* Collection, storage, retrieval, and analysis of data generated by the supply chain organizations as well data about the environment.
- *Decision making:* Decision making regarding the use of resources and materials, shipments, storage, pricing, etc.
- *Physical aspect:* Handling and processing of materials (raw materials, parts, and finished products) in each organization and across the supply chain.

The first part deals with the collection, storage, retrieval, and analysis of data such as data from a customer's new order that enters the system or new data created by transactions. New data collected from a variety of internal and external sources is the "raw material" of the information system.

(a)   (b)   (c)

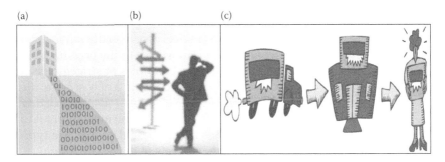

FIGURE 4.3
Three parts of a supply chain: (a) data, (b) decisions, and (c) physical aspects.

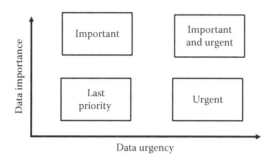

FIGURE 4.4
Data classification by importance and urgency.

As depicted in Figure 4.4, data can be classified according to importance and urgency.

Data that is both important and urgent must be processed and analyzed quickly, and immediately made available to decision makers. An example of such a situation might be a quality problem that may cause safety issues and require a recall. Other data can wait for a weekly or a monthly report, for example, regular financial data regarding payment by customers, but a situation that may necessitate a recall cannot wait for the data to be collected, processed, and make its way up through different levels of management. Nonurgent but important data should be stored until it is time to retrieve it for further processing and analysis. It is advisable to set regular times for dealing with such data and give it priority over other less important data, as this data may be pertinent to high-priority decisions. Unimportant data should not be collected in the first place. Technical aspects of data collection and retrieval are discussed in detail in books dealing with information systems (see, e.g., Laudon and Laudon, 2011).

4.1.1 Use of Information to Support Decision Making

Three types of decisions are made by management or other employees in most organizations and are described as follows:

- *Strategic decisions:* These are long-term decisions usually having significant economic impact and high importance to the organization. These decisions are made infrequently. Examples of strategic decisions are: the decision to develop a new product or a new service, the decision to build a new plant or add a production line in an existing plant, and the decision to enter a new market. Implementation takes a long time and the cost involved is high. These decisions are typically made by the senior management levels of the organization.
- *Tactical decisions:* These are medium-term decisions; for example, the decision to change the supplier of a certain raw material, or

TABLE 4.1

Characteristics of Decision Types

Type of Decision	Level of Decision Makers	Percentage of this Type of Decisions from Total	Cost of Mistake
Strategic	Top management	Very small	Very high
Tactical	Medium level	Small	Medium to high
Operational	Low level	Very large	Low to medium

the decision to stock finished units of certain product during low-demand periods, to avoid the shutdown of a production line; and to sell these products during high-demand periods when production capacity may not be sufficient to supply all the demand.

- *Operational decisions:* These are short-term decisions. Most decisions in the organization belong in this category; for example, a decision when to process a certain product unit on a specific machine, or how much raw material to order next week. These decisions are made very frequently and usually at lower levels of the organization.

This classification is described in Table 4.1.

When designing an information system, it is important to distinguish between two types of decisions:

- *Routine decisions:* These are usually lower level (operational or tactical) decisions. Because these decisions are made routinely, it is possible to use past data and to develop a decision process that can be supported by the data or even automated (automated means that it is performed by the information system without human intervention). Decisions relating to payments to suppliers are an example of such decisions. If there is a long-term contract with a supplier and an agreed-on cost for the products supplied, and the quantity supplied and the cost per unit are all known, the decision of how much to pay and when to pay is routine and the computer can do it automatically. The money can be transferred electronically or a check may be sent by the information system.

  In many supply chains, there are routine operational decisions that can be automated. For example, when the inventory level drops below a certain level, predefined by the user, the system can order a predefined quantity, directly from a supplier with whom a long-term supply contract was signed. The decision in this example is an operational decision. In general, operational decisions are easier to automate.

- *Nonroutine or ad hoc decisions:* Such decisions are difficult or impossible to automate. Often the decision-making process also relies on the

intuition and experience of the decision maker. Examples of nonroutine decisions are decisions to stop production of a specific product,
to start development of a new product, to change the list price of a
product or to offer a special discount to a one-time customer. Strategic
decisions and some tactical decisions belong in this category.

It is very difficult or impossible to automate strategic decisions. Strategic
decisions are based on many different considerations and on partial information, and therefore, involve risk. The strategic decision process can be
supported by computerized information systems such as simulation and
forecasts, as well as data and information regarding the organization and its
environment. It is not, however, usually automated.

However, there are many tactical and operational routine decisions that
could and should be automated. A poor supply chain design does not employ
automated decision making in its day to day operations. In some extreme
cases, every decision requires a separate discussion, and senior management's attention. As a result, senior management devotes a lot of time to
an endless stream of routine decisions. Since senior management's time is
limited, decisions are delayed, causing long lead times in the supply chain.

In a good supply chain design, routine decisions are automated as much as
possible. If automation is not possible for some reason, an effort to instruct
those who participate in the process regarding how to make routine decisions is needed, along with an information system that provides the right
information to the right employee at the right time.

In a well-planned process, most of the time senior management deals with
strategic and nonroutine decisions. Senior managers can devote their time
and energy to monitoring the environment, the competition, the customers,
and the performance of the supply chain, to identify as early as possible the
need for changes and to decide what to change and how to do it. To support
this monitoring and control activity, it is important to track the performance
of every process and every organizational unit, so that management can be
alerted when their attention is required. This is especially true in a competitive,
rapidly changing environment, where a high degree of uncertainty is present.

An important step in process planning is to decide which decisions to automate, how to do it, and how to monitor and control the process. The purpose of
monitoring and controlling is to identify problems or poor performance as early
as possible. Once the necessary data is collected and processed, the information
is used to support decision making. To ensure success in the implementation of
decisions, workers participating in the process should get the information when
it is needed, and in a form they understand and know how to use. Training is
an important part of the operating and use of information systems.

Information sharing within the organization and between organizations is
the basis of supply chain management and a key success factor in the success
of supply chains. In the past, the typical functional organization had legacy
information systems for each function (e.g., shop-floor control software was

unrelated to financial auditing software and made by a different developer using different hardware and software tools and techniques). Each function had its "own" data and its "own" information.

In modern integrated information systems, such as those used in supply chains, data and information are shared among all those engaged in the process. The supply chain management approach is based on sharing information among the chain's organizations or links, which together provide hardware, software, information, and services. Organizations in the supply chain collaborate and share information in order to maximize quality, minimize cost, shorten response time, and maximize flexibility of the whole supply chain. The benefits are shared among the organizations forming the supply chain and their customers, so everyone benefits.

The planning of a supply chain is a strategic undertaking, given that the selection of the supply chain's organizational members and the decision about the contribution that is expected of each member are long-term decisions that will have significant impact on the organization. Some tactical issues should also be addressed when designing a supply chain; for example, whether to hold stock of a particular component and how many employees to hire before the holiday season, when demand is expected to rise. Finally, there are a large number of operational decisions such as how much inventory to order next week and where to hold this stock or when to schedule the preventive maintenance during the coming week which have to be resolved during the planning process.

The basis of supply chain management is the availability of advanced information systems that can share information among supply chain partners. Each partner must install and operate such an information system, and share the information with the other partners.

4.1.2 Data Handling

Information systems are based on data that (1) accurately represent the real situation, (2) are timely, and (3) are readily available. To achieve this, several steps are required:

1. *Data collection:* Data can be collected in many ways. In the past, the common way to collect data was to do it manually and to key the data into the system. Due to inefficiencies inherent in it, the high rate of errors and the high cost of labor associated with it, other methods have been developed to collect data using advanced technologies. Technologies such as pattern recognition, OCR, handwriting recognition, speech recognition, and artificial intelligence systems along with various types of sensors enable gathering of immense amounts of data. Data can be collected using portable terminals/scanners. Once collected, the data has to be transferred to a computer using technologies such as electronic data interchange (EDI), Wi-fi,

Bluetooth, and infrared. To avoid the need for repeatedly collecting the same data, technologies or devices such as barcodes, RFID tags, and magnetic stripes are printed on or attached to packaging. Thus, wherever these goods go, their information can automatically be gathered by barcode readers, magnetic tag readers, and RFID antennas. These developments have enabled data to be collected quickly, accurately, and at low cost.

EDI systems create a direct link between different parts of the organization in different locations, and between organizations in the supply chain. For example, data collection systems used by retail chains collect and distribute real-time sales data to warehouses, factories, and suppliers electronically instead of via paper reports. Thus, participants in the supply chain know the level of inventories in different locations along the supply chain and can decide when to renew a specific inventory, without the need to manually transfer data on inventory status within the organization or between organizations.

2. *Data storage:* Data are stored in databases. Data are also stored on electromagnetic media such as magnetic disks (hard disk), electro-optical media, or solid-state devices such as flash memory that allows users to write, erase, and rewrite. The Internet allows fast retrieval of data stored in remote locations as well their backup from anywhere around the world. Companies that specialize in data-related services offer enterprises storage and data handling services on their remote server farms (also known as the cloud). Storing data on the cloud enables organizations (and individuals) to have backup data in several remote places around the globe. This protects the data against any local catastrophic loss.

3. *Data retrieval:* Information systems have to retrieve the data and present it to decision makers before and after analysis and processing, to support the decision-making process. Presentation of data is done via electronic means such as a tablet, smartphone, computer monitor or projector, as well as through printed reports that present data in tables or graphs.

4. *Data analysis:* Even in medium-sized companies, the amount of data that accumulates is gigantic. It also contains an immense amount of minute details, all of which must be processed and summarized to make some sense. Data analysis must work through a large variety of data types, with the objective being sometimes to summarize and other times to also show trends, correlations, or patterns. For example, carriers such as FedEx, UPS, and United States Postal System (USPS) gather data on several millions of deliveries every day. The records of each package help management when deciding on a general load plan for their fleet of trucks. Records on each delivery are also gathered at several points along the way; the vehicles that

carried the package, and the times it was unloaded from one vehicle and loaded on to the next one are documented. The data have to be processed to enable analysis of quantities and efficiencies.

There are many tools for data processing. Some popular tools are as follows: (1) Standard Query Language (SQL), which is used in relational databases; (2) Online Analytical Processing (OLAP), with its multi-dimensional cube, for quick query execution by pivoting; (3) the R programming language, which is convenient for dealing with many files and big data; and (4) Apache Hadoop, which is an open-source software framework for storage and large-scale processing of data sets. Data processing generates useful information from raw data and is the main component of data analysis.

At the end of the process described previously, the data become information that supports decision making by humans, helps in monitoring and control, or is used for both automatic and managerial decision making.

4.2 Components of the Information System

Several components of information systems support the management of supply chains:

1. *The transaction processing system:* This system receives and records transactions, such as sales order entries in the sales department, inventory transactions, and deliveries to customers. This is the core data collection, enabling reliable real-time data base. Such systems produce reports for managers, summarizing the transactions over a period of time, to support the decision process.

2. *The management information system:* This system serves the areas of planning, monitoring, and control. Such systems help low- and medium-level managers in making decisions via either fully automated processes or by providing relevant information when needed.

3. *The decision support system:* This system integrates data and analytical models to support semi-structured decision making. Such systems are designed to help managers make strategic and tactical decisions. They do not provide a decision that can be applied directly, but present different types of information and recommendations on request to provide decision makers with insight into the situation, while the decision remains in their hands.

These three systems are based on a database and a model base as explained in Section 4.2.1.

4.2.1 Database Systems

All three systems described previously exploit data from internal and external sources. An important component of a well-planned information system is the database where data are stored and managed.

Database systems are designed to handle large amounts of data. The system provides a physical storage for the data on electronic media, electro-magnetic media or electro-optical media and a mechanism for retrieving it, which is implemented using the database management system (DBMS). The DBMS is a collection of programs that enable easy access to the data in order to update it and process it into useful information. The database system provides the user with the possibility to store and retrieve data in an easy to use and efficient environment. This is done by separating the management of the data's physical location from its logical location. The user does not need to manage the physical location on the magnetic or optical media. He or she uses a logical location, and the DBMS translates it into the physical location. In addition, the system takes care of protecting data against technical failure and unauthorized access.

Database system users are not necessarily computer experts. To enable efficient and beneficial use of the system for all its users, these systems are designed to hide the complexity of physical data storage and retrieval using three levels of abstraction (Silberschatz et al., 2002):

1. *The physical level:* This is the lowest level of abstraction of data. It describes how and where data are actually stored, for the purposes of programming. Physical level data structures are the most detailed, including the physical location of data (on the magnetic optical or other type of storage).

2. *The conceptual or logical level of the data:* This is the next higher level of abstraction. It describes what data are stored in the database, and the relationships between the types of data. For example, this includes the list of columns/fields and the data type of each column/field. It also includes tables describing the various relationships of data types. At this level, the physical location of data in the system is not specified. The logical level is used mostly by database administrators who decide what information will be stored in the database and how.

3. *The observation level:* This is the highest level of abstraction. It is developed for serving the end users, which must see the data in a simplified way. Despite the simplified presentation of the structures at the logical level, some of the complexity of the system still remains at this level, due to the size of the database and interconnection between the stored data elements. In most database systems, users need only part of the database, and thus their interaction with the system is simplified at the observation level. The system provides a

variety of different ways to view the data from the database to provide users with the specific information that they need.

An important feature of the DBMS is the ability to separate the logical
aspect from the physical aspect of the data, so that users can retrieve data
without having to pay attention to the physical location of the data in the database. The physical aspect, as far as the end user is concerned, is transparent.

4.2.2 Queries and Structured Query Language (SQL)

Silberschatz et al. (2002) explained that the query language is a language
through which the user seeks information from the database. Query languages are classified into procedural languages and nonprocedural languages.
In procedural languages, the user tells the system to perform a sequence of
operations on the database for the purpose of obtaining the desired result. In
a nonprocedural language, the user describes the information needed without giving detailed instructions on how to get the information. Modern database system query languages provide elements of both approaches.

A common query language is the structured query language (SQL). This
language is user-friendly and very common in database management. IBM
developed the original version of SQL in its research labs in San Jose. Originally,
this language was called Sequel, and it was developed as part of the R system
in the early 1970s. The language has evolved since then and was renamed SQL
(structured query language). SQL is considered a standard language for data
queries.

It is important to note that although SQL is referred to as a query language, it
can perform additional actions such as setting up data structures, implementing changes in the database information, and defining security constraints.

4.2.3 Data Flow Diagrams

A data flow diagram (DFD) is a model that shows the flow of data in a computerized system. DFDs are used early on in systems analysis to help define
the following: (1) the required data, (2) its sources, (3) the operations performed on it, (4) where it is stored, and (5) the output it is used to create. The
basic building blocks of a DFD are as follows (Whitten and Bentley, 2005):

1. A bubble that represents a process or a function in the information
   system.
2. A rectangle that represents an external entity that provides input or
   receives output.
3. Two parallel lines that represent a database or a file where data are
   stored.
4. An arrow that represents flow or data transfer between other parts
   of the diagram.

The DFD chart models data flow and data contained in the database. The DFD model is based on the following simplifying assumptions:

- There is no reference to the timing or frequency of the different elements in the diagram.
- Decisions are not modeled (e.g., it is impossible to show that data flows in one direction if a certain condition is true and in another direction otherwise).
- Errors are not represented in the DFD model.
- Material flow is not represented in the DFD model.

DFD charting conventions:

- Every process has at least one input flow and one output flow of information.
- There must be a process on at least one side of the flow of information.
- There is no direct information flow to the entity or the entity's database repository.
- There is no flow of information from a process back to itself.
- Each database has at least one input flow and one output flow.
- An external entity on the left-hand side represents input and on the right-hand side represents output.

Figure 4.5 illustrates the general elements of the DFD.

DFDs are usually constructed in a hierarchy such that the highest level is the least detailed and each level details its upper level. The charts are numbered in the ascending order with respect to the level of detail. The upper level is least detailed and is called DFD zero. At the upper level, bubbles represent the main processes and the DFD shows the relationship with the environment. Each process and its inferences are further detailed in the next lower level. Lower levels of the DFD show more details of the information system and its processes.

A DFD data dictionary accompanies many diagrams and is divided into two parts:

1. Dictionary of the DFD components: It contains a detailed description of each component on the DFD chart.
2. Dictionary of the DFD data elements: It contains attributes/fields used by the databases and the data streams.

4.2.4 Model Base

The data in the database have little value for most decision makers. Although it can be retrieved easily and presented, in most cases, it must be processed

Legend:

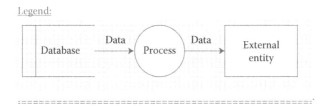

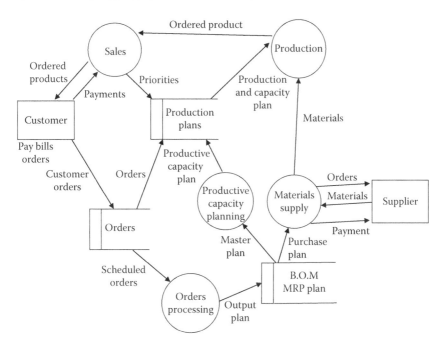

FIGURE 4.5
An example of data flow diagram (DFD).

in order to support specific decisions. As mentioned earlier, three types of decisions are supported by the information system:

- *Routine:* Well-structured decisions that frequently can be fully automated. An example of such decisions is the decision to pay a supplier following the delivery on time of the correct quality and quantity of raw material ordered from the supplier. Based on information in the contract with this supplier, a proper model can be used to issue a check or to transfer the payment electronically, in accordance with the agreed-on price and terms of payment.

- *Nonroutine:* Ad hoc decisions that cannot be automated. Nevertheless, partial insight into the situation can be provided by applying some general-purpose models to the relevant data. A very simple example is an Excel spreadsheet constructed for a specific problem in order to analyze it. Investment in a new production facility is a typical example of its nonroutine decisions that can be supported by an Excel cash flow analysis (typically, calculating the net present value [NPV] or the internal rate of return [IRR] of several alternatives).

- *Monitoring and control models:* These models are designed to identify deviations from planned or normal progress. In particular, they are designed to identify out-of-control situations. For example, a statistical process control (SPC) model uses information from samples of units produced on a production line to identify as early as possible out-of-control cases or trends that may lead to the production of defective items. Based on the SPC analysis, a decision may be made to temporarily stop production and readjust the production process.

By making simplifying assumptions and using data from the database, models can help decision makers solve difficult problems. A good model is simple enough to allow understanding and analysis of the real problem, and yet close enough to the real problem so that the solution obtained from the analysis of the model can be implemented on the real problem. In many cases, a sensitivity analysis is performed to ensure that despite the simplifying assumptions of the model, the solution will withstand various unexpected changes in the real problem.

Selecting decision-making suitable models is a typical task of the industrial engineer. To perform this task, the industrial engineer must be familiar with the decisions that the information system should support and the models that can be used to support these decisions. Therefore, the curriculum of industrial engineering programs focuses on different types of problems and the related models that can support the decision makers dealing with these problems.

In many cases, the same model can be used to solve a variety of problems, by applying the respective relevant data in the database to each problem. Accordingly, separation of data from models is important. An important aspect in designing an information system is the integration of the model base and the database to ensure that all the necessary data for the selected models are in the database and can easily be retrieved and used by the appropriate model.

In many organizations, the industrial engineers have to choose the models and the DBMS that best fit their organizational needs. Development and maintenance of an information system for a particular organization is very expensive, and therefore, most organizations prefer to buy off-the-shelf information system modules and adapt them to their needs when possible. Examples of information system modules that are available off-the-shelf are as follows: (1) customer relationship management (CRM) modules, (2) supplier relations management (SRM) modules, and (3) human resource management (HRM) modules. These modules must be compatible and integrate with the organization's enterprise resource planning (ERP) system. The cost of acquisition and maintenance of off-the-shelf information systems is usually lower than the cost of development. On the other hand, off-the-shelf information systems are usually less flexible. Therefore, it is important to analyze and prioritize the requirements from the information system, and only then select the most appropriate information system product. This is the task of many industrial engineers. If this task is not done properly, the information system could impose limitations on the decision makers, and restrict the competitiveness of the organization.

Let us consider two types of organizations that need a new procurement system (non-repetitive vs. repetitive). There are procurement systems that deal with research and development of new products. They belong to organizations that are characterized by dynamic purchasing requirements and raw material consumption that is not repetitive. Therefore, the procurement system must support flexible short-term (or one time) supply contracts. These procurement systems are very different from procurement systems designed for organizations that deal with repetitive production. Examples of repetitive production include an industrial plant, a car assembly plant, or an engine manufacturing plant. In these cases, the overall raw material consumption is regular and repetitive. In other words, the same raw materials are purchased again and again over time.

In the organization dealing with research and development, procurement requirements are generated by development engineers or researchers engaged in the development process. They initiate the process that creates a new purchase order (PO) in the information system. There might be an approval process requiring the electronic signatures of other members of the organization. After approval, the purchasing agents turn to candidate suppliers for quotes, which are fed into the system's database. These quotes are evaluated based on parameters such as price, quality, and delivery time. The preferred supplier is selected and a purchase order is sent to the supplier. Supplier selection may be supported by a proper model in the database models that compares the various quotes received.

In the production plant, procurement requirements are generated automatically by the information system that monitors the inventory levels continuously. The system issues a purchase order for the required raw material to preselected suppliers (with whom a long-term contract was signed) when the inventory level drops below a predetermined point (called the reorder point). In some cases, the supplier is selected from the list of alternative suppliers listed in the system's database. For each item listed in the database, the order-point, and suppliers from whom it can be purchased from are also in the database. In many cases, for each alternative supplier, the prices of each part and raw material are stored in the database along with the promised lead time (based on the organization's agreements with its suppliers). Once a supplier is selected and the order approved, it will be sent to the selected supplier.

In both examples, the information system can support the procurement process, if it is well defined in the system's model base. In both cases a sequence of actions, which are a result of a decision supported by data from the database and models from the model base, is required to issue a purchase order. There are clear differences between the two organizations described previously. For example, in the first case, bids are required; in the second example, the decision is based on the organization's long-term agreements with suppliers. It is the task of the industrial engineer to decide on the models applicable to his or her organization and to select the right information system that contains these models and satisfies the requirements of the organization.

4.3 Quality of Information

In Figure 4.4, we classified data as important and not important. We now introduce another classification based on the quality of the data. Information obtained from processing data has value only if it is based on high-quality data that represents reality accurately. The quality of information generated from processing the data is dependent on the quality of the data, as well as on the user's ability to understand the information correctly. Any information system is subject to the famous garbage in, garbage out (GIGO) rule (i.e., poor data quality leads to poor information quality).

The designer of information systems must ensure that the information is understandable, valid, relevant, accurate, and complete. The only way to ensure this is to perform a thorough analysis to understand what information is needed (and for what purpose) and to develop, at a reasonable cost, a system that supports these needs.

The processing of data into useful information takes many forms. Statistical analysis is used to summarize large quantities of data and to estimate attributes of a population based on samples. Operations research models are used to find the best solution to complex problems that can be presented by quantitative models, and data mining technology is used to identify patterns and relationships between variables based on past data collected and stored in the database.

One aspect is common to all these processes: The effort to use past data to support decisions regarding the future. To understand this aspect, the next section focuses on models used for forecasting future values of important parameters, such: as future demand for a product, or future cost of a commodity.

4.4 Forecasting

Good information systems provide high-quality information—information that is understood, valid, relevant, accurate, and complete. In reality, there is always some information that is not available, and at best, it can be estimated. This is especially true when dealing with future events. Information about future interest rates, future inflation rates, and future exchange rates is important for investment decisions, information on future demand and competition is important for operating and marketing decisions, and information on future absences of employees and future breakdown of machinery is important for production planning. Since this future information is not available, if no major changes are expected, it is customary to use accumulated data to project future quantities. Therefore, many information systems use models to forecast such information based on past data. In this unit, we will present a forecasting model that is used to support the forecasting of future demand in supply chains. The model is based on time series analysis—the analysis of a sequence of data points, measured at successive points in time, spaced at uniform time intervals. The consumer price index (CPI) is an example of such an analysis. The assumption is that it is possible to predict the future based on data from the past. It is important to estimate the accuracy of such forecasts, and therefore, models for measuring the quality of forecasts are presented as well.

We distinguish between the future value of a time series, called a forecast, and the past values, called observations. In discussing forecast methods and determining the quality of the forecasts, we use the following notation:

F_t is the forecast for period t

O_t is the observation for period t

4.4.1 Moving Average Model

The simplest forecast form is the moving average (MA): For each point in time, the moving average gives a forecast value of the mean of the last n periods. The number of periods (n) is the major parameter of the moving average method, and therefore, it is customarily written as MA(n).

For example, a moving average of two periods (MA(2)) generates a prediction (F_t) for each time period (t) based on the actual (O_t) quantities of the last two periods: $F_t = [O_{t-1} + O_{t-2}]/2$. For MA(3): $F_t = [O_{t-1} + O_{t-2} + O_{t-3}]/3$, and so on. In this case, n represents the relevant history. The larger the n, the more weight is spread among longer periods of time.

In addition, the moving average has an attenuating effect and is used to smooth fluctuations. This is illustrated in Figure 4.6.

4.4.2 Estimating the Quality of Forecasts

The mean absolute deviation (MAD) measures the average error in absolute values (the error is the difference between the forecast for a period and the observation for the same period). If each forecast is exactly the same as the

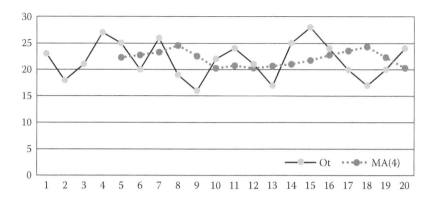

FIGURE 4.6
Example of data and a moving average of four periods—MA(4).

observation for the respective period, MAD is equal to 0. The larger the value of MAD, the lower the forecast accuracy. Mad averages the absolute values of the errors so that negative errors will not cancel out positive errors.

$$MAD = \frac{\sum_{i=1}^{N} |F_i - O_i|}{N}$$

Another measure that can be used is the mean squared error (MSE), which measures the average of squared error values. MSE averages the squared error values so that negative errors will not cancel out positive errors.

$$MSE = \frac{\sum_{i=1}^{N} (F_i - O_i)^2}{N-1}$$

4.4.3 Exponential Smoothing Model

To demonstrate the use of MAD and MSE, consider the exponential smoothing forecasting model. The following notation is used in this model:

F_t is the forecast for period t

F_{t-1} is the forecast for period $t-1$

O_t is the observation for period t

O_{t-1} is the observation for period $t-1$

α is a constant between 0 and 1, which could either be regarded as the weight for the most recent error or the weight for the most recent observation.

The exponential smoothing model is based on the following equation:

$$F_t = F_{t-1} + \alpha(O_{t-1} - F_{t-1}).$$

The forecast in period t is equal to the forecast in period t–1 plus a correction term based on the error in period t–1. The error term is multiplied by a factor α that takes the values between zero and one. The logic is straightforward: If in the previous time period the forecast was equal to the observation, use the same forecast in the next time period; otherwise, correct it by adding a fraction (or all) of the error term.

If we rewrite the model equation:

$$F_t = \alpha O_{t-1} + (1-\alpha)F_{t-1},$$

the same model can be used (recursively) for period t–1:

$$F_{t-1} = \alpha O_{t-2} + (1-\alpha)F_{t-2}.$$

By substitution, we get

$$F_t = \alpha O_{t-1} + \alpha(1-\alpha)O_{t-2} + (1-\alpha)^2 F_{t-2}.$$

If we continue this process for t–2, t–3, and all past periods, we get

$$F_t = \sum_{i=0}^{i=\infty} \alpha(1-\alpha)^i O_{t-i-1} = \sum_{i=0}^{i=\infty} a_i O_{t-i-1}.$$

In this equation, all past observations are taken into account and each one is multiplied by a factor (a weight) equal to

$$a_0 > a_1 > a_2 > \ldots, a_i = \alpha(1-\alpha)^i.$$

This is classic exponential series of fractions and the sum of these weights is equal to 1:

$$\sum_{i=0}^{\infty} a_i = \sum_{i=0}^{\infty} \alpha(1-\alpha)^i = \alpha \sum_{i=0}^{\infty} (1-\alpha)^i = \frac{\alpha}{1-(1-\alpha)} = 1.$$

Thus, the exponential smoothing model is a weighted average of all past observations where each weight is an exponential function of α.

TABLE 4.2

Forecasts with Different Alpha Values

Period	0	1	2	3	4	5	6	7	8	9	10	11	Forecast
Actual	100	105	103	107	106	108	110	107	105	107	109	110	
Alpha = 0.1	N/A	100	100.5	100.8	101.4	101.8	102.5	103.2	103.6	103.7	104.1	104.6	105.095283
Alpha = 0.3	N/A	100	101.5	102	103.5	104.2	105.4	106.8	106.8	106.3	106.5	106.3	106.080543
Alpha = 0.8	N/A	100	104	103.2	106.2	106	107.6	109.5	107.5	105.5	106.7	106.7	106.636697

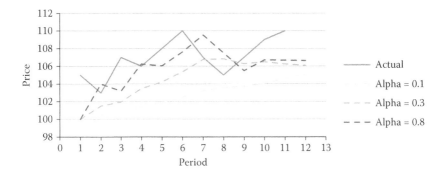

FIGURE 4.7
Example of exponential smoothing with various alpha values.

The value of α changes the forecast as well as the values of MAD and MSE (Table 4.2). See equations at the end of Section 4.4.2. The following are MSE and MAD values for three levels of alpha.

MSE $(\alpha = 0.1) = 24.35$
MAD $(\alpha = 0.1) = 4.63$
MSE $(\alpha = 0.3) = 10.84$
MAD $(\alpha = 0.3) = 2.87$
MSE $(\alpha = 0.8) = 7.38$
MAD $(\alpha = 0.8) = 2.41$

It is clear that in this time series, $\alpha = 0.8$ is the best model as it minimizes both MAD and MSE. This is an important observation as we can improve the quality of the forecast by adjusting the value of α to best fit the time series. The resulting model is called the adoptive forecasting model (Figure 4.7).

References

Laudon, C.K. and Laudon, P.J. 2011. *Management Information Systems: Managing the Digilul Firm*. 12th edn. Englewood Cliffs, NJ: Prentice-Hall.

Silberschatz, A.F., Korth, H., and Sudarshan, S. 2002. *Database System Concepts*. 4th edn. Singapore: McGraw-Hill.

Witten, J.L. and Bentley, L.D. 2005. *System Analysis and Design Methods*. Irwin, NJ: McGraw-Hill.

5

Supply Chain Management: The Interface with the Customer

Educational Goals

Supply chains are composed of a collection of organizations, many of which are customers of other organizations in the supply chain and at the same time are suppliers of their own customers in the chain. This chapter presents the issues concerning the design and operation of organizational interfaces within the supply chain. In most supply chains, there are multiple such interfaces as each organization in the supply chain (a single link in the chain) has interfaces with one or more organizational customers. These interfaces are the points of contact with the customers, where information on customer needs and expectations is collected and used as input, to support the decisions of that link in the chain regarding its future activities. The design of these interfaces includes considerations related to the flow of material, the flow of information, and the decision-making process. Customers choose their suppliers from a variety of alternative suppliers. The chosen supplier has actually won a competition in which its superiority over alternative suppliers was determined. This superiority may stem from the quality, cost, or the timeliness of its products and services. In short, dealing with customers requires being competitive, and continuously managing the competition.

This chapter presents the following aspects of the customer interface design:

- Introduction: The need for the customer interface design and relevant considerations
- Selecting a customer interface policy
- The master production schedule
- Lead time and time-based competition
- Quality-based competition
- Cost-based competition
- Flexibility-based competition

The chapter is one of the three chapters that deal with the supply chain building block—the single organization (or the single link) in the chain. Following the discussion in these three chapters focusing on the single link of the chain, integration of these links into the whole chain is discussed. The next chapter deals with suppliers and outsourcing, while the chapter that follows it deals with the scheduling of operations within each link of the supply chain.

This chapter explains the complex interface with the customers, relevant policies, the flow of information through the interface, as well as the models developed to support decisions by using that information and the resulting flow of material.

5.1 Introduction to the Customer Interface and Its Design

Supply chains are a collection of organizations linked to each other by customer–supplier relationships. Each organization in the chain supplies goods or services along with information to its customers, while providing the information to its own suppliers in order to get the goods and services it needs from them. Thus, the supply chain has many interfaces between customers and suppliers, and the efficient design and management of these interfaces is a key to the success of each link in the chain, and to the success of the whole chain.

The relationship with customers may take many forms. On one extreme is a one-time order for specific goods and services to be supplied in a given period of time and in specific quantities. On the other extreme are long-term relationships with a customer with contracts specifying periodic deliveries on request either in prespecified dates or when needed.

The interface with the customers is very complex; in addition to needing to share information among different information systems, it includes the management of customer orders, as well as activities that influence customer demand such as advertising, special sales, quantity discounts, etc. These activities are based on strategies that take into account the competition in the market.

As mentioned in earlier chapters, competition has several dimensions: time, cost, flexibility, and quality. Each organization has to decide how to compete in each of these dimensions, and these decisions affect the competitiveness and long-term survival of the organization. In a competitive market, low cost, high quality, and a short supply time may play a major role, as well as a fast reaction to changing needs of customers and changing market conditions (i.e., flexibility). Changing market conditions and competition force organizations to continuously seek ways to improve their performances in competitive dimensions, in order to assure their long-term success.

The starting point of the customer interface design is the "voice of the customer" (VOC)—understanding the needs and expectations of the customers.

It is also important to understand alternative ways to satisfy those needs and expectations, and the way competition is trying to satisfy these needs and expectations. The analysis is an effort to determine the "right" price that the customers are willing to pay and hence the target cost or the "cost objective" of products and services supplied by the organization. Since profit is the difference between the prices the customer pays and the cost to supply the goods and services, cost analysis and cost reduction are very important parts of the decision process and are the cornerstone of cost-based competition. However, cost is only one of four dimensions of business competition. Like the cost dimension, the delivery time, the quality, and flexibility are subject to customer expectations. These expectations should be analyzed and used as input for the customer interface design. The interface design focuses on the flow of information, the flow of material, and decision making. An example where these flows are essential is the connection between external customer orders and internal work orders. The design of that connection includes the flow of information and materials and the use of organizational resources triggered by work orders. A related example is deciding how to link orders from customers with procurement decisions. For example, customers may require a certain quality of materials or components acquired from the suppliers of the organization. The design of the customer interface is, therefore, a major factor in the design of the supply chain.

As an example of customer interface design issues, consider a market in which time-based competition dominates. In some markets, the customers are not willing to wait and want to get the goods and services they need instantaneously. For example, a customer of a supermarket expects to get groceries off the shelf, and in most cases will not be willing to wait for a delivery truck to come. In such cases, safety stock decisions are based on forecasts of future demand. The supermarket uses forecasting models such as those discussed in Chapter 4 and based on those forecasts, orders from its suppliers. Since forecasts are subject to forecasting errors, a major design issue is how to protect the system against these errors in forecasting or, in other words, how to ensure that the customer will find what he or she needs on the shelf when he or she needs it, without holding excessive amounts of stock. A possible decision is to order quantities larger than the forecast to protect (or to "buffer") against uncertainty. The result of such a decision is the extra cost of holding excessive goods and the risk that they may not be sold, especially if they are perishable. Another result of such a decision is limited flexibility to demand changes. For example, if the demand changes and some products are no longer popular, the organization still holds a stock of these products and usually will not dispose of it until the market trend is clear.

The trade-off between time-based competition and cost-based competition that is between the required short customer delivery time and the extra cost of inventory is just one aspect of customer interface design.

A different design problem is presented when the process is triggered by orders from customers, namely the case where only on receipt of a customer order is the delivery process initiated. In this case, the promised delivery time to the customer may include: (1) the time needed to process the order, (2) the time to deliver the finished products, (3) the time needed for the purchase and delivery of raw materials, (4) the actual manufacturing time, (5) the time for assembly and packaging, and (6) the delivery time of the products ordered (Rogalski, 2011).

In some supply chains, a great deal of flexibility and a large variety of products is the basis of competition. When the customer order requires product design, the delivery time also includes the product design time, as well as the time to acquire raw materials and to process them. For example, kitchen cabinets can be ordered from a carpentry based on customer specifications. In this case, delivery time will include the following: (1) the time for translating the order and its design to production plan, (2) the time to order raw materials and parts, (3) the time to acquire the raw material, (4) the time to manufacture the product and package it, and (5) the time to ship it to the customer.

Most organizations are using mixed strategies. Managers can decide to purchase parts and raw materials on the basis of forecasted demand, but production and assembly of finished products begin only after receiving orders from customers. Several examples of linking procurement orders and work orders to customer orders are as follows:

- Purchase orders and work orders are issued based only on confirmed orders from customers.
- Purchase orders and work orders are issued based on demand forecasts.
- Purchase orders of raw materials and parts from external suppliers, which have a relatively long lead time, are based on forecasts. Purchase orders and work orders for short lead time items are based on confirmed orders from customers.

In many supply chains, a number of different alternatives are used simultaneously, as each product, part, and raw material may be managed differently based on the trade-off between delivery time, flexibility, quality, and cost.

The above discussion highlights the need for a well-designed customer interface that is integrated with the rest of the supply chain. Marketing, production, and procurement considerations must be incorporated into the design process, and trade-off analysis is required.

One example of a mixed policy is based on the "supermarket approach." This approach is based on comparing the total lead time required to supply a product (including the time to order raw material and parts, manufacturing time, assembly time, and time needed for the inspection of packaging and shipping to the customers). If the supply time required by customers is as long as the time required for the fabrication and delivery process, there is no need to carry any inventories and the supply process can be triggered by the customer order. However, if the supply time required by the customers is shorter than the time of the delivery process, some stocks will be required to compensate for the difference between the required delivery times and the time it takes to supply the goods or the services. In this case, an incessant effort must be made to shorten delivery process time to minimize or to eliminate the need to carry stocks.

The logic that determines when the selected work orders and purchase orders will be issued is known as "time phasing," and it is discussed in depth in the following chapters. For now, if the policy is, for example, to minimize the work in process inventory, an effort to shorten production times will be required (accelerating the transformation of work-in-process into final products). This can be done by reducing the time needed to switch production resources from one operation on one product to another operation on the same product, or on a different product (known as setup time). Another approach may be an effort to reduce the duration of different activities, and perform activities in parallel and not in sequence. If the policy adopted is to issue work orders and purchase orders based on forecasts of demand, it will be necessary to develop a proper forecasting model.

The discussion thus far has focused on when to issue work orders or purchase orders. A related issue is to determine the size of each order. Economy to scale may suggest that larger orders are preferred, but large orders may create unnecessary inventories. This is the case, for example, when orders are shipped in large containers, and the shipping cost of a container is fixed. (The cost does not depend on the weight or volume of the goods inside it.) When shipping cost is based on the number of containers and it does not depend on the contents of the containers, it seems more economical to order quantities that fill the entire volume of the container, but such a practice creates excess inventories and reduces flexibility.

5.2 The Impact of Inventory

5.2.1 Forecast-Based Orders

As discussed earlier, one way to reduce the lead time to customers is to issue work orders and procurement orders based on forecasts of future demand. When using this policy, inventories of finished goods that can satisfy future demand are created. This policy reduces the delivery time to customers at the cost of holding inventories.

The following are the policy benefits:

1. It is possible to offer customers a short delivery time, if a large enough inventory is carried to cover future demand.

2. Standard products can be produced in large quantities and benefit from economies of scale that enable high efficiency and low cost.

Policy shortcomings:

1. Inventory holding cost may be substantial—instead of earning interest in the bank, the money is invested in inventory (when loans are taken, the cost of holding inventory is even more obvious).
2. There is a risk that stocked items will be damaged, stolen, perish, or even become obsolete (so that they cannot be sold).
3. Stock of standard products reduces flexibility.
4. Inventory creates costs associated with storage, maintenance, etc.

For example, many food and beverage products are produced to inventory. A customer requiring a bottle of soda is not willing to wait and wants an immediate supply from the shelf. However, cocktails are made to order at the bar. Only the raw materials are kept in inventory.

5.2.2 Production to Order

Production-to-order policy is based on issuing a purchase order or a work order when a customer order is received. In some cases, the product design is also done at the request of the customer (design to order); this happens when there are no available "off the shelf" product alternatives. Usually, these are tailor-made components or products. For example, a manufacturer may order a conveyor belt for a production line including the design according to given specifications; and the conveyor belt installation company may order special parts to be designed for the given specifications. Some organizations such as NASA, the Air Force, Army, and Navy are famous for being the customers of production to order.

Advantages in production-to-order are the following:

1. Saving costs associated with holding inventory.
2. The flexibility to accommodate special requests—a unique product can be produced on request.

Disadvantages in production-to-order are as follows:

1. Long lead time, including the time required for procurement and delivery of raw materials, production time, assembly time, packaging, and transportation to the customer.
2. Production is typically done in small quantities, so there is no economies of scale, resulting in a relatively high cost.

This policy is common when the product is unique, and therefore, it is impossible to hold a stock of finished products. Ships, for example, are produced to order. Each ship has its special design, navigation system, and engines. Therefore, shipyards may carry stocks of metal that are used for the construction of ships, but it will not carry a stock of engines.

5.2.3 Assemble to Order

Assemble to order compromises between the two extreme policies as discussed previously. In this case, certain types of raw materials, parts, or subassemblies are kept in inventory but the assembly of final products begins only after an order is placed by the customer.

Personal computers are frequently assembled to order. There are predefined modules that the customer can choose, such as hard drives, motherboards, and graphic cards. The most popular components are kept in inventory and assembled according to customer order to form the desired product. The assembly time is short, and a large number of combinations are possible, so there is no point in having all possible combinations in stock. The resulting flexibility is high. If a customer is interested in a component that is not kept in inventory (e.g., a unique graphic card), delivery time will be longer due to the need to order and receive this unique part. This approach was adopted by DELL Computers, which sells personal computers through the Internet, and delivers them to the customer in a very competitive delivery time.

5.2.4 Customer Interface Policy Considerations

The decision to choose a customer interface policy and how to implement it depends on the product and on the market. The same product may, therefore, be produced in some markets based on demand forecasts to ensure fast delivery time; while in other markets, the same product may be produced on the basis of actual customer orders. There are situations where certain product sales to retail customers are made off the shelf, but the production of large quantities for wholesale customers is based on actual customer orders.

Organizations that provide services are using a similar approach in designing their customer interface. For example, fast food restaurants prepare standard food items and keep them in inventory to achieve very short delivery time and relatively low cost as a result of economies of scale. Upscale restaurants cook the food exactly according to customer requirements, and the customer knows that he or she must wait to get the food served exactly as he or she likes it.

Technology can play an important role in this decision. If a single raw material is used to produce a variety of end products, and the supply time of the raw material is relatively long, it may be best to maintain a large stock of this raw material. This is the situation in some process industries, such as refineries, where crude oil is the main raw material from which the refinery produces a variety of end products.

An important consideration in this analysis is the shelf life of the final products, raw materials, parts, and components. If the shelf life of the finished product is short while the shelf life of raw materials is long, it may be better to stock raw materials and not finished products, to avoid getting into situations of spoiled finished goods while keeping short delivery times.

5.3 Bill of Materials

The information on the product structure, such as what parts, raw materials, and subassemblies are needed to manufacture the product, is organized in the form of a bill of material (BOM). BOM is a schematic description, kept as a computer file, which describes the various product components in a hierarchy of levels of detail. For example, a bicycle has a frame, two wheels, crankset (including the chain and pedals), a gear set, and a seat. Each assembly or subassembly is also detailed. For example, a bicycle wheel assembly consists of its central axis, the spikes, a tire, and the inner tube. In the same way, the frame is made up of welded steel pipes. In addition, the bike seat, pedals, and gear as well as the handlebar, brakes, headlight, and bell are added to the BOM. Eventually, the BOM describes the logical relationship between the final product and its components. Figure 5.1 describes the BOM for the bicycle example above.

The structure of the BOM can help in the design of the customer interface. Consider the following examples:

- The final production link in the supply chain manufactures a single product or a very small variety of end products. These products are made up of a large number of different raw materials. In such cases, it might be advantageous to produce these end products from stock, especially when the market requires a short delivery time, and the final products have a long shelf life (Figure 5.2).

- The final production link in the supply chain produces a large number of different end products from a single type of raw material. For example, oil refineries produce many kinds of oils and fuels from crude oil. In this case, keeping a stock of the raw material may be worthwhile because it increases the level of flexibility as many end products can be produced from that stock. The raw material can be used for the production of a specific product according to customer order, while shortening delivery time because the raw material is in stock (Figure 5.3).

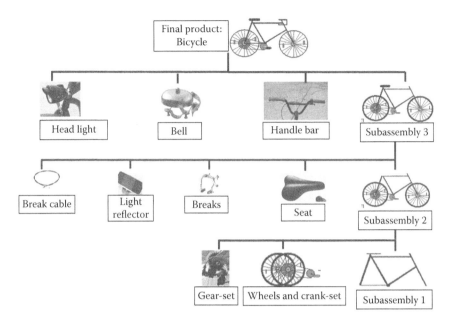

FIGURE 5.1
Schematic example for a bicycle bill of material.

FIGURE 5.2
A bill of materials with many raw materials and few end products.

FIGURE 5.3
A bill of materials with many end products and few raw materials.

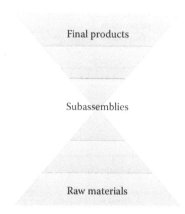

FIGURE 5.4
A bill of materials with many end products many raw materials and few subassemblies.

- The final production link in the supply chain produces a family of similar final products consisting of a variety of subassemblies. The number of possibilities for the final products is great, as well as a variety of raw materials used, so it is better to stock the relatively small number of subassemblies, and to assemble to order the specific final product ordered by the customer. For example, the PC producer mentioned earlier will store subassemblies such as hard drives, monitors, motherboards, etc. and will assemble the required computer at the request of the customer. No inventory of assembled computers will be stored because computers will be assembled on demand from the stored subassemblies (Figure 5.4).

5.4 Master Production Schedule

The master production schedule (MPS) is a list detailing future gross requirements of end products (products sold to customers) scheduled for production in each of the future periods within the planning horizon. This list serves as a plan that specifies the requirements for each type of final product, and the quantity that the company plans to manufacture in each period in the planning horizon. To implement this plan, the needed production capacity (resources used for production) should be available, as well as the necessary quantities of raw materials, parts, and subassemblies whether these are self-produced or provided by subcontractors and third parties. The MPS interface encompasses marketing, purchasing, and operations. Marketing people work with customers to decide the number of end products of each type required in each period. The MPS is the output that results from their work and drives the production and procurement activities in the supply process.

The MPS is based on future market demand in the form of customer orders, in the form of forecasts, or a combination of the two. When the MPS is based on customer orders, the goods produced are already assigned to specific customers. When forecasts are used, some future requirements can be supplied from the inventories of end products, and customers can enjoy a short delivery time. When the MPS is based on forecasts, some end products are said to be available-to-promise (ATP). In Chapter 8, we discuss how to calculate the amount of each end product ATP in each period in the future.

The management of the MPS is an important task in the delivery process because the MPS reflects future commitments to customers and the allocation of production capacity to different products in future time periods. The management of the MPS consists of the introduction of new requirements and the resulted adjustments to the MPS. The MPS is continuously updated to reflect changes in demand, new customer orders, and changes in existing orders. Integrated management of the MPS requires a joint effort of marketing, inventory management, manufacturing, and purchasing.

When the management of the MPS is not well integrated, it is very likely to create conflict between marketing, purchasing, and production. The reason is that marketing in many organizations strives to provide customers with a short delivery time and a quick response to the changing needs of customers, while production is focusing on the efficient use of resources and purchasing is trying to minimize cost of purchased parts and raw materials. Thus, marketing people are interested in large quantities of ATP and its flexibility to change the MPS according to the changing needs of the customers. Production likes to reduce the ATP to save inventory cost and both production and purchasing like to minimize changes in the MPS, especially in the near future, as such changes create inefficiencies and difficulties to manage the system.

To avoid conflict, the management of the MPS should be an integrated joint effort (of marketing, purchasing, and production) in each link of the supply chain, based on common goals and performance measures that apply to the whole delivery process. In a team of marketing, production, and purchasing, people managing the MPS together is the cornerstone of successful supply chain management.

The management of the MPS has several aspects:

- Introduction of new requirements based on customer orders or forecasts
- Updating of existing requirements in the MPS
- Monitoring and control of actual production and delivery compared to the MPS plans and taking corrective actions when needed

TABLE 5.1

Simplified Illustration of the MPS

Month/End Product	1	2	3	4	5	6
End product 1	100	90	80	100	70	80
End product 2	60	60	60	60	55	50

Each of these should be based on an integrated approach. For example, the introduction of new requirements to the MPS should take into account: (1) the needs of customers (marketing), (2) the available production capacity (production), and (3) the available inventory of raw materials and parts supplied by suppliers (purchasing). In a similar way, any change to the MPS must take these factors into account, while monitoring and control should eliminate unnecessary deviations between the MPS plan and actual production and delivery to customers.

In Table 5.1, the MPS presents two end products—end product 1 and end product 2. The planning period (also known as the time bucket) is 1 month and the planning horizon is 6 months. Gross requirements for each of the next 6 months are presented, but it is not clear if these requirements are based on customer orders or on forecasts.

In Chapter 8, other MPS formats are presented, in which the source of each requirement (forecast or customer order) is presented along with the number of units available to promise in each period.

5.5 Delivery Time and Time-Based Competition

Time plays an important role in supply chains because it is one of the dimensions of competition in the market. As explained in the previous section, ensuring shorter delivery time to customers improves the competitiveness of the organization and its ability to obtain new business (Zhang, 2003).

Delivery time of the single link in the supply chain (delivery lead time) is defined as the time from the moment a customer issues an order for a product until the moment the required product is delivered to the customer. The total delivery lead time is composed of several components, such as order processing time, time required to purchase parts, components, and raw material (supplier lead time), time required for the production process (production lead time), and the time required to ship the goods to the customer. In time-based competition, the goal is to shorten the lead time as much as possible (Nicholas, 2010).

The delivery process is a sequence of actions performed over time. The time required for purchasing production, assembly, and shipping affects the length of the delivery process. The organization that strives to improve its performance in the time domain has to examine the activities performed during the delivery process, including activities related to the flow of information, the flow of material, and the decision-making processes (Blackburn, 1991).

When the delivery process is well designed and well executed, marketing people can commit to a due date knowing that the required products will be supplied on time, to ensure customer satisfaction and loyalty. A critical factor in the development and maintenance of the organization's competitive advantage is a constant effort to shorten the lead time of the delivery process and supply in the time promised to the customer.

The lead time of the delivery process is determined by the time required to perform the three parts of the process, namely, data processing, decision making, and processing and physical transport of materials, parts, and finished products. This process is a collection of activities. Throughout the process, avoiding delays prior to performing activities, reducing the duration of activities, and performing activities in parallel are key factors affecting the lead time of the delivery process. Therefore, in order to shorten the delivery time, organizations try to examine and improve every activity related to data processing, decision making, and physical processing of materials.

This effort concentrates on the following four targets:

1. Eliminating unnecessary activities
2. Shortening the duration of necessary activities
3. Minimizing delays before, during, and after each activity
4. Minimizing dependencies between activities so that activities can be performed in parallel

5.6 Avoiding Unnecessary Activities

A critical examination of all the activities in the delivery process is a good starting point. Why is this activity performed? Is it really needed? The most important question is, "What is the added value of each activity?" A simple rule of thumb is that an activity has an added value only if the customer is willing to pay for it. It is not always possible to eliminate activities that do not add value to the customer. For example, some activities in the food and pharmaceutical industries are required by law (while having no clear value to the customer). In many cases, activities related to transportation and storage are not generating any customer value and should be reduced to a minimum. The goal is to reach the minimum possible transportation, material handling, and storage of materials (Rother et al., 2003).

An example of an efficient delivery process with respect to transportation, material handling, and storage is the system developed by the Walmart company that uses it in its retail operations across the United States. The system is based on a method called cross docking (Cohen and Keren, 2009). In this method, the company's logistics center is a node—a meeting place for trucks that bring goods from the factories and trucks that deliver goods to stores. Goods are unloaded from the trucks that bring goods from the factories and immediately loaded onto the trucks that deliver goods to the stores. Therefore, there is no need to store goods in a logistics center, and the transport time from the factory to the store is very short.

An example of unnecessary activity is multiple data input (e.g., typing the same data more than once). Typing data is inefficient and doing it multiple times (e.g., in different locations along the process) has no added value and it is a waste. Moreover, data typing is time-consuming, costly, and creates data entry errors, which reduces the quality of data and information (as explained in Chapter 4), so it is advisable to avoid this kind of activity altogether. Automatic data acquisition, such as barcode readers, RFID readers, OCR, and pattern recognition, should be used for data collection.

Another way to reduce the lead–time is to improve the process and its flow in order to eliminate some activities. For example, developing stable and reliable processes, that do not create defects, enables the elimination of quality control activities. This could be achieved by integrating quality control activities within each processing operation. In this way, the operator that performs each activity in the process takes responsibility for the quality of his or her work.

5.7 Shortening the Duration of Value-Added and Necessary Activities

After identifying all nonvalue added activities that can be eliminated and taking care of them, the next step focuses on value added activities in an effort to shorten their duration and reduce their cost, for example, by using computers to automate some of these activities or to support the person performing these activities. Time and motion studies are used to perform this analysis, and the proper way to perform such studies is taught in industrial engineering programs at universities around the world.

Data processing and decision-making activities are also analyzed, for example, an effort to improve and automate the activities of data collection (by using RFID, speech recognition, or barcode readers) or using workflow engines that manage the flow of information while monitoring and controlling the process. Development of appropriate information systems and decision support systems, and integration of these systems with databases and model bases allow further reduction in the duration of the delivery process.

5.7.1 Reducing Delays in the Process

Delays occur in different parts of the delivery process. There are three common types of delays: (1) delays in transport, (2) delays in operations (activities), and (3) delays in data processing and decision making. These types are detailed below.

5.7.1.1 Several Reasons for Transportation Delays

1. *Transportation batch sizing:* When transportation is performed in batches such as a whole container, the first product in each batch has to wait until the last product in the batch is ready. This delay can be avoided if the transfer is performed for a single product. This kind of delay may exist on the shop floor when transfer from one workstation to the next is performed in batches. In this case, the machine that is scheduled to perform operations on the batch may be idle while waiting for the whole batch to be transported.

2. *Availability of transportation resources.* For example, if several transport processes require a forklift, but only one forklift is available, the forklift becomes a bottleneck, and the waiting time for transportation increases. Another example of limited transportation resource is an employee transferring crates or a carts of products between various workstations.

3. *Poor monitoring and control.* This happens when processing is complete and the products are ready for transportation, but the workers that should transport these products do not get the required information in a timely manner.

5.7.1.2 Delays in Operations Are Created for Several Reasons

1. *Setup time:* Setup time is the time required to switch a machine or other resource from performing one operation to another, or to switch from production of one item to another. When the setup time is long, there is a tendency to increase batch sizes to reduce the number of setups, and this causes a delay in the delivery process.

2. *Process batch sizing:* When several operations are performed on each unit of the product, each batch is delayed until the completion of all units on each workstation. Thus, the time delay for a batch at each workstation is the setup time plus processing time per unit multiplied

by the batch size. To avoid this delay, an effort to minimize setup time and consequently minimize the batch size is needed. When the setup time is short, there is a lesser need for large batches and delay is avoided. By reducing setup times, reducing the size of batches is less costly and by reducing the transfer-batch delays, the delivery process can be minimized.

3. *Bottlenecks in the process* are also a source of delays. Bottlenecks are created when the load on a resource like a forklift or a machine is higher than the maximum production capability (called its available capacity). In this case, if the rate of production activities that feed the bottleneck exceed the production rate of the bottleneck, parts awaiting processing are accumulated before the bottleneck in the form of in-process inventories and the process is delayed. To prevent this type of delay, the MPS and the production schedule must be synchronized with the production rate or capacity of the bottleneck. The scheduling method called Just In Time (JIT) is based on these principles, and is discussed in Chapter 8. It is also possible to use methods based on finite capacity scheduling that take into account the available capacity of the resources on the factory floor, as explained in Chapter 8.

5.7.1.3 Delays in the Decision-Making Process Are Created for Several Reasons

1. *Uncertainty:* Delay occurs in cases where the decision makers are waiting for more information to increase the certainty of the decision results. The execution of actions often depends on such decisions and is delayed as long as the decision is not made.

2. *Serial decision making:* When data and information are transferred serially to users, decisions are made in sequence (one after the other). Even when these decisions are independent of each other, they cause delays in operational processes. Several situations lead to sequential flow of data and decisions:

   a. In some places, part of this sequential flow is a result of sticking to outdated documentation habits of a paper trail. This mode of communication of passing printed pages from table to table and via internal mail belongs to a bygone era. Computer technology should be used to transfer the same data to any number of users in parallel and to accelerate the decision making.

   b. Another reason for a sequential decision-making process occurs when information is transmitted between hierarchical levels, and different organizational functions in the form of documents, and a formal decision-making process requires the signing of documents by executives.

c. The situation is worse when decisions are made following lengthy discussions and presentations. Discussions always cause delays in decision making. The use of electronic transfers of data and electronic signatures can avoid such delays. Moreover, the establishment of a combined treatment of any supply process and the use of communication networks will eliminate many of the bureaucratic delays in the functional structure of decision making.

d. Organizations that do not have a predefined policy regarding routine decisions and a proper MIS suffer from delays in the decision-making process. In this case, routine decisions are treated as ad hoc decisions, requiring the involvement of managers. When the number of these decisions is high, decisions (or the lack thereof) can delay the entire process.

e. Manual control in the processing of data and its distribution are a common source of delays. When the control is done by regulating the flow of information, it impedes the flow of information and forces a serial process of decision making.

5.8 Quality-Based Competition

5.8.1 What Is Quality?

To satisfy customers' needs and expectations, suppliers of goods and services must be able to listen to the voice of customer (VOC) and to translate it into a product or a service that satisfies the customers' needs and expectations. Quality in its broadest sense is the ability to satisfy customers. Whether it is a product, a service, or a process, its quality is related to meeting and even exceeding expectations. Hence, quality-based competition is an ongoing effort to translate the VOC or what the customer "wants" into a design of a product or a service that satisfies the customers' needs and expectations.

Customers are satisfied when their expectations are fulfilled. Garvin (1987) suggested eight dimensions or quality performances that are frequently required or desired by the customers:

1. *Performance:* Performance measures how the product or service meets the purpose for which it was created. This refers to the main features of the product or service. For example, a car's performance is related to its driving comfort, acceleration, deceleration, safety, and fuel economy. A computer's performance is related to its processor speed, its memory storage, and its screen resolution. Understanding customers' needs and expectations is the basis of performance

analysis in the design of a service or a product. The design is always aimed at exceeding the required level of performance, as a key factor in quality-based competition.

2. *Additional attractive features:* While being secondary aspects of performance, these features are contributing toward exceeding customer expectations. These are the "bells and whistles" of a product or a service. In many cases, different customers like different bells and whistles, and in these cases good design lets the customer choose the options out of a list of such features that contribute to the quality of the product or service.

3. *Reliability:* Reflects the probability of malfunction or failure of a product or a service within a specified time. It is also reflected in the cost of maintenance and downtime of the product.

4. *Compatibility (conformance):* The extent to which a product or service complies with prevalent standards. The more the product or service shows conformity with advanced standards, the better is the quality of the product or service. Some conformances are mandatory, for example, a computer should comply with the voltage and frequency of the electrical network in the country where it is sold, as well as the standards for communications such as Wi-Fi network. A car fueling interface must comply with the gas-station nozzle size.

5. *Durability:* The operational and technical service life duration of a product is part of its quality. Durability refers to the amount and duration of use that a product provides, until the need arises to replace it (because of technical or economic considerations).

6. *Serviceability:* Reflects the speed, courtesy, and professionalism of service provided in case of failure of the product. Product reliability and serviceability are complementary. Reliable products rarely go bad and in the rare occasions that the product fails, fast and inexpensive service is provided so that the product will serve its owner again.

7. *Aesthetic:* Aesthetic products are always preferred to unaesthetic products of a similar nature. Service given in aesthetic surroundings is preferred to the same service in a less attractive environment. This dimension is related to a subjective impression of the product or service through the senses: touch, taste, sight, and smell, and reflects personal preferences.

8. *Perceived quality:* A subjective measure (reputation) related to the product or service. Perceived quality is mostly related to reputation. Reputation can be based on past experience and on partial information, but in many cases, customer decisions are based on perceived quality and aesthetics, especially when there is no accurate information about the six first performance metrics mentioned previously.

To ensure that customer needs and expectations are satisfied in a consistent and persistent way, a systematic method should be used to translate the VOC into a design of the product or service. One such method is based on the model called quality function deployment (QFD), as discussed next.

5.8.2 Quality Function Deployment Model

The QFD model, which was published by Akao (1994, 2004) supports the identification and implementation of customer requirements at every stage in the development of a new product or service. The model supports a structured design process, which can be applied in an integrated engineering environment, that is, an environment where the development process is carried out by a team representing the various functions in the organization: development, engineering, production, purchasing, marketing, maintenance, operation, and so on. Development of a new product by an integrated team of this type is known as concurrent engineering. The development team is responsible for the entire development process. The advantage of the concurrent engineering approach is that the various product requirements are analyzed early on for the whole life cycle of the product. The design, production, operation, and maintenance aspects are all considered early on, and in an integrative way in the design process. The accumulation and integration of information saves changes and modifications in the later stages of the product life cycle and ensures a design that balances the requirements of different stakeholders.

The QFD model is based on analysis of customer requirements. In its simplest form, the QFD model is a matrix that shows basic customer requirements in each row and engineering or technical parameters in each column. Each cell in the matrix represents the correlation between the corresponding customer requirement and the technical parameter. In a more detailed QFD model, a comparison to competing products in the market is also presented as well as the correlation between the technical parameters. Extension of the model that combines a number of matrices is known as the "House of Quality."

The steps to construct a simple QFD model are as follows:

1. Identify customer requirements for the product or service by actively listening to the VOC. Listening to the VOC requires active pursuit of interface with the customers. Examples are conducting a written questionnaire, a quick telephone survey, using undercover customers to talk to other customers, and simply communicating with customers asking their input on possible improvements. Summarizing the VOC yields the customer requirements. These requirements, also known as "what the customer wants," define the product features. For instance, in the example displayed in Table 5.2 (taken from the original Mitsubishi publication), the product is a car door, and

TABLE 5.2

Customer Requirements

Customer Requirements for Car Doors
Easy to close
Stays open on a hill
Easy to open
Does not leak in rain
No road noise

the following requirements are listed: The door is easy to close, it remains open even when the car is on a slope, it is easy to open, and it prevents the penetration of rain and noise into the car.

2. Determine the relative importance (to the customers) or weight of each requirement—on a scale of 1–10, with the most important requirements having the highest weight, and the least important having the lowest weight. In the example, the highest weight is assigned to "Easy to close the door," and the lowest weight is assigned to "No road noise." There may be a number of features with the same importance level, so the weight will be the same. For example, features "Easy to open" and "Does not leak in the rain, weigh the same in Table 5.3.

3. Identify engineering characteristics (parameters) of the product, which are supposed to meet the customer's requirements–, and are to be determined during the design process. These features are described in the columns in the matrix. In the example, there are six characteristics: (1) the energy required to close the door, (2) door seal resistance, (3) the force required to maintain an open door, (4) the energy required to open the door, (5) acoustic transmission of the window, and (6) water resistance. This is illustrated in Table 5.4.

4. Evaluate the relationships between customer requirements (the "whats") and technical characteristics—these relationships are recorded in the corresponding cells in the matrix. That is, a cell

TABLE 5.3

Weights Assigned to Customer Requirements

Car-door Customer Requirements	Importance
Easy to close	7
Stays open on a hill	5
Easy to open	3
Does not leak in rain	3
No road noise	2

TABLE 5.4

QFD Matrix with Customer Requirements and Engineering Characteristics

Customer Requirements	Importance to Customer	(1) Energy to Close Door	(2) Door Seal Resistance	(3) Check Force on Level Ground	(4) Energy Needed to Open Door	(5) Account Trans Window	(6) Water Resistance
Easy to close	7						
Stays open on the hill	5						
Easy to Open	3						
Does not leak in rain	3						
No road noise	2						

describes the relationship between the customer requirements in the corresponding row and the engineering characteristic in the corresponding column. Graphic symbols are used to describe these relationships as shown in Figure 5.5. In the conventional legend, a single circle describes a medium connection (the power is 3 on a scale from 1 to 10). A triangle describes a weak connection (the power is 1 on a scale from 1 to 10), and a double circle describes a strong connection (the power is 9 on a scale from 1 to 10). For example, the relationship between ease of closing a door and the energy needed for closing the door is strong (double circle), while the relationship between ease of closing the door and the door seal resistance is medium (single circle). However, there is no connection between the ease of closing the door and the energy required to open the door, so the appropriate cell in the matrix is empty (Figure 5.6).

5. Determine the importance of each engineering characteristic as a function of the relative importance of customer requirements—the process of calculation is based on multiplying the relative weight of each customer's requirement by the relative strength of its connection to the engineering characteristic whose relative importance is calculated, and a summary of each of these multiples for the engineering characteristic in question. For example, the total importance of the

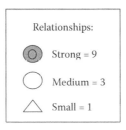

FIGURE 5.5
Legend: Relationship strength annotations.

	Importance to cust.	Engineering characteristics	Energy needed to close door	Door seal resistance	Check force on level ground	Energy needed to open door	Accoust. trans window	Water resistance
Customer requirements								
Easy to close	7		◎	◯				
Stays open on a hill	5				◎			
Easy to open	3			◯		◎		
Doesn't leak in rain	3			◎				◎
No road noise	2			◯			◯	

FIGURE 5.6
Cells in the QFD matrix.

"energy needed to close the door" is 63 since the relative weight of the requirement "easy to close" is set to 7, and its relationship strength to "energy needed to close the door" is 9. Thus, 7×9 gives the total value 63. The same value was calculated for "door seal resistance." In this case, the value is a sum of the following relationships (Figure 5.7):

$$(3 \times 7) + (3 \times 3) + (9 \times 3) + (3 \times 2) = 63$$

6. The relative weight of the engineering characteristics enables engineers to focus on the characteristics strongly related to the most important requirements of the customer.

	Importance to cust.	Engineering characteristics	Energy needed to close door	Door seal resistance	Check force on level ground	Energy needed to open door	Accoust. trans window	Water resistance
Customer requirements								
Easy to close	7		◎	◯				
Stays open on a hill	5				◎			
Easy to open	3			◯		◎		
Doesn't leak in rain	3			◎				◎
No road noise	2			◯			◯	
Importance weighting			63	63	45	27	6	27

FIGURE 5.7
QFD matrix with importance weighting.

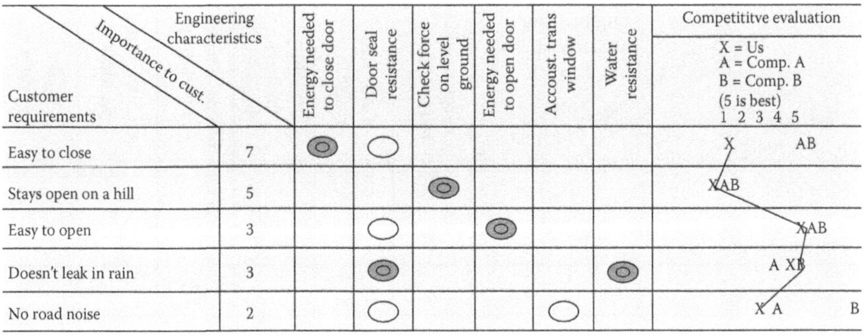

Customer requirements / Importance to cust. / Engineering characteristics		Energy needed to close door	Door seal resistance	Check force on level ground	Energy needed to open door	Accoust. trans window	Water resistance	Competititve evaluation X = Us, A = Comp. A, B = Comp. B (5 is best) 1 2 3 4 5
Easy to close	7	◎	○					X AB
Stays open on a hill	5			◎				XAB
Easy to open	3		○		◎			XAB
Doesn't leak in rain	3		◎				◎	A X B
No road noise	2		○			○		X A B

FIGURE 5.8
QFD matrix with competitive evaluation.

7. Comparison to competitors—the last column in the matrix presents an estimate of how successful our design is with respect to each customer's requirements compared to competitors' designs. The best solution gets a score of 5, while the worst gets a score of 1. The comparison is illustrated in Figure 5.8 and helps to identify the best design of competitors and learn from it. In Figure 5.8, X in the right column indicates our current product design, where A and B indicate competing products in the market. Each of the customer requirements is compared on a scale from 1 to 5, where 5 is the highest score possible. This comparison helps the design team focus on changing engineering characteristics relating to technical features in which the competition is better.

   For example, with respect to the issue of noise penetration, competitor's product B is much better than our product, which is marked with an X, so it is advisable to study the design of B in order to learn from it and to improve our design.

8. The target values of the engineering characteristics are set to meet customer expectations. The bottom row in Figure 5.9 shows the target value for each of the engineering characteristics.

9. The relationships among the engineering characteristics of the product are recorded at the top of the matrix that looks like a roof (Figure 5.10). This is done to describe the relationship between the different engineering characteristics. A detailed discussion of these relationships is part of advanced engineering design courses.

The QFD model enables engineering teams that develop new products to focus on the engineering characteristics that affect the most important customer requirements and ensure that the new products meet customer needs and expectations, and can thrive in competitive markets.

Engineering characteristics / Customer requirements	Importance to cust.	Energy needed to close door	Door seal resistance	Check force on level ground	Energy needed to open door	Accoust. trans window	Water resistance
Easy to close	7	◉	◯				
Stays open on a hill	5			◉			
Easy to open	3		◯		◉		
Doesn't leak in rain	3		◉				◉
No road noise	2		◯			◯	
Importance weighting		63	63	45	27	6	27
Target values		Reduce energy level to 7.5 ft/lb	Maintain current level	Reduce force to 9 lb	Reduce energy to 7.5 ft/lb	Maintain current level	Maintain current level

FIGURE 5.9
QFD with target values for engineering characteristics.

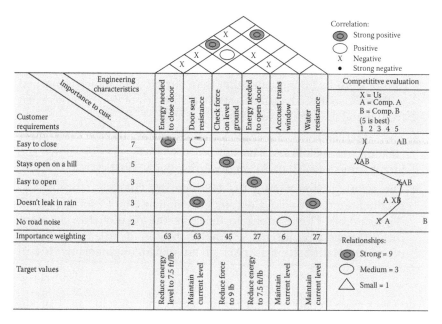

FIGURE 5.10
Relationships between engineering characteristics of the product.

Quality-based competition requires the organization to understand the needs and expectations of its customers by listening to the VOC and to produce quality products that satisfy those needs and expectations and to design and manage quality processes. A product that is supplied and supported by faulty processes will suffer from poor reputation and poor perceived quality.

In addition to QFD, several other models are used for similar purposes in quality-aware organizations. The role of these models is to make sure that products meet the customer-required quality level. These models fall into two categories: quality control models and quality assurance models.

Quality assurance is an approach based on the concept that quality should be designed and built into the product and process to prevent defects. It is worthwhile to work hard in the planning and designing phase and to train employees to prevent any damage to product and process quality. Under this approach, the organization adopts policies aimed at zero defects and should therefore avoid the need for detecting and fixing problems.

Quality control is based on a continuous effort to monitor against deterioration and failures and to prevent shipping poor quality products to the customer. Problems should be avoided by proper design and quality assurance. However, if there is a problem, it should be detected as early as possible when the product is still within the plant, by appropriate tests.

In recent years, there is a tendency to reduce quality control importance by better designs and better quality assurance. This trend is the result of a desire to minimize the number of defective products. Poor quality products cause a waste of time, money, resources, and poor reputation.

5.9 Cost-Based Competition

5.9.1 Effect of Cost on Profit

Profit is a critical competitive measure, and cost directly detracts from the profit. Therefore, cost is a major competitive performance measure. Since profit per unit of product is defined as the difference between the selling price of a product and the cost of that product, an organization can increase its profits by either raising prices or reducing costs (or both). The selling price in a competitive market affects the demand, and therefore, the volume of sales and total revenue. Although it is difficult to predict the exact relationship between the price and sales, in general raising the price tends to result in lower demand and lower sales. Therefore, raising prices reduces the volume of sales, and the resulting profit may not go up, or it may even decrease.

A safer way (usually the recommended way) to increase profits is to carefully reduce the cost of products while keeping high-quality products and processes. Reduced costs can lead to higher profit per unit if the selling price remains constant; it can even be used to reduce prices and increase demand and sales volume. In both cases, the key to increasing profits is lower cost.

To reduce costs and compete successfully in the market, the effort should focus on maximizing the value to the customer while minimizing waste (Shtub et al., 2004). In its simplest form, value is anything for which the customer is willing to pay (e.g., whatever satisfies the needs and expectations of the customer) while waste is any use of resources, money, or time that does not add to customer satisfaction, and for which the customer is not willing to pay.

Proper planning of the product and the production process should focus on the elimination of features and activities that do not generate value to the customer—for example, the transport activities of products and parts. It is interesting to note that efforts to reduce costs and shorten time bring similar results in many cases. For example, the prevention of transport operations reduces time and cost simultaneously.

Focus on creating value for the customer is the basis of Lean manufacturing (Lean production). In this approach, the focus is on creating value for the customer by eliminating waste. That is, preventing any action for which the customer does not see its value and therefore is not willing to pay for it (Nicholas, 2010).

To implement the Lean manufacturing approach, a mapping method has been developed (value stream mapping—VSM). The VSM method maps the flow processes in the organization from the order received by the customer until the product reaches the customer. The mapping process is performed in a wide range of topics: manufacturing cycle times, inventory, the movement of materials, information flow, etc. The purpose of VSM is to identify and to eliminate activities in the process that do not provide value-added to the product or service that the organization provides (Rother et al., 2003).

Reducing costs requires a combined effort of all participants in the process. The participants should focus on the following: the handling of materials, procurement of quality and cheap raw materials, minimizing inventory costs, and avoiding situations where inventory becomes unusable (dead stock). Production personnel should focus on utilizing the resources at their disposal to minimize the cost of production.

A combined effort that focuses on the different dimensions of competition is the key to organizational success. Since organizations operate in a dynamic environment, they must introduce flexibility along with attempts to increase quality, reduce time, and reduce costs.

5.10 Flexibility-Based Competition

In the modern economy, customer needs, economic conditions, demand, and competitors in the market are changing frequently. Therefore, it is very important to develop flexibility—the ability to adapt products and processes to changing requirements. The flexibility of the process, a product, or a service is expressed in the ability to change the process, product, or service quickly and at low cost; for example, to change a predetermined delivery date on request of the customer by phone, or change the type of products delivered, or change the quantities thereof. The importance of flexibility stems from the need to allow adjustment of the product or service provided to the changing market requirements. Flexibility is the ability to introduce a change in a product or process according to customer requirements or the changing environment.

The flexibility is achieved through proper process design that allows inexpensive and rapid changes. Flexibility is also achieved by designing products that enable quick changes to the product, without harming quality or increasing cost. There are various ways to achieve flexibility. One example is a continuous effort to train employees to perform a large variety of different activities. Another example is investment in flexible equipment such as computer-controlled machines and robots, with which it is possible to perform a wide variety of manufacturing operations, assembly, and transportation at low changeover cost and high quality.

In the past, products designed according to customer requirements were generally more expensive, and many factories were designed for mass production of a uniform and cheap product (mass production). An extreme example was the production line of the Ford Model T, which provided identical cars, all of them in black. Similarly, based on standardization, some fast food restaurants are serving dishes that are completely uniform, for example, MacDonald's which managed to significantly decrease costs while maintaining the quality of the process and the product, but had to give up flexibility. In the past, it was difficult to achieve a simultaneous rise in flexibility and decrease in cost. That is, improving flexibility would involve a higher cost. Modern computers allow flexibility when the means of production are controlled by a computer and changes are done by software, for example, the rapid construction of a prototype (rapid prototyping) of some products can now be done by a three-dimensional printer. Another example is the use of robots and machines operated by a computer (this is often called computer integrated manufacturing—CIM). Computer-controlled resources allow the processing of a large variety of different products by loading the appropriate software onto the computer (this is often called flexible manufacturing systems—FMS). Manufacturing flexibility is not only the possibility to change the characteristics of the products, but also the ability to increase and decrease the volume of production in accordance with market requirements and the ability to modify deliveries as per customer requirements. This capability is achieved through a skilled workforce, which means deploying a training program so that each worker possesses a wide variety of skills that can easily be directed to a new required type of work.

Development of computing power and the decline in the prices of computers, forms the basis for the flexibility in production and services. For example, mass customization based on computer technology allows Dell to assemble custom-made computers. Another example is the flexibility of Toyota mixed-model car assembly line. This line can easily adjust to fluctuations in product mix and total demand. The combination of high flexibility, high quality, low turnaround time, and competitive price was made possible by focusing on customer needs, the prevention of waste, the creation of value, and the use of intelligent computing and automation.

5.11 Summary

Organizations achieve their goals by providing products and services to their customers. Customers (including organizational customers) are always free to choose their suppliers. It is therefore imperative for businesses to gain competitive advantage to survive. Competition dimensions are time, cost, quality, and flexibility. While each dimension is very important, being competitive means having achieved some degree of each dimension. In addition, the interface between the organization and the customer is of great importance. It is crucial to understand the customers' needs and expectations and to translate the VOC into products, services, and processes that meet these needs and expectations. One of the tools that assists engineers to align their work with customer preferences is the QFD model. In this model the VOC and its requirements are translated into engineering characteristics of a product. Another important tool is the MPS. By introducing new orders and updating existing orders in the MPS, the delivery process can link to market demands. Determining the appropriate policy for management of the MPS and deciding how to link demand to production have a decisive influence on the organization's ability to succeed in the competitive field, maintain profits, and survive.

References

Akao, Y. 1994. *Development History of Quality Function Deployment. The Customer Driven Approach to Quality Planning and Deployment.* Minato, Tokyo: Asian Productivity Organization, p. 339.

Akao, Y. 2004. *QFD: Quality Function Deployment–Integrating Customer Requirements into Product Design.* New York: Productivity Press.

Bengtsson, J. 2002. The impact of the product mix on the value of flexibility. *Omega*, 30(4), 265.

Blackburn, J.D. 1991. *Time Based Competition*. Homewood, IL: Business One–Irwin.

Chan, L.K. 2002. Quality function deployment: A literature review. *Eur. J. Oper. Res.*, 143(3), 463.

Cohen, Y. and Keren, B. 2009. Trailer to door assignment in a synchronous cross-dock operation. *Int. J. Logist. Syst. Manage. (IJLSM)*, 5(5), 574–590.

Garvin, A.D. 1987. Competing on the eight dimensions of quality. *Harvard Business Review*, November–December.

Nicholas, J. 2010. *Lean Production for Competitive Advantage: A Comprehensive Guide to Lean Methodologies and Management Practices*. Boca Raton, FL: CRC Press.

Rogalski, S. 2011. *Flexibility Measurement in Production Systems: Handling Uncertainties in Industrial Production*. Berlin, Germany: Springer.

Rother, M., Shook, J., Womack, J., and Jones, D. 2003. *Learning to See: Value Stream Mapping to Add Value and Eliminate MUDA*. Cambridge, MA: Lean Enterprise Institute.

Shtub, A.F., Bard, J.F., and Globerson, S. 2004. *Project Management: Processes, Methodologies, and Economics, 2/E,* Irwin, NJ: Prentice Hall.

Zhang, Q. 2003. Manufacturing flexibility: Defining and analyzing relationships among competence, capability, and customer satisfaction. *J. Oper. Manage.*, 21(2), 173.

6

Interface with Suppliers
and Subcontractors

6.1 Procurement and Outsourcing for Gaining
Competitive Advantage

To successfully compete in today's markets, organizations must adopt a strategy that fits the market conditions, the organization's goals, and its capabilities—or develop the capabilities deemed essential. A successful strategy must provide an edge to the company in the different dimensions of competition, which are as follows: time based, cost based, quality based, and flexibility based. Frequently, some key capabilities considered by the organization as critical to its success will be missing. When these needed capabilities must be acquired, and it is possible to develop them in house (which may not always be economical or even feasible, as we will see later), devoting the time and effort to such development may not be practical or economical, or in line with the organizational strategy. Consequently, the organization may prefer to use external sources available from suppliers and subcontractors that can provide the required capabilities. It is simply impractical to expect organizations to develop all the capabilities they may need to succeed or even just to maintain their business (Barney, 2010).

Most enterprises prefer to focus on their core business and core competencies and not to develop capabilities that are external to this core. For example, most large enterprises do not see any advantage in self-operation of a cafeteria or lunch services for their employees yet they may consider these services to be a very important part of their business. Therefore, organizations commonly contract with a catering supplier that has the knowledge, experience, and, therefore, the competitive advantage in this area. In most organizations, it is possible to improve competitiveness by purchasing products and services that are not directly relevant to the organization's field of expertise from outside experts. Suppliers and subcontractors can often provide a competitive advantage by providing a product or service at a lower cost, higher quality, and higher flexibility in a relatively short time. This happens, for example, when a supplier specializes in the mass production of some components that the purchasing organization needs in small quantities, and thus, the supplier benefits from economies of scale that the purchasing organization cannot achieve (due to the small quantities it needs). Another example relates to an organization's possible limited resources (facilities, equipment, personnel, etc.) that may result in limited capacity (measured by the maximum output level for a given period). In many organizations, overtime or extra shifts can be used to add capacity in the short term. In the long term, an increase in capacity can be achieved by adding resources such as employees and equipment. All these remedies are costly and must be compared to the alternative of subcontracting or purchasing. When demand fluctuates, it may not be practical to make frequent changes in the level of resources or in resource utilization, especially when the cost of such changes is high. Changes can create quality problems and cause instability in the supply process.

The use of suppliers and subcontractors as external sources of capacity or expertise is known as outsourcing, and is common for all of the above reasons. Very few organizations do not use any outsourcing. In addition to outsourcing, most organizations get some of the materials, parts, and services they use from external sources, for example, suppliers and service providers. The process of contracting with external organizations for obtaining resources is also known as procurement, purchasing, or acquisition.

Buying products and services from an external source may take many forms. In its simplest configuration, it would be a one-time purchase of material or parts. The procurement of standard raw materials and components used for the organization's manufacturing or service processes is a repeated process, typically covered by contracts signed with the suppliers. These contracts establish a stable flow of reliable supplies required as input to the firm's operations. Another form of buying external resources is the acquisition of land, buildings, and facilities for the organization's locations. A smaller form of acquisition may involve procuring furniture and equipment needed for the operation of the organization's offices. Purchasing services could either take the form of routine support of the organization's operation (such as catering services, cleaning, and security services), or the form of contracting projects that are not in the organization's core business or area of expertise. All these examples can help increase the organization's effective capacity without requiring an investment in facilities, machinery, recruiting, and training of employees.

Purchasing from an external source may be useful in the short term, the long term, or both—and is beneficial as long as it enhances the organization's competitive advantage in terms of cost, time, quality, and flexibility. The organization strives to select the best sources and decides on the nature of the ongoing relationships with the suppliers it selects, in the aim of increasing its competitive abilities.

To ensure that purchasing from an external source is justified and will improve the organization's competitiveness, it is necessary to assess the following points carefully:

1. Rigorously compare the "do it yourself" alternative to the purchasing of any materials, goods, and services and decide whether it is more economical to "make or buy."
2. If the decision is to "buy," carefully choose the suppliers from all possible suppliers based on a clear set of criteria.
3. Decide on what quantities to buy, when to buy, and how to manage the logistics of transportation and storage.
4. Decide on the type of contract to sign with the supplier and how to manage this contract during its life cycle.
5. Decide on the process and tools to monitor and control suppliers' performance and the quality of supplied goods and services.

In the following sections, we will elaborate on each of these points.

6.2 Purchasing from an External Source: The "Make or Buy" Decisions

In many cases, it is feasible to produce materials or components or even routine services "in-house"—that is, within the organization. Even though the organization's competitors may already be producing these materials and components, their selling prices may be higher than production "in-house." The only time one should consider "making" when it is not cheaper, is for control and quality considerations. The following model is based on the fixed cost per purchased product, as well as the (different) fixed cost per produced product. This model (sometimes called "the Break-Even Model") is described in Figure 6.1.

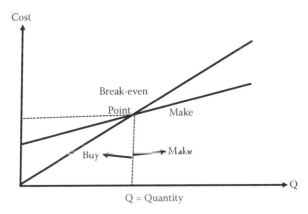

FIGURE 6.1
The break-even point model.

In Figure 6.1, it is clear that for quantities below the break-even point, it is cheaper to buy, while for bigger quantities it is better to make. Apart from price, several other factors should be considered when moving toward "make or buy" decisions.

6.2.1 Capacity Considerations

An external source of additional capacity may be needed when demand fluctuates, resources are hard to get, or when the organization does not want to invest in producing additional capacity in-house. In the short term, using an external source to acquire additional capacity can help deal with a sudden increase in workload and provide a temporary source of capacity, which may be better than using overtime and extra shifts. In the long run, the organization may choose to increase its capacity by procurement from an external source because the cost of adding capacity is too high or the resources required are not available. In services, service capacity must be much higher than demand for avoiding excessive wait and long waiting lines.

6.2.2 Know-How Considerations

When technological or commercial know-how is not available and developing such know-how requires considerable investment in the organization or may be uneconomical, impractical, or impossible (e.g., due to patents that belong to other organizations that are the potential suppliers), purchase from an external source is the best way to go.

6.2.3 Core Technology Considerations

Technologies that are at the core of the organization's products or service and are the basis of its competitive ability are irreplaceable assets. Companies go to great lengths to develop these even when it is hard to find an economic justification. Mastering such technologies usually has a long-term impact and also has the potential for future advances. Proprietary technologies are typically protected by the organization that tries to keep the knowledge on which these technologies are based under control. Therefore, organizations tend to avoid outsourcing such technologies in an effort to minimize or prevent their competition from stealing or learning to use these technologies.

6.2.4 Economies of Scale Considerations

Typically, establishing production or service capabilities involves an initial investment that makes the product or service cost a decreasing function of volume (known as economy of scale). This situation motivates outsourcing when quantities are too low. For example, when expensive special purpose facilities, machinery, and equipment are needed and the required volume of production is too low to justify the investment, other alternatives are required. In such cases, purchasing may be the preferred solution. Accumulating experience, ongoing learning, and continual improvement also decrease the unit cost and give advantages to organizations that produce (or serve) large quantities. When the market has other organizations that sell much larger quantities of a certain item, it might be uneconomical to produce this item when the quantities are too small. For example, car producers such as Ford, GM, and Toyota buy their tires from suppliers such as Goodyear or Michelin that produce tires for the whole automotive market.

6.2.5 Quality Considerations

Any product or service requires some input, and its quality depends on the quality of this input. When the quality of the components, materials, or services is high—good overall quality can be achieved in-house. Product or service quality also depends on the quality of means and processes. For example, when the provider has advanced machinery and advanced measuring and testing machines, it increases the quality of its products. However, the cost of such machinery is justified only for high enough volume production (see break-even point Figure 6.1). Thus, using the expertise and the high quality of external suppliers can be a way to improve the competitiveness of a quality-based organization.

6.2.6 Control Considerations

Procurement increases the organization's dependence on other organizations that are not under its direct control. As a result, the organization is less effective in dealing with uncertainty and risk. A strike by the supplier's workers, for example, can be destructive to the client organization. Also, the quality of components and materials is not under the organization's direct control. Monitoring and control of quality is an essential tool in the management of risk and is used as a basis for corrective actions when needed. However, when the items are purchased—that is, not produced in-house— the required corrective actions are beyond the reach of the organization. One remedy is signing a strong and binding contract. However, to ensure compliance, supplier monitoring and management is necessary, and is one of the most important tasks when outsourcing is used.

6.2.7 Accessibility Considerations

Supply logistics is an important factor that affects both costs and lead times and must be taken seriously. Distances and transportation means give an advantage to the more accessible supplier. There are times when purchasing from an outside source depends totally on accessibility constraints. For example, when the US military had a large presence in Iraq, many of its acquisitions were made in Iraq or neighboring countries only because of geographical accessibility. Another example is, when raw materials are found only in some specific locations, purchasing from foreign countries may be a necessity. Holding local inventory storage is frequently used in such cases to provide accessibility. Accessibility constraints are not always geographical. For example, the production of hazardous materials may be legal in some countries and illegal in other countries. In many cases, such materials must be purchased from countries where authorized suppliers can operate.

Decisions to buy or make are made at the strategic level (the level of policy making), as well as at the tactical and operational levels. At the strategic level, decision tends to focus on which parts of the product or process have to be made in-house. At the lower levels, ad hoc decisions such as to maximize the use of the organization's production capacity and some subcontractors when demand is high may be needed. In both cases, if the organization decides to purchase from an external source, the contract and supplier management are important issues to be considered.

6.3 Introduction to Suppliers Management

Companies get their revenues from sales while their suppliers and workforce comprise their expenses. The cost of purchases by certain companies is a substantial percentage of the amount of their sales revenue. Burt et al. (2011) found that many companies spend 30%–50% of the revenues they get from sales on purchasing. In extreme cases, the amount of purchasing reaches 80% of sales. Thus, supplier management is a central activity in many companies.

Supplier management can be divided into the following three subactivities:

1. Finding potential suppliers by requests for quotations, bids, or proposals
2. Selecting the right suppliers and negotiating the contracts
3. Managing the process of executing the contract with the selected supplier

6.3.1 Finding Potential Suppliers by Requests for Information, Quotations, Bids, or Proposals

When new technology or a new field of expertise is considered, issuing a request for information (RFI) is a popular way to find out the main players and the best practices in the new field. An RFI is published in main media channels such as popular newspapers and invites organizations to send professional information on the new subject. Organizations that specialize in this field have incentive to participate and establish their position as a leading candidate for being a supplier. Answers to an RFI assist the issuing organization in writing the specifications of the requested product or service, which is used in the request for quotation (RFQ) or request for proposal (RFP). An example of an RFI is given in Figure 6.2.

In most outsourcing cases, no new technology or expertise is involved so that the specification of the product or service required can be written directly (without RFI) as part of the request. However, pricing may not be obvious and, in many cases, there is not even a basis for its estimation. In cases where the company knows fairly well what the alternatives are, but does not know the prices, an RFQ is a powerful tool through which to get this information. An RFQ includes detailed specification of the needs, but requests potential suppliers to give price quotes on achieving the specifications. However, an RFQ is not a fully binding contract. The final contract has to be signed only after issuing an RFP and getting offers from suppliers. The RFP provides information on at least four major issues: (1) Product or service

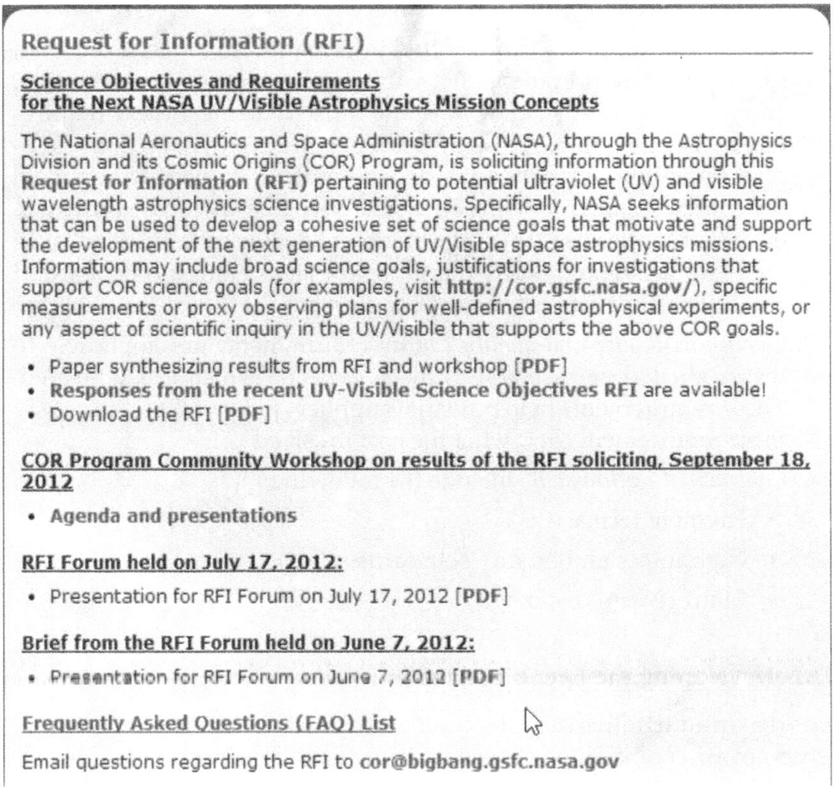

Request for Information (RFI)

Science Objectives and Requirements
for the Next NASA UV/Visible Astrophysics Mission Concepts

The National Aeronautics and Space Administration (NASA), through the Astrophysics Division and its Cosmic Origins (COR) Program, is soliciting information through this **Request for Information (RFI)** pertaining to potential ultraviolet (UV) and visible wavelength astrophysics science investigations. Specifically, NASA seeks information that can be used to develop a cohesive set of science goals that motivate and support the development of the next generation of UV/Visible space astrophysics missions. Information may include broad science goals, justifications for investigations that support COR science goals (for examples, visit **http://cor.gsfc.nasa.gov/**), specific measurements or proxy observing plans for well-defined astrophysical experiments, or any aspect of scientific inquiry in the UV/Visible that supports the above COR goals.

- Paper synthesizing results from RFI and workshop [**PDF**]
- **Responses from the recent UV-Visible Science Objectives RFI are available!**
- Download the RFI [PDF]

COR Program Community Workshop on results of the RFI soliciting, September 18, 2012

- Agenda and presentations

RFI Forum held on July 17, 2012:

- Presentation for RFI Forum on July 17, 2012 [**PDF**]

Brief from the RFI Forum held on June 7, 2012:

- Presentation for RFI Forum on June 7, 2012 [PDF]

Frequently Asked Questions (FAQ) List

Email questions regarding the RFI to **cor@bigbang.gsfc.nasa.gov**

FIGURE 6.2
An example of an RFI issued in 2012 (http://cor.gsfc.nasa.gov/RFI2012/).

requirements; (2) delivery process requirements; (3) quality assurance and quality control requirements; and (4) contractual requirements.

1. *Product or service requirements and specifications* focus on the required functional, physical, and technical characteristics. The supplier's response is crucial information that enables the organization to judge whether the supplier can provide the product or service as required.

2. *Delivery process requirements* include information such as
   - Required delivery times
   - Required delivery method (land, sea, or air transport or a combination thereof), as well as the required quality check before delivery is confirmed
   - Where the goods or services should be delivered
   - Number and size of shipments
   - Shipping arrangements and frequency of delivery
   - Required flexibility in the delivery process

   Potential suppliers can use this information to judge if they can satisfy these requirements, to estimate the cost of transportation, and to assess whether they have the capacity to meet these requirements. The flexibility required relates to issues such as will the supplier be willing to change the delivery date, the quantity supplied, the characteristics of the product or service that should be provided, etc. The supplier uses this information to decide how far in advance the customer should notify the supplier that a change is needed, and what the estimated cost of such changes is.

3. *Quality assurance and quality control* requirements are applicable to the product or service and to the process by which it is delivered. These requirements help potential suppliers judge if they can satisfy these requirements and what the cost involved is.

4. *Contractual requirements* include the following:
   - Payment terms
   - Guarantees and quality standards
   - Management of change

6.3.2 Developing the Specifications

Consider an automotive manufacturer of cars that decides to contact a manufacturer of air-conditioning systems during the early conceptual design stage of a new car and invites the manufacturer to participate in the design process of the air conditioning system for the prototype of the car. At this stage of design, only functional requirements may be known, and the expertise of the supplier is a very important factor in developing these functional requirements into detailed specifications.

A very different example is the manufacturer of printed circuit boards that receives detailed designs of a printed circuit board required for the control system of the air conditioning unit. In this case, the printed circuit board manufacturer does not design the board and only has to manufacture it based on the drawings the circuit board manufacturer gets from the manufacturer of the air conditioning system (this process is known as built to print—the production is based on a design prepared and supplied by the customer).

A third example is a wholesaler of electronic components that supplies standard components "off-the-shelf," to be installed on printed circuit boards. In this case, the manufacturer of the air-conditioning system uses the wholesaler's catalog to select the proper components and orders the components based on the catalog terms and conditions.

Each of the above cases requires information on the product or service to be provided, but the type of information and level of detail are considerably different. The information is needed to help the suppliers assess their potential ability to provide the products or services required, and to estimate the associated time and cost for the required deliverables.

6.3.3 Delivery Issues

The second issue, delivery process requirements, helps potential suppliers assess their ability to meet the delivery time and quantities. Two important questions are whether the required capacity will be available to supply the amounts required at the time they are required, and whether the allotted lead times fit the supplier's delivery process. In some cases, the supplier may have to adopt an inventory policy of manufacturing to stock to ensure timely delivery.

The logistics associated with the delivery process include issues such as the means of transportation by which the goods will be transferred to the buyer (ship, rail, truck, etc.). Examples of common terms are Free On Board, which means the seller pays for transportation of the goods to the port of shipment, plus loading costs. At this point, the buyer assumes responsibility for any risk and pays all other costs such as the cost of marine freight transport, insurance, unloading, and transportation from the arrival port to the final destination. The passing of risks occurs when the goods are loaded on board at the port of shipment. Another term is cost, insurance, and freight (CIF), which means the seller pays all costs and insurance to bring the freight to the destination.

6.3.4 Quality Issues

The third issue is related to the quality of the product or service and the requirements of the supplier's quality system. In the past, product quality was measured on the basis of the percentage of defective products with correspondence to an acceptable level, that is, the assumption was that some items shipped from supplier to customer (usually a small percentage) may be defective and that this is acceptable. Special acceptance tests based on random samples were used to assess the percentage of defective items. When the percentage of defective items in the samples was higher than the maximum specified as acceptable, the whole shipment was rejected and returned to suppliers (this has a severe effect on the supplier's delivery cost).

This approach, which is heavily dependent on acceptance tests, based on the assumption that some items are shipped defective reduces the organization's ability to compete in the areas of cost, quality, and delivery time. Tests increase the cost and extend delivery time, and a random sample does not preclude the possibility that some poor quality items will not be discovered and that this will impact the quality of product supplied to customers.

The modern approach defines the quality of products supplied in terms of their characteristics and how we use them, that is, fitness of use. Today, the accuracy and consistency of all stages in current production processes has gone from 3 sigma to 6 sigma, and recently even to 9 sigma—meaning that each process has a very small variation and defects are less than one per million. Proper adherence to quality performance metrics eliminates the need to check incoming shipments, as the effort is not on detecting defective items but rather on locating the source of problems in the process, fixing these problems, and avoiding almost completely defective production. As a result, purchasers can authorize their supplier to transfer the goods directly to the point of use in the production process, without the need for testing and storage—thereby improving the quality performance and delivery process cost and time.

In some cases, teams of experts from the purchasing organization examine the supplier's processes, including design, manufacturing, testing, packaging, etc. Such involvement helps the supplier achieve better quality, meet special customer needs, and fix problems. It may also form part of the supplier certification for transferring products directly to the point where they are used in the customer's/purchaser's production process, without the need for testing and storage. This gives the customer some control over the input to its processes.

6.3.5 Contractual Issues

The fourth issue deals with the contractual relationship between the suppliers and the buyer. There is a wide range of contractual agreements. Some are designed for one-time engagement through a bidding process, and some are designed for long-term engagement. For example, ordering the estimated quantity of the product for a year but leaving the size of the exact timing of shipments open for coordination between the buyer and the supplier.

6.4 Selecting the Right Suppliers

When the requirements are defined, the supplier selection process can start. The first step is to identify potential suppliers that can fulfill the requirements for the product and process. Information on possible suppliers is available from a variety of sources including the Internet, trade magazines, and trade associations. The objective is to find the best suppliers to contact. The traditional approach was to select the best suppliers and use the competition between suppliers to reduce the cost through a bidding process. In some cases, parts of the order are split between several suppliers, assuming that multiple sources reduce dependence on a single source; and thus reduce the risk of: a sudden increase in cost, a sudden shortage due to a strike at the supplier's premises, a longer shortage due to supplier bankruptcy, or late delivery due to an unexpected load on the carrier.

A different approach was taken by some organizations in Japan. A single supplier was selected for each part, and partnership relations were built between the supplier and the client organization. The supplier was selected after careful evaluation, focusing on the supplier's ability to improve the competitiveness of the client organization. The dependency and loyalty of these Japanese firms were well preserved because each supplier (to ensure their continued livelihood) strove to keep the customer organization satisfied all the time.

Many organizations adopt an intermediate approach, and do business with a small number of suppliers for each item purchased. The small number of suppliers allows certification processes, and the creation of long-term relationships; and competition among suppliers allows reasonable prices to be maintained.

Newman (1988) lists seven areas in which the potential buyer has to assess the potential supplier:

1. *Process capabilities:* The quality level of the process by which the supplier produces the required parts; assessment tests determine whether the quality level is adequate.

2. *Quality assurance:* Keeps the required supplier quality level over time; this is done using quality assurance practices.

3. *Financial stability of the supplier:* Assessment of financial stability includes investigating what is the risk involved in business with a supplier. In other words, the likelihood that the supplier will go out of business for economic reasons is evaluated.

4. *Cost structure:* The supplier's actual costs of materials, labor, etc.; high costs and low profit foreshadow future problems. The supplier must make some profit from the deal; otherwise, the contract will not last for a long time.

5. *The ability to perform value analysis:* The supplier's ability to professionally examine the technical and economic value of the products in the eyes of the client indicates the degree of understanding that the supplier has about customer needs regarding product characteristics and their relative importance.

6. *Production scheduling:* The ability of the supplier's system to enable efficient planning of resource utilization along the timeline. Efficient planning includes the ability of production to adapt to changes and deliver within the required delivery time. This supplier ability is crucial to improve the purchasing organization's time-based competition and flexibility-based abilities.

7. *The ability of the supplier to comply with the agreements and his/her record in this area:* Meaning, timely delivery of the contracted items and services meeting specified quality and quantity. This performance is documented and accumulated, and used for evaluating supplier performance over time. The purchaser's information systems should make it possible to monitor the performance of suppliers and their degree of compliance with the terms of contracts.

To assist in the process of selecting suppliers, the ISO-9000 series of standards provides a lower threshold (with stress on the quality of processes and quality of management at the suppliers' premises). Many companies require that their suppliers comply with ISO-9000 standards. Another standard system used in relation to suppliers is the supply-chain operations reference-model (SCOR). SCOR is used within the context of a supply chain, and spans the range from the suppliers of the direct suppliers of an organization, to the customers of its direct customers. For example, an SCOR for an organization that builds engines would include its direct supplier of the main body block and the next supplier—a metal mining supplier. The SCOR also include customer: the car assembly plant, and its customers: wholesalers and dealers.

6.5 Managing the Process: Contract Management

When a supplier is selected, the next step is to decide on the type of contract to be signed.

There are different types of contracts. Some examples are as follows:

- *Fixed-price contract:* This is a contract where payment is fixed and agreed on between the parties. In this type of contract, the risk of cost overrun is on the side of the vendor (the supplier).

- *Cost plus fixed-fee contract:* This is a contract where the buyer promises to cover all the costs plus a predetermined fee that was agreed on at the time of contract signing.
- *Cost plus incentive fee contract:* This is a contract where a larger fee is awarded if the vendor/supplier meets or exceeds performance targets, including cost savings.
- *Cost plus award fee contract:* This is a contract where the buyer pays a fee based on the vendor/supplier's work performance.
- *Cost plus percentage of cost:* This is a contract where the buyer pays a fee that rises as the vendor/supplier's costs rise.

The selection of the right form of a contract is based on the degree of uncertainty in the definition of the product or service, the degree of uncertainty in assessing the cost, and of course, market forces, such as supply and demand in the existing product or service, and the relative merits of the contract viewed by each of the sides (the supplier and the purchaser).

After signing the contract at the end of the selection process, the focus shifts to the management of the ongoing relationship with the supplier or contract management. Pence and Saacke (1988) define three categories of relationships between buyers and suppliers, focusing on quality issues:

1. *Testing:* The buyer uses product testing to prevent defects. This is usually done in situations of one-time procurement, when the buyer does not know how good the supplier's processes and products are, and relies on acceptance testing of the goods.
2. *Prevention:* The buyer uses his or her expertise to teach the supplier how to achieve the required product and process quality, with the aim of building a foundation for a long-term relationship; the buyer helps the supplier develop defect-free manufacturing processes in an effort to build cooperation with the supplier.
3. *Partnership:* A long-term relationship between the buyer and the supplier is built, based on teamwork between the two parties in the design, manufacture, and supply of the goods and services as required.

    For example, Toyota helps suppliers set up factories close to its plants. The purpose of this is to reduce inventory and transportation costs and improve the delivery process. As a result of physical proximity between supplier plants and Toyota assembly plants, shipping costs are reduced and the relationship between Toyota and the supplier is carefully maintained. When this type of close partnership is achieved, the buyer and supplier can align their information systems to improve the flow of information and, consequently, improve the decision making. Such capabilities enable the supplier to send parts directly to the assembly line, to the point where they are needed (Ihara, 2007).

The level of cooperation between the buyer and supplier can take different forms. The traditional approach is based on a bidding process where the winning bidder receives a one-time contract that defines the required goods, quantities, due date, cost, and terms of payment. In case of nonstandard items, a technical specification is added. This approach usually requires testing. This means that products are shipped to a reception area (sometimes a special warehouse), where they are tested and stored until they are needed for the production process. The added cost of handling components or materials, holding them, testing them, and shipping them is quite significant.

A different approach is to sign a long-term (e.g., annual) contract with a supplier, based on a cost per unit, the estimated total requirements and the promised delivery time. The customer's (buyer's) planning and control system periodically issues supply orders directly to the supplier's planning and control system. When this form of cooperation is used, there is no need to negotiate the price and terms of each order during the agreement period, and the two sides build long-term relationships. When both parties are using information systems, such as a material requirements planning (MRP) system or an enterprise resource planning (ERP) system, it is possible to link the systems and transfer requirements for periodic shipments electronically.

A third approach is based on the idea of "just in time" (JIT). In this case, the supplier sends frequent shipments (sometimes several times a day) directly to the assembly line or production line. This is possible when there is an established supplier, and a long-term system based on a relationship of trust. In such cases, the supplier is a true partner in the production process, and therefore, executes quality assurance testing before making the shipments. This form of partnership can develop in situations where the buying organization transfers a whole process or a whole function to one of its suppliers, which is known as outsourcing (Chapter 6).

6.6 E-commerce and Supplier Management

E-commerce is commerce via computer networks and information communication technology (ICT). In case of standard products or orders, the client can purchase online using the supplier's web portal. In other cases, client and supplier may communicate through the Internet in various ways such as email, Skype, chat, etc. This effectively eliminates transaction and communications costs.

E-commerce is divided into two types:

1. *B2C (Business to Customer):* This is the case when the customer is located at the end of the supply chain. This type includes businesses such as Amazon and Dealextreme online and shopping sites that sell products directly to the customer.

2. *B2B (Business to Business):* B2B is a key component in many supply chains supporting purchasing and supplier management. B2B business agreements are made between businesses over the network. Proponents of B2B procurement systems claim that the costs of the procurement process can go down significantly.

As early as 1994, CommerceNet was a pioneer in promoting the use of the Internet for commercial purposes. The organization's pioneering activities are reflected in several initiatives that we use even today:

- Development of a secure system for making online transactions
- Development of open trading network online
- Establishment of a regulatory agency
- Conducting surveys for understanding the demographics and needs of users using the Internet for e-commerce

Turban et al. (2000) describe the effect on General Electric Lighting Company when switching to an online electronic procurement system. In 1996, GE arrived at the conclusion that its procurement system was not efficient enough. The procurement process included a number of activities that took a long time and required resources. Before the change, the procurement departments in GE's many plants would individually send hundreds of daily bids (RFQs) for mechanical low-value items. For each request, the process included a number of activities: retrieving the item from the database programs, locating the requirements, processing, aggregation of requirements, preparing envelopes, and sending mail. This paper-handling process took about seven days. The process consumed labor resources at a level that allowed GE to send only two or three suppliers tender packages in parallel.

In an effort to improve the procurement system, the company decided to switch to an electronic procurement system based on the Internet. In this system, the purchasing department receives the bid requirements electronically and sends packages of tender information to suppliers around the world via the Internet. The system pulls the appropriate data from the database automatically. Two hours after the start of the procurement process, all the necessary information is forwarded to potential suppliers via email or fax and gives them a period of seven days to bid. Suppliers return the bids electronically, and the contract can be signed on the same day.

The advantages of this procurement system are as follows:

1. The labor invested in the procurement process has decreased by 30%. At the same time, material costs were reduced by 5%–20%.
2. More time became available to spend on the procurement of strategic items, instead of spending it on paperwork.

3. Before the change, locating suppliers and preparing tenders took between 18 and 23 days; after the change, this process took 9–11 days.

4. All transactions are handled electronically, changes are immediately updated in the system. This creates uniformity of information contained in the information system and reliable documentation of the procurement process.

5. GE's purchasing departments around the world share their information on the best suppliers. In 1997 alone, GE Lighting found seven new suppliers via the Internet, including one whose bid was lower by 20% compared to the next lowest bid in the tender.

Hofacker (2001) points out several features of purchasing through e-commerce:

- Internet technology reduces the operational problems that come with locating suppliers, negotiating with suppliers, contract monitoring, and handling of financial arrangements.

- The Internet allows several companies to band together virtually and take advantage of their group size to achieve discounts and better conditions for future purchases from suppliers.

- The Internet reduces the costs of searching for suppliers, making it easier and faster to find alternative suppliers, while providing more information.

- The negotiation process is less time-consuming because the Internet increases the number of possible modes of communication: email, chat, web pages, and more.

6.6.1 Inventory Management: Cost/Benefit Considerations

As discussed in previous chapters, organizations can use inventory of raw materials, parts, and finished products to enhance their competitive advantage in time-based competition. The example discussed showed how to shorten delivery times by carrying inventory of finished products that the customer can buy off the shelf. This is one reason to carry inventories of one type. However, organizations carry other types of inventories as well.

For example,

- Raw materials inventory
- Inventory of parts and components needed for the finished product
- Work in progress inventory
- Inventory of materials that are not used as part of the finished goods

The decision to carry inventory and how much to carry should be based on a proper trade-off analysis. Trade-off between the benefits of carrying inventories and its cost is an essential part of inventory management. To carry out this analysis, the industrial engineer has to understand the different types of inventory, their benefits, and associated costs. Trade-off analyses answer the questions regarding the types of inventory needed, the right quantities, and where these inventories should be stored. This is discussed in the following sections.

6.7 Benefits of Inventories

Because of their costs, maintaining inventories should be considered only if they enhance the competitive advantage of the organization. The benefits of having an inventory on hand come at a cost, which must be evaluated. It is important to perform cost–benefit analysis to decide what types of inventory to carry and in what quantities. In this section we focus on the benefits, while in Section 6.8, we discuss inventory costs.

The major benefits are as follows:

1. Time-based competition
2. Cost-based competition
3. Technological consideration

6.7.1 Time-Based Competition

This issue was already discussed in Chapter 5. Inventories can bridge the gap between a short delivery time required by customers and a longer lead time dictated by various constraints. It is possible to shorten the time required to manufacture a product by maintaining inventories of raw materials parts, subassemblies, or even finished products. When finished goods inventories are carried, products can be delivered to the customer instantaneously from inventory. When the customer does not require instantaneous delivery, it might be better to store subassemblies in inventory if assembly time is not too long compared to the delivery time required. When the allowed lead times are long enough, it may suffice to hold only raw material inventories that save procurement and transportation time. In such cases, the lead time includes the time to process these raw materials into parts and components of the finished product. The decision on what inventory to hold depends on the market competition. For example, when the market demands to be supplied "off the shelf," the only way to compete is to keep inventories of finished products. The result, however, is not only additional cost but also loss of flexibility as only the variety of products kept in inventory can be supplied.

Dealing with uncertainty: Demand uncertainty is part of any B2C organization. In a B2C organization, no one knows exactly how many customers will come tomorrow or how much he or she will buy. So without inventory, these demand fluctuations cannot be met. The demand in B2B organizations is different because the demand fluctuations are already attenuated by the B2C inventories. Therefore, in B2B organizations, the fluctuations are less of an issue, but trends and seasonality are much more of an issue. The forecasting error may cause two types of problems:

1. The forecast is higher than actual demand and excess production creates surplus inventories.
2. The forecast is lower than actual demand and shortages are created.

Avoiding shortages: When the cost of shortages is high—lost sales and loss of reputation—the decision is usually to protect against shortages. The protection could be maintaining a constant inventory, or overproducing a product (producing more than the forecasted demand), which if continued may cause accumulation of inventories. This is known as a buffer inventory as it buffers or protects against shortages.

Reducing dependence (decoupling): In the supply process, when the demand is seasonal with maximum demand in one season and a substantially lower demand in other seasons (e.g., demand for swimming suits or ski equipment) and changes in production rate are expensive, it might be beneficial to produce at a constant rate, accumulating inventories at low demand periods and using this inventory in seasons of peak demand.

Alternatively, the need to decouple may arise when a relatively slow process in the production sequence is fed by a much faster production process. The slow production process is a bottleneck that limits the total output of the system. In this case, a decision to run the bottleneck process in a second shift or as overtime, while the rest of the production process is operated in a single shift at regular time, may solve the problem. In resolving the problem this way, however, inventories are created during regular time, and there will be parts waiting to be processed at the bottleneck juncture during the second shift or overtime.

6.7.2 Cost-Based Competition

Inventory may be the result of trying to get discounts and reduce costs while supporting cost-based competition. For example, consider that the organization has the opportunity to purchase raw materials at low prices in the commodity market. The price of some commodities is subject to significant fluctuations and by purchasing large quantities of a commodity needed for the production process, the organization can benefit from substantial cost savings, and use these inventories in the future. Another example is when savings are possible due to economies of scale, when the per-unit cost is a function of the quantity purchased. The larger the order, the lower the cost per unit. In such cases, there may be justification for purchasing large quantities, storing them, and using the items over a period of time (such a decision must consider the resulting inventory holding cost). Another example is the case when a large fixed cost is associated with each order. In this case, large orders divide the fixed cost over a large number of units and thus reduce the cost per unit.

In addition, the cost of shipping can be lowered at times by purchasing and transporting large quantities. For example, the cost of shipping by containers is calculated based on the number of containers and not by weight, so to save on shipping costs, companies often purchase raw materials and parts in quantities that fill containers even when this creates inventories of these raw materials and parts. Similarly, finished goods that are shipped by containers are shipped in quantities that utilize all the space in the container.

6.7.3 Technological Considerations

In some production processes, setup or changeover cost is substantial. This cost is a function of the time and material needed to switch from one operation to another or from one product to another. When setup time (and cost) is high relative to the time it takes to process a single product, it may be possible to save on the number of setups and the associated cost by running large batches after each setup. Processes whose cost is about the same for a single product unit or several units are another example of leveraging inventories by exploiting technology. For example, a substantial component of the cost of heat treating in a furnace is the cost of energy. Filling the furnace to its capacity is an obvious way to save energy costs. However, such a practice creates a parts inventory.

The above examples of benefits associated with inventories show that inventories can help organizations in developing and maintaining competitive advantage, but these benefits must always be understood as being only one side of the cost–benefit analysis. The costs of inventory are discussed next.

6.8 Costs Related to Inventory

Inventory costs can be classified into three categories:

1. *Capital cost:* This is the cost of maintaining the inventory, as well as the facility and equipment used for holding and handling the inventory. The capital cost includes the interest that the money spent

on inventory, facilities, and equipment could make if invested in a bank and the stock market or, alternatively, the interest paid to the bank (or to issued bonds) when money is taken as a loan. The inventory cost includes the costs of all materials, components, and work hours invested in the inventory. The facility cost used for holding the inventory includes the area, the development cost of this area (e.g., paving access roads or paths to and within this area, and laying foundation for water, electricity, and communications), the value of the building used to house the inventory (e.g., storerooms), the adjustments for holding inventory (such as installing shelves, bins, or material handling equipment, e.g., fork lifts or conveyor belts).

2. *Operational costs:* This is the cost to operate the inventory system including salary costs of employees who manage and operate the inventory system (receiving, testing, storing, maintaining the information, retrieving, and ordering). It also includes the cost of energy required for air conditioning and lighting, and the cost of fuel to operate material handling equipment. On top of these, operational costs are overhead costs such as taxes that apply to facilities, etc.

3. *Costs related to risks:* Inventories can be damaged or lost due to fire, water, theft, etc. These costs or the cost of insurance against these risks are part of the cost of carrying inventories.

The decisions of what and how much inventory should be carried should be based on the analysis of the costs listed above and the benefits that the inventory generates. While in the past studies have focused on inventory cost minimization, the current trend is to examine the issue in a broader perspective and to take into account the disadvantages of inventories whose cost is hard or impossible to estimate. For example, Toyota developed the approach known as Just in Time (JIT) (discussed in detail in Chapter 8), which is based on the philosophy that the organization should strive to expose and solve problems in its operations. These problems (e.g., machine failure) generally cause short breaks in the production flow but the presence of inventory may hide these flow fluctuations and the problem remains hidden. The JIT recommendation is to expose such problems by gradually reducing the inventory to a minimum level. By exposing the fundamental problems and then solving them, the organization gains a competitive advantage without having to bear the costs resulting from holding inventory.

Inventories of raw materials, work in process, and finished goods exist at different stages of the supply process and may differ in various ways from one another. The different inventories should be managed by proper models and governed by the right policies. However, such models and policies are not necessarily the same for all inventory items. Accordingly, for each inventory item, models and policies should be examined and determined or developed. For example, there are items with a very long shelf life of several years, while there are other items that can be stored for only a few days at most (perishable items). Obviously, one policy cannot fit all the different items in stock, and usually, a method for classifying different items for the purpose of inventory management is needed. A common method for classifying inventory items is ABC analysis or Pareto analysis.

6.8.1 Pareto Analysis (or ABC Analysis)

Vilfredo Pareto was an Italian economist who developed the Pareto principle based on his observation in 1906 that 80% of the land in Italy was owned by 20% of the population. To his surprise, he found this ratio in many other phenomena. This observation led to the development of the Pareto chart, which contains both bars and a line graph, where individual values are represented in descending order by bars, and the cumulative total is represented by a line.

In inventory management, the vertical axis is the percentage of value of inventory items out of the total value of inventory. The right vertical axis is the cumulative percentage of the total cost. Because of the decreasing order of item value, the cumulative function is a concave function. A Pareto chart is used to highlight the most important (costwise) inventory items.

The Pareto principle is also known as ABC analysis, in which inventory items are classified into the following three categories:

1. *Type A:* The highest value items—about 20% of the inventory items whose cumulative value amounted to about 80% of the cumulative inventory value.
2. *Type B:* About 30% of the inventory items whose cumulative value amounted to about 15% of the cumulative inventory value.
3. *Type C:* About 50% of the inventory items whose cumulative value amounted to about 5% of the cumulative inventory value.

ABC categorization is very popular in inventory management and typically results in different inventory management approaches. Type A items are the most important, and therefore, get more attention. While detailed optimization is typically applied for Type A items, Type C items such as screws, nails, etc. are usually held with quantities that may last for several months and are replenished periodically (e.g., once every one or two months). Type B items are usually treated in a way that strives to optimize their cost via generic models such as the economic order quantity (EOQ) (discussed in Section 6.9.1).

An alternative approach to using the Pareto principle or the ABC analysis is using the annual usage value (AUV). Instead of using the inventory value as the basis for the ABC analysis, one can use the AUV, that is, the product of the annual demand for an item and the cost per unit. AUV is widely used in the retail sector. For example, in the supermarket, the most important items (Type A items) are determined by multiplying the number of units sold in each period by their unit cost; whichever item gives the largest number is a Type A item.

Analysis of inventory items based on cost or on the basis of annual usage value allows the industrial engineer to develop an inventory management policy for each group. Typically, such a policy tends to limit the quantity of inventory items A and have supply items on hand exactly at a time when they are needed. Type C items, on the other hand, can be ordered in large quantities to achieve savings through economies of scale.

An example of a very simple model of inventory management is the approach for managing inventories of Type C items, the two-bin model. The capacity of each bin is big enough to cover the largest possible demand during the lead time required for the replenishment of this item. Whenever the current bin is emptied, the inventory in the second bin is used, and a shipment to fill the empty bin is ordered. This method was widely used before the era of computerized inventory management. It is simple, inexpensive, and easy to apply. It provides an inventory buffer at a reasonable cost, and protects against uncertainty regarding Type C items.

Consider the manufacturing process of passenger aircrafts. Companies such as Boeing purchase many parts in order to manufacture the planes. For example, engines are a Type A item. The engines, which cost millions of dollars each, are managed in a JIT fashion to minimize their inventory holding cost. They are ordered to arrive on the day of their installation in the aircraft. The organization avoids keeping engines in stock because the cost of holding the engines in inventory is high (due to the high cost of these items).

On the other hand, rivets that cost a few cents are Type C items. It would be inconceivable to stop an airplane assembly for lack of rivets. Therefore, a company in this industry has a bin for each type of rivet. Each full bin should last for at least three months and the company fills all the bins once a month. In this way, no shortage ever occurs. For bigger Type C items, two bins are held for each part type. Each bin holds enough parts for a few weeks. When one is empty, the second bin is opened and a new bin is ordered.

6.9 Inventory Management Models and the Assumptions on Which They Are Based

Inventory management models provide a recommendation for when to order an inventory item and how much of it to order. As any other model, these models are based on simplifying assumptions and it is possible to classify inventory management models according to the assumptions on which they are based. Next, we list some main assumptions used in inventory models:

- *Dependencies between inventory items:* Simple models assume that inventory items are independent (e.g., the demand for these items is not dependent on the demand for other items). Complex models assume that inventory items are dependent on each other (e.g., they share the same shipping container) or the demand for one inventory item is dependent on the demand for other items (e.g., the demand for ski boots is correlated with the demand for other ski equipment).
- *Number of inventory items managed simultaneously:* Simple models treat inventory items separately (based on the assumption that inventory items are independent of each other). Complex models treat all the inventory items that are dependent on each other simultaneously. For example, there may be inventory items that are manufactured on the same machine, and a setup time is required when switching from one type of item to the next. The model for this case will take into account all these items and will try to find out how much to order of each item and when to order it based on data for all relevant items.
- *Uncertainty:* Simple models assume that all the information the decision maker needs is available and perfectly accurate. Complex models assume that some information may not be available, and at best, it is possible to forecast this information—which is subject to forecasting error (e.g., information about future demand may not be known as well as information about future cost of inventory items or the cost to carry these items in inventory).
- *Optimality:* Simple models can be solved to optimality (the best possible solution for the model is found). Complex models, especially those that take uncertainty into account, provide good feasible solutions but not necessarily the best possible (optimal) solutions.

Industrial engineers learn when and how to use the different models. In the following section, a basic model is illustrated and discussed.

6.9.1 Economic Order Quantity Model

A simple inventory management model is the economic order quantity (EOQ) model. More sophisticated models can be found in Hugos (2011).

The EOQ model was developed by Harris (1913) and is based on the following assumptions:

1. Demand rate (number of units per time period) is known and is constant; that is, it is spread evenly over time.
2. Immediate delivery to customers is required; that is, no shortages are allowed.

3. Delivery lead time is constant and known (it does not depend on the size of the order or the capacity of the delivery process of the supplier).

4. The cost per unit is constant and known (it does not depend on the size of the order; there are no quantity discounts and it does not change over time).

5. The fixed cost per order is constant and known (it does not depend on the size of the order; there are no quantity discounts and it does not change over time).

6. Inventory holding cost per unit per period is constant and known. It can be estimated based on the average inventory level.

7. The objective is to minimize the total cost per period.

Based on these assumptions, the inventory level along the timeline is depicted in Figure 6.3.

Based on these assumptions, the following notation is used to formulate the model:

c = Purchase price per unit or unit production cost in a production process

Q = Order size (decision variable)

Q^* = Optimal size of order

D = Demand per period (e.g., annual) in units

K = Fixed cost per order; the cost includes the transportation, handling, and administrative work per order and/or setup cost per order in a production process

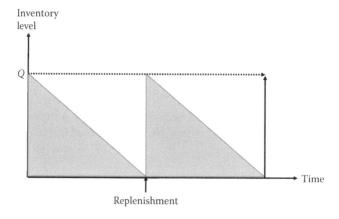

FIGURE 6.3
Inventory-level timeline according to EOQ with order quantity Q.

h = Inventory holding cost per unit per period (e.g., annual holding cost per unit)

Based on the assumptions, the different cost components can be expressed in terms of the above notation:

- Purchase or production cost of units per period is $c \times D$
- Ordering cost (and/or setup cost) per period: This cost is composed of the fixed cost K of placing and handling the order, times the number of orders per period D/Q. Ordering (or setup) cost per order = $K \times D/Q$
- Inventory holding cost per period: $h \times Q/2$ (inventory holding cost per unit per period, times the average number of units in inventory)

The total cost (TC) per period is the sum of these three components:

$$TC = c \times D + K \times \frac{D}{Q} + h \times \frac{Q}{2}$$

If no discount is offered for increasing the order, then the direct cost ($c \times D$) is a constant, so optimizing the order quantity includes only the holding cost and the ordering cost.

These two cost components and their sum are illustrated in Figure 6.4.

To determine the minimum total cost per period, we partially differentiate the total cost with respect to Q and set the derivative to 0:

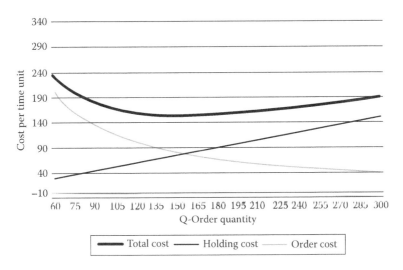

FIGURE 6.4
The holding cost, order cost, and total inventory cost for optimization.

$$0 = -\frac{DK}{Q^2} + \frac{h}{2}$$

Solving for Q gives Q^* (the optimal order quantity):

$$Q^{*2} = \frac{2DK}{h}$$

$$Q^* = \sqrt{\frac{2DK}{h}}$$

6.10 Summary

Chapter 6 presented important points in the interface with suppliers and subcontractors. The use of external sources of knowledge and capacity to enhance the competitive advantage of the organization is very important as organizations try to focus on their core competencies and to buy everything else to reduce cost, increase flexibility, reduce lead time, and improve quality. Each link (organization) in the supply chain must decide on what to focus and use external sources as a way to improve its competitive position. In areas that are totally new to the company, a RFI should be issued to acquire initial information about the subject and the market. A RFQ is a tool for establishing or improving cost estimation. Finally, a RFP is the real binding tender, and it is from the responses to the RFP that the best supplier's offer is selected.

A competitive edge may be attained in different ways depending on the selected dimensions: cost, time, quality, and/or flexibility. The management of operational processes, logistic operations, quality procedures, inventory policy, and outsourcing policy are the keys factors in achieving a competitive advantage in one or more of the competition dimensions. These factors and their relationship to the competitive dimensions are presented and discussed deeply in this chapter.

References

Barney, J. 2010. *Gaining and Sustaining Competitive Advantage.* 4th edn. Englewood Cliffs, NJ: Prentice Hall.

Burt, D., Petcavage, S., and Pinkerton, R. 2011. *Proactive Purchasing in the Supply Chain: The Key to World-class Procurement.* New York: McGraw-Hill.

Harris, F.W. 1913. How many parts to make at once. *Oper. Res.*, 38(6), 947–950.

Hofacker, C.F. 2001. *Internet Marketing*. 3rd edn. New York: Wiley.

Hugos, M.H. 2011. *Essentials of Supply Chain Management*. 3rd edn. New York: Wiley.

Ihara, R. 2007. *Toyota's Assembly Line: A View From the Factory Floor*. Melbourne, Australia: Trans Pacific Press.

Newman, R.G. 1988. Single source qualification. *JSCM* 24(2), 10.

Pence, L.J. and Saacke, P. 1988. A survey of companies that demand supply quality. *ASQC Quality Congress Transactions*, Milwaukee, IL, USA, pp. 715–722.

Turban, E., Lee, J.K., King, D., and Chung, H.M. 2000. *Electronic Commerce: A Managerial Perspective*. Englewood Cliffs, NJ: Prentice-Hall.

7

Scheduling

In this chapter, we discuss the time dimension in operations planning of an organization, with focus on the operations at hand (daily, weekly, and monthly operations). The chapter does not discuss strategic scheduling, supply chain scheduling, or project scheduling (see Chapters 3 and 11). Here, the focus is on the day-to-day operations of a single organization. We think of the organization as a typical link in the supply chain and analyze its internal operations. These operations, along with the flow of material through the organization (the link in the chain of organizations), the flow of information, and the decision-making processes, transform the link's input (raw materials, parts, subassemblies, or even finished goods from suppliers) into the link's output, that is, the goods and services that the organization/link provides to its customers.

We discuss general concepts as well as specific scheduling tools and models for scheduling. We start with a single-machine scheduling and then move to the two layouts discussed in Chapter 2: the job shop and the flow shop. Finally, we discuss the just in time (JIT) scheduling approach and the theory of constraints (TOC). These are general scheduling philosophies that can be used in any type of organization or link of the supply chain.

7.1 Introduction to Operational Scheduling

This section introduces the topic of operations scheduling and provides a broad overview of this topic. Typically, operations scheduling comes after product/service quantities are determined. This means that capacity planning and material requirements planning (MRP) (and frequently even lot size optimization) occur prior to operational scheduling. Therefore, operational scheduling deals mainly with the time dimension of the given activities.

This focus shall remain on the industrial side of scheduling. However, the service side of scheduling is broad and has its own scheduling models. Some examples for service models are timetabling, reservations, scheduling in entertainment and sports, workforce scheduling (and rostering), crew scheduling, and priority scheduling. In general, service processes apply mostly to people, and the order of service is the order of arrival (first come first served—FCFS).

Each scheduling model is based on assumptions. These assumptions simplify the real-life situation regarding the layout of the production system, the presence of uncertainty, the type of jobs processed, the organization's goals, constraints governing the organization, and the environment in which it is operating.

The solution to the job shop scheduling problem is a list of operations and the job (work order) they belong to, along with the planned start time and the planned end time of each job on each machine that has to process it. Since each job has a unique routing, the number of possible combinations grows rapidly with the number of different jobs, the number of different operations, and the number of different machines. This creates a difficult combinatorial problem.

Some basic elements of machine scheduling include the simple Gantt chart, flow time, due dates, and delays.

7.1.1 Simple Gantt Chart

Gantt charts are composed of resource time-lines along with the graphical division of the timeline of each resource into intervals. Each interval is dedicated to a specific activity. The name or identifying number of the operation or activity done in each interval appears clearly within or above the interval.

7.1.2 Flow Time

Flow time is the time that the order, part, or product (or customer) spends in the system; this is measured from the time of arrival to the time of departure, including wait and processing.

7.1.3 Due Dates

Due dates are delivery deadlines usually imposed by customers (through their order), but in some cases may also be artificially imposed by a company's policy. The due date for each process is used to determine the process earliness or tardiness.

7.1.4 Delays

If an activity or process is finished after its due date, it is considered "late" or "tardy." Its tardiness is the positive difference between the finish time and its due date. The general delay is the sum of the amount of all delays.

These basic terms are now illustrated using a small example. The example data are summarized in Table 7.1.

Note that the completion time is a cumulative sum of the process time and waiting time. It is the figure to designate the start of the next process, and the end of the previous process in the Gantt chart, as shown in Figure 7.1.

TABLE 7.1

Data of Seven Tasks on One Server/Machine, in an FCFS Schedule with Completion Times and Tardiness

Task No. i	Process Time (Days) t_i	Due Date d_i	Completion Time (Days) c_i	Delays T_i
1	3	16	3	0
2	5	18	8	0
3	1	18	9	0
4	2	9	11	2
5	4	12	15	3
6	3	15	18	3
7	5	19	23	4

The basis for scheduling decisions is the goals the organization sets. Possible examples of goals are as follows:

1. Timely completion of all jobs (work orders), according to the master production schedule (MPS).
2. Completion of all jobs (work orders) as soon as possible.
3. Minimization of process inventory (jobs waiting processing) in the shop floor.
4. Maximization of the utilization of resources by shortening the periods in which they are idle or being adjusted for the next job (setup time of machines).
5. Minimization of cost by using less-expensive resources (such as using regular business hours, and avoiding the use of overtime and additional shifts).

The schedule must be feasible, that is, satisfy the relevant constraints. Examples of typical constraints are as follows:

- Each machine can perform one operation on a single product unit at a time.
- Only one machine at a time can process a product unit.
- Jobs must be processed according to their predetermined routing.

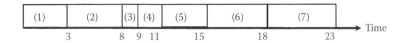

FIGURE 7.1

Gantt chart for the example of Table 7.1 (single server).

- A machine operating on a product unit cannot be stopped to process another unit and then complete processing of the first unit.
- Setup of machines may be required when switching from one operation to another operation.
- Setup time depends on the next operation as well as the previous operation performed on this machine.

An important constraint is imposed by the limited capacity of the job shops' machines. For example, if the number of hours required for processing all the jobs on any machine during the planning period is larger than the number of hours available on this machine during the planning period, a capacity constraint is imposed. Furthermore, the number of hours required for processing all the jobs on any machine during the planning period may not be more than the number of hours available on this machine during the planning period. However, if setup time is required on this machine, the total processing time and setup time may exceed the machine available time during the planning period. Consequently, a capacity constraint is imposed.

A possible solution in the case of capacity constraints with setups is to plan groups of jobs so that each group contains a large number of identical items with no need for setups between them (known as a batch). The large batch size minimizes the number of setups and saves setup time. Processing large batches of the same product on the machines saves time but it results in a relatively high in-process inventory.

7.2 Single-Machine Scheduling

The simplest approach to scheduling is to focus on the single machine and schedule it according to one priority rule. The appropriate rule is selected according to the desired target that fits the specific production environment and the relevant corporate priorities. Priority rules based on a single parameter are known as simple rules. Priority rules based on a combination of a number of parameters are called complex rules. We start by introducing simple priority rules as described next.

7.2.1 Simple Priority Rules

1. *First in first out (FIFO):* This rule is based on the order of arrival of jobs to the specific job shop machine. The jobs that arrive first are processed first. When the current job is finished, the job in the queue in front of the machine that arrived first is processed first. The

advantage of the FIFO rule is in its simplicity. In service systems, the rule has another advantage as it generates a "fair" plan so a person that arrived first to the service system will be served first. Another advantage has to do with limited life of materials. When perishable materials (e.g., fresh food) are considered, the FIFO rule tends to minimize waste. Therefore, this rule is common in the service and food industries. In the service sector, it is known as first come first served (FCFS), and it promises "justice" to customers.

2. *Last in first out (LIFO):* This priority rule is based on the cartridge principle: the job that arrives last will be processed first. This priority rule is used when jobs are stuck on top of each other in the in-process inventory and the last one is on top of the stack, and therefore, is the easiest to retrieve. An example is the unloading a truck, or unloading of containers from a ship.

3. *Earliest due date (EDD):* Jobs are scheduled by their due dates. The earlier the due date, the higher the priority of the job. The rule can minimize delays under some special conditions, for example, it minimizes the maximal delay time in a job shop with a single machine. Table 7.2 illustrates the effect of EDD: only one late delivery with 4 days delay, while FCFS in Table 7.1 yields four late jobs with total delays of 12 days.

4. *Shortest processing time (SPT):* This rule attempts to minimize the waiting times and work-in-process (WIP) inventory awaiting a machine. By processing the shortest jobs first, a large number of short jobs get processed quickly, and the number of jobs waiting for processing on the machine is rapidly reduced. The problem with this rule is that jobs that require long processing time may be delayed and eventually be late, while jobs with a short processing time will be finished earlier than needed. Table 7.3 illustrates the effect of SPT on the data in Table 7.1.

TABLE 7.2

The Effect of an EDD Schedule on the Example in Table 7.1

Task No. i	Process Time (Days) t_i	Due Date d_i	Completion Time (Days) c_i	Delays T_i
4	2	9	2	0
5	4	12	6	0
6	3	15	9	0
1	3	16	12	0
2	5	18	17	0
3	1	18	18	0
7	5	19	23	4

TABLE 7.3

The Effect of SPT Schedule on the Example of Table 7.1

Task No. i	Process Time (Days) t_i	Due Date d_i	Completion Time (Days) c_i	Delay T_i
3	1	18	1	0
4	2	9	3	0
1	3	16	6	0
6	3	15	9	0
5	4	12	13	1
2	5	18	18	0
7	5	19	23	4

5. *Current job:* This priority rule minimizes setup time. When the currently processed job is finished, the machine operator searches the WIP inventory of jobs in front of the machine for another job with the same items that require the same processing on the machine. If such a job exists, it is processed next so that no machine setup is needed and setup time is saved.

   From the above discussion, it is clear that the selection of a priority rule may have a significant impact on the performance of the job shop. Furthermore, there is no one rule that is always the best. The selection of the right rule depends on the goals and constraints facing the specific job shop as well as on the mix of jobs and their processing requirements on the specific machines in the job shop.

7.2.2 Complex Priority Rules

Some priority rules are based on several parameters, and priority is determined by a formula in which the selected parameters are the independent variables.

1. *Slack time remaining (STR):* This rule is based on the difference between the time remaining until the due date and the time required to finish processing the job. For example, if the remaining job-processing time is 8 days and the delivery date is in 10 days, then the value of the STR is 2 days. The priority of a job is higher as the value of the STR decreases.

2. *Critical ratio (CR):* This rule is based on the ratio between the time remaining until the due date (numerator) and the time required to finish the job processing (the denominator). The job with a lower CR receives a higher priority.

3. *Slack time remaining per operation (STR/OP):* This rule is based on the ratio between the STR and the remaining number of operations to complete the job. Jobs with higher STR/OP values have lower priority.

The transformation process that converts input from suppliers and contractors into goods and services provided to the customers can take many forms depending on a variety of environmental factors, the technology used, the markets, and the organization performing the transformation. In previous chapters, we discussed two examples of layouts of the machine and equipment on the shop floor—the job shop and the flow shop. We shall now discuss scheduling in each of these environments to expand our knowledge of these important disciplines.

7.3 Scheduling the Job Shop

The job shop environment is suited to producing multitudes of different products. However, the layout of job shops varies widely based on the process complexity and the quantity produced. In small shops producing a variety of simple products, there are several different machines scattered throughout the floor space and the flow of products is nonrepeating and has no general pattern. As production quantity increases, more capacity is required and the number of machines increases. It is often the case that capacity requirements dictate having multiple machines of the same kind. In such cases, it is customary to cluster or group together similar machines and resources. For example, in a metal processing job shop, all the grinders are clustered together in one department or work center, while the presses are clustered together in another department or work center. The pooling of similar machines and resources has the effect of promoting learning and increasing professionalism in performing each type of operation. It also gives the manager (of a group of similar machines) an opportunity to manage the efficiency of the machines. While scheduling each group of machines is relatively simple, the overall scheduling becomes extremely difficult, as we will see later in this chapter.

Job shop scheduling deals with the specific order in which the variety of items scheduled for processing during a given time (or planning horizon) are processed by the job shops' resources (machines, tools, humans). It deals with the allocation of resources (e.g., machines and people) to specific operations on specific items at specific times. The term used to describe a group of identical items processed together in the same way by the job shop resources is a "job" and this is the source of the name job shop. Typically, each job is accompanied by an operations list document called a "work order" or a "route card" manifest. It specifies the processing activities required by the job, the type of items to be processed as a group in this job, the number of items to be processed, the specific order in which the processing activities should be performed (known as routing), and the operations to be performed by each resource for each item. Sometimes a due date is also specified for the work order—the date the order should be finished and the items should be ready for delivery to a customer or to another facility.

Scheduling is a decision regarding when each job shop machine specified in the work order routing will process each job. The layout or physical arrangement of the machines on the production floor has a significant impact on scheduling decisions. In the functional-oriented layout (groups of similar machines), a large variety of items can be processed, and each job may have different processing and routing requirements. This flexibility does not come free of charge: it makes tracing the operations of a specific order extremely difficult. In turn, this makes job scheduling very difficult (due to the large number of possible combinations of job-processing orders). Figure 7.2 illustrates the job shop layout.

There are several approaches to developing schedules for the job shop. Ideally, the goal is to develop optimal solutions. An optimal solution is based

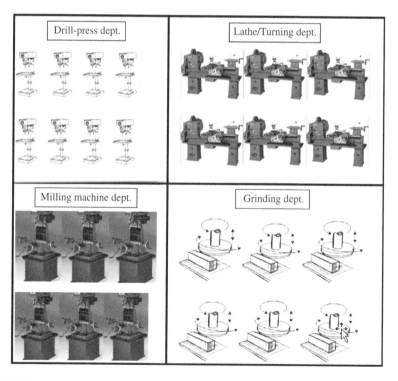

FIGURE 7.2
Example of functional layout for job shop processing.

on a mathematical model that considers all types of products, all types of machines and their capacity, all the necessary work orders, including quantities and routing of each job represented by the corresponding work order. Due to the complexity of such models, this approach works only on relatively small problems (Pinedo, 2012), that is, a small number of machines and a small number of jobs. Another approach is to divide the job shop scheduling problem with n different machines into n problems, each with one machine. This is known as problem decomposition. This approach simplifies the problem but it leads to local optimization, and frequently the combination of the local optimum for each machine is not necessarily the global optimum for the whole job shop scheduling problem.

The local optimization of a scheduling problem with one machine is often based on priority rules that dictate the order in which jobs awaiting processing in front of the machine are processed. The quality of the solutions produced by these models when solving real problems is limited because the models are based on a limited local view of a broader problem.

7.3.1 Single-Machine Scheduling in a Job Shop

The simplest approach to scheduling job shops is to focus on the WIP of jobs in front of each machine, ignoring all other jobs. This local optimization provides a simple solution to a very complex combinatorial problem, and therefore, the solution is usually not the best for the whole job shop. In its simplest form, a supervisor will instruct the machine operator how to select the next job for processing whenever the job currently processed by the machine is finished. In this case, new jobs arriving from machines preceding the one being discussed are considered for processing when the job processed during their arrival is finished. It is possible to use priority rules to prepare a plan that schedules all the jobs waiting for processing in front of a machine; however, this schedule must be updated when new jobs arrive. The job routing determines the order in which the work is carried out on the different job shop machines. Priority rules based on a single parameter are known as simple rules. Priority rules based on a combination of a number of parameters are called complex rules. Because it is difficult to predict which priority rule is the best for a given case, several rules might be tested and the best plan selected. In an effort to predict *a priori* what priority rule is best for a given situation, simulation studies are conducted to assess the relative efficacy of various rules regarding a variety of job shops.

7.3.2 Use of the Gantt Chart as a Job Shop Scheduling Aid

The simple priority rules discussed in the Section 7.3.1 for scheduling the single-machine job shop can be used in a job shop with several machines by decomposing the problem into several single-machine problems. The following job shop example, where three machines (machine 1—M1, machine

2—M2, and machine 3—M3) are processing three jobs (A, B, and C), illustrates this situation. These three jobs arrive at the job shop in the following order: job A arrives first, job B arrives second, and job C arrives last. The assumption is that each machine in the job shop is scheduled locally by its operator using a priority rule. A Gantt chart is used to model the schedule of each machine and the three Gantt charts integrate three schedules into a schedule for the whole job shop.

At time 0, the three jobs await processing by the machines in the job shop. The first job consists of a single unit of product A, the second job consists of a single unit of product B, and the third job consists of a single unit of product C. Table 7.4 presents the routing of the three jobs on the three machines, the processing time, and the due dates.

FIFO rule. The three Gantt charts represent the processing of the jobs on the three machines (Figure 7.3).

The Gantt charts show when each job starts on each machine and the time it finishes. It is easy to see on the Gantt charts that each machine is processing no more than one job at a time and at most one machine in any period processes each job. A quick look at the routing column of Table 7.4 shows that job B can start processing on M2, while both job A and job C are set to start on M1. In line with the FIFO rule, job A—which arrives first (before job C)—will be loaded first on M1 and will be finished at time 2. Job C will be loaded on M1 at time 2 and will finish at time 3. Job B will be loaded on M2 (this is the first machine in its routing) at time 0 and will finish at time 3. Job A will be loaded on M2 at time 3 (when job B is finished on this machine) and will be finished at time 6, while job B is loaded on M1 at time 3 (just after its processing on M2 is finished) and will be finished at time 6. Job C will start processing on M3 at time 3 and will finish at time 6. At that time, job A will be loaded on M3 and will finish processing at time 8. Finally, job B will be loaded on M3 at time 8 and will finish processing at time 9.

The due dates and actual finish times of the three jobs are presented in Table 7.5.

LIFO rule. Using the LIFO rule, the highest priority job is C and the lowest priority job is A (Figure 7.4). The actual finish times and job statuses are illustrated in Table 7.6.

TABLE 7.4

Data for the Job Shop Example

Job Arrivals	Routing (Processing Sequence)	Processing Time on Machine 1	Processing Time on Machine 2	Processing Time on Machine 3	Job Due Date
A	M1, M2, M3	2	3	2	8
B	M2, M1, M3	3	3	1	7
C	M1, M3, M2	1	4	3	8

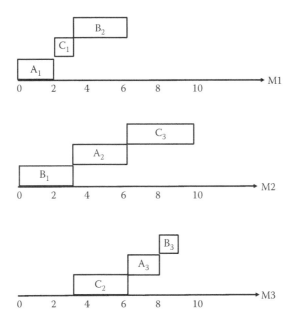

FIGURE 7.3
Three Gantt charts of the FIFO three-machine example.

EDD rule. Using this rule, job B has the highest priority, while jobs A and C have the lowest priority (Figure 7.5). The actual finish time and job status are illustrated in Table 7.7.

SPT rule. The three Gantt charts in Figure 7.6 represent the processing of the jobs on the three machines.

Jobs A and C start on M1. Since job C takes less time, it is loaded first on M1. The processing is finished at time 1. Job A is loaded next and the processing is finished at time 3.

Job C is loaded on M3 at time 1 after processing on M1 and processing is finished at time 4.

Job B is processed by M2 first and finishes processing at time 3. It is then loaded on M1 at time 3 and finished at time 6.

TABLE 7.5

Actual Finish Time and Job Status of the Three Jobs in the FIFO Example

Job	Due Date of Job	Actual Finish Time	Job Status
A	8	8	On time
B	7	9	Late
C	8	10	Late

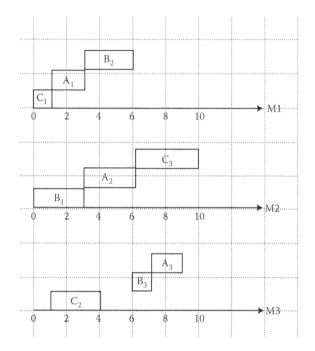

FIGURE 7.4
Three Gantt charts of the LIFO three-machine example.

After finishing processing job B, M2 processes job A, starting at time 3, and finishes at time 6. M2 then processes job C, finishing at time 10.

M3 is idle during the first period as there is no job that needs to be processed on it first. After processing job C at time 4, M3 is again idle until M2 finishes processing job A and M1 finishes job B. Since job B has a shorter processing time on M3, it is processed first, from time 6 to time 7, and job A is the last one on M3, from time 7 to time 9.

The due dates and actual finish times of the three jobs are illustrated in Table 7.8.

The above example shows how the selection of a priority rule affects the job shop schedule. In all four schedules, job C is finished at time 10, hence it is not sensitive to the priority rule selected. Jobs A and B are sensitive; job A

TABLE 7.6

Actual Finish Time and Job Status of the Three Jobs in the LIFO Example

Job	Due Date of Job	Actual Finish Time	Job Status
A	8	9	Late
B	7	7	On time
C	8	10	Late

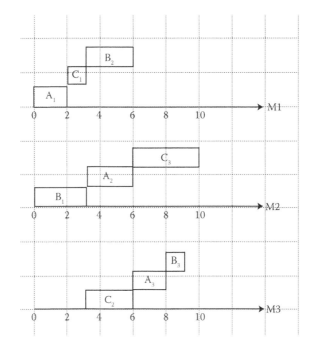

FIGURE 7.5
Three Gantt charts of the EDD three-machine example.

is finished on time when FIFO and EDD are used, while job B is finished on time when LIFO and SPT are used.

The use of simple priority rules for scheduling does not take the capacity of the job shop machines into account. The assumption is that the machines have enough capacity during the planning horizon to process all the jobs by their due date. This may not be correct. To demonstrate this point, the total processing time on each machine is added to the previous example (Table 7.9).

It is clear that if all three jobs have due dates of 8 or less days and the total processing time of the three jobs on M2 is 10 days, it is impossible to finish all three jobs by their due date.

Assume that the goal is to finish as many jobs as possible by their due date. Further, assume that when a job is late, it does not matter by how much time

TABLE 7.7

Actual Finish Time and Job Status of the Three Jobs in the EDD Example

Job	Due Date of Job	Actual Finish Time	Job Status
A	8	8	On time
B	7	9	Late
C	8	10	Late

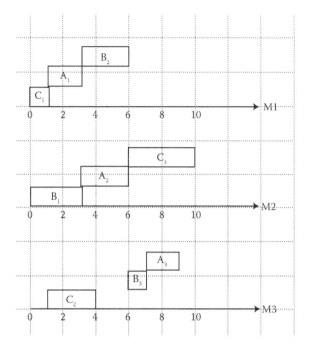

FIGURE 7.6
Three Gantt charts of the SPT three-machine example.

TABLE 7.8

Actual Finish Time and Job Status of the Three Jobs in the SPT Example

Job	Due Date of Job	Actual Finish Time	Job Status
A	8	9	Late
B	7	7	On time
C	8	10	Late

TABLE 7.9

Summary Table for the Example Data

Machine Job	Routing (Processing Sequence)	Processing Time on Machine 1	Processing Time on Machine 2	Processing Time on Machine 3	Job Due Date
A	M1, M2, M3	2	3	2	8
B	M2, M1, M3	3	3	1	7
C	M1, M3, M2	1	4	3	8
Total processing time on the machine		6	10	6	

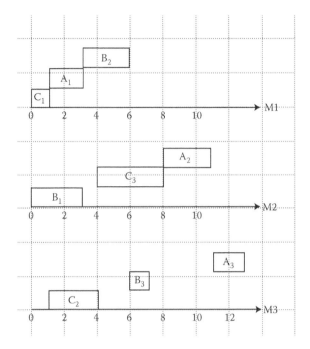

FIGURE 7.7
Three Gantt charts of the three-machine example with two on-time jobs.

it is late. Figure 7.7 and Table 7.10 present a schedule that finishes two of the jobs on time.

The observation that machine capacity is an important factor in scheduling is the focus of the theory of constraints (TOC) developed by Eli Goldratt (1983). The idea behind this theory is that the scheduler should focus on the bottleneck machines—the machines with the tighter capacity constraint compared to the load generated by the processing time of all jobs on this machine during the planning horizon (Goldratt, 1999). Goldratt developed a scheduling technique known as the drum buffer rope (DBR). While the DBR's primary focus is on the flow shop (for a more detailed discussion, see Section 7.7), its ideas are adaptable to the job shop environment

TABLE 7.10

Actual Finish Time and Job Status of the Three Jobs in the "Two On-time Jobs" Example

Job	Due Date of Job	Actual Finish Time	Job Status
A	8	13	Late
B	7	6	On time
C	8	8	On time

(Goldratt and Cox 2012). The DBR assumption is that a good schedule for the bottleneck machine is a good schedule for the whole job shop. Machines that feed the bottleneck (machine) with jobs should be scheduled in a way that best ensures that the bottleneck is busy and gets the jobs it needs to process according to its schedule. Machines that come after the bottleneck should be scheduled so that the jobs already processed by the bottleneck will be finished as soon as possible.

7.4 Schedule Control

The scheduling problems discussed thus far are based on the assumption that the job shop is static and deterministic.

1. *Static.* All jobs to be processed in the relevant period are ready to start processing. No new jobs arrive until all the current jobs are finished.
2. *Deterministic.* There are no uncertainties. The exact duration of processing of each job on each machine is known as well as machine capacity (i.e., machines are operating, operators are not absent, etc.).

In many real situations, the scheduling problem is dynamic and stochastic—new jobs arrive while other jobs are in process. In some instances, the job's arrival time as well as the type of items to be processed and their quantities are not known in advance. Furthermore, unexpected events such as machine breakdown and absent operators are quite common. To deal with such situations, monitoring and control systems were developed. These systems continuously collect information on the current situation in the job shop and compare it to the planned schedule (the planned schedule is known as a baseline for comparison or, simply, a baseline). In many shops, data collection is automated by using a barcode reader or RFID (radio-frequency identification) technology. When deviations from the planned schedule—the baseline—are detected, corrective actions are considered and the baseline is updated accordingly.

Corrective actions could be working overtime (to catch up on late jobs), outsourcing (part of the job is done by another company), or increasing the workforce (and the pace of work). Another typical example of a corrective action is expediting. This is the case when a job is late, for example, due to a machine breakdown, and consequently, it might miss its due date. To fix the problem, the job is assigned a higher priority and is processed earlier on the machine for which it is waiting. Expediting is a modification by management of the job shop's planned schedule. When this method is used correctly, it can help deal with unexpected events such as a machine breakdown or a customer request to change the promised due date of a job and to deliver it earlier than planned according to the baseline.

A simple monitoring and control system is based on the in-process inventory of jobs waiting for processing on each job shop machine. When too much inventory accumulates in front of a machine, this might signal that the machine is a bottleneck and it does not have enough capacity compared to the load generated by the jobs that require processing on it. A possible corrective action for the short term is to operate this bottleneck machine during overtime or in a second/third shift. If the situation persists in the long term, the better options may be to buy another machine of the same type or use external sources of capacity such as a subcontractor. Another corrective action may be to invest in upgrading the machine to increase its available capacity.

Excessive bottleneck starvation is another indication that a corrective action might be needed. Starvation occurs when a bottleneck machine is idle while waiting for jobs that are not ready yet. These jobs have preceding operations (according to the specific job routing of each job) that are either waiting in a line for a busy machine or in process on a machine. In such cases, the corrective action may be to expedite jobs waiting for other machines, whose next operation is on the bottleneck machines.

The last two examples deal with the idea of bottleneck machines, which due to lack of capacity should be carefully scheduled, monitored, and controlled. This is the essence of the TOC and its DBR scheduling logic, a logic discussed later in this chapter.

Another layout is the flow shop, which is based on clustering products sharing similar routing or processing requirements. In the flow shop, higher efficiency and easier scheduling compensate for the limited flexibility due to a fixed routing. Although it is possible to schedule the jobs in a flow shop with the same tools and techniques used in the job shop, special models developed specifically for the flow shop perform much better in this environment.

7.5 Flow Shop Scheduling

Recall that in the flow shop, all jobs share the same routing, that is, the same processing sequence on flow shop machines applies to all the jobs. When the flow shop is processing a few different products, the system is called a flow line. When all the items processed in the shop are identical, the system is called a production line. In the same manner, complex products such as cars are assembled on assembly lines. When a variety of different product models are assembled on the same line, special models for line balancing are used for scheduling. In the following section, we will discuss the case where each job has a different processing time on each machine.

TABLE 7.11

Data for the Example of a Two-machine Flow Shop

Job Name	Processing Time on the First Machine 1	Processing Time on Machine 2
1	3	3
2	4	6
3	7	5
4	2	9
5	2	3
6	1	4

Consider, for example, scheduling six jobs on the two machines of a flow shop. All jobs must be processed by M1 first and then on M2. Processing times are presented in Table 7.11.

Assuming that the scheduling goal is to finish all six jobs as soon as possible, two steps should be undertaken: (1) the time the second machine is idle (waiting for the first machine to complete the first job) should be minimized and (2) the time the first machine is idle (waiting for the second machine to process the last job) should be minimized. This idea is implemented by the Johnson algorithm.

7.5.1 Johnson Algorithm

1. Create two lists: "available" is showing the availability of all the jobs that have not been scheduled yet and "scheduled" is showing all the jobs that are already scheduled. The scheduled list is ordered so that processing starts on the leftmost jobs in the list while the rightmost jobs are processed last.

   The scheduling process starts with an empty scheduled list as no jobs are scheduled and all the jobs appear in the available list.

2. Select from the available list the job with the minimal process time on either the first or the second machine and delete it from the list. If there are two or more such jobs, arbitrarily select one of these jobs.

3. If the minimal time of the selected job is on the first machine, add this job after (to the right of) the last job on the left side (toward the beginning) of the scheduled list. If the minimal time of the selected job is the second machine, add the job before (to the left of) the last job on the right of scheduled list.

4. If the available list includes only one job, go to step 5. If not, go back to step 2.

5. The last job remaining is added between the two parts of the scheduled list, that is, it is added after the last job added on the left

side of the scheduled list and before the last job added on the right of the scheduled list so the two lists are merged into a single schedule.

7.5.2 Scheduling the Example Problem

Job 6 has the minimal time (one unit of time) and it is on the first machine; accordingly, job six is assigned to the first location on the left-hand side of the scheduled list:

6					

Job 6 is deleted from the available list, which now includes jobs 1, 2, 3, 4, and 5.

The minimum processing time of jobs in the available list is 2, and it is the processing time of jobs 4 and 5 on the first machine. We arbitrarily select job 5 and it is scheduled next on the left-hand side of the scheduled list.

6	5				

The available list now includes jobs 1, 2, 3, and 4; the minimum processing time is 2. This is the processing time of job 4 on the first machine. Job 4 is scheduled next on the left-hand side of the scheduled list.

6	5	4			

The available list now includes jobs 1, 2, and 3; the minimum processing time is 3. This is the processing time of job 1 on the first and second machines. We arbitrarily select the second machine. Job 1 is scheduled last on the right-hand side of the scheduled list.

6	5	4			1

The availability list now includes jobs 2 and 3; the minimum processing time is 4. This is the processing time of job 2 on the first machine. Job 2 is scheduled last on the left-hand side of the scheduled list.

6	5	4	2		1

The available list now includes only job 3. This job is now placed between the left-hand side of the list and the right-hand side to complete the schedule:

6	5	4	2	3	1

The final schedule (from left) is presented on a Gantt chart (Figure 7.8).

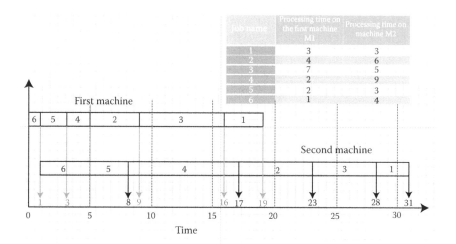

Job name	Processing time on the first machine M1	Processing time on machine M2
1	3	3
2	4	6
3	7	5
4	2	9
5	2	3
6	1	4

FIGURE 7.8

The Gantt chart for the final schedule of the two-machine flow shop example.

This schedule minimizes the time it takes to finish all the jobs on both machines; in other words, it minimizes the time it takes to finish the last job on the second machine.

7.6 Applying the JIT Philosophy in Scheduling

To maximize value and minimize waste, Toyota, led by Taichi Ohno (1988) developed the JIT philosophy. It results from the simple observation that

$$\text{Profit} = \text{Revenues} - \text{Costs}$$

Briefly, it espouses the pull idea, that is, material is pulled by the next operation in a process only when this material is needed. Thus, to minimize in-process inventory, jobs on the shop floor are processed and moved to the next operation only when the machine that performs the next operation is ready to process these jobs. The book *The Machine that Changed the World* (Womack et al., 2007) describes the JIT philosophy and its implementation.

Organizations (here we focus on for-profit organizations) can increase their profits by increasing revenues, by reducing costs, or by doing both. The revenues are a function of the volume of sales (number of product units sold) and the selling price (price per product unit). The organization's willingness to increase the selling price (per product unit) is limited as this may reduce the market demand. In a competitive market, demand is a function of price and the former tends to decrease as the latter increases. Consequently, the volume of sales (the number of units sold times the price per unit) may decrease or remain the same when unit price is increased. Furthermore, increasing revenues by raising prices can be risky, as a high price may attract competitors to the market. For these organizations, the key to increasing profits is cost reduction.

The focus on reducing costs (by reducing all kinds of waste) is the foundation of the JIT approach. In a very simple way, JIT focuses on maximizing the value the customers get (for our simple discussion, value is anything that the customer is willing to pay for) while minimizing waste (any cost or time that does not add value to the customer). According to the JIT philosophy, everything that does not add value is a waste that the organization should reduce to a minimum. Managing the supply chain in order to maximize the value to the customers and to bring waste to a minimum is also known as *Lean management*.

Waste can take several forms, for example:

1. *Waste associated with defective product quality.* This is the waste of materials and work invested in defective products. An ongoing effort to continuously improve quality is a cornerstone of the JIT approach. Line stopping is another keystone and it means that each worker is required to stop the production process if a defect is detected so that additional defects will not be created. This is frequently done by automatic means that sense a problem and stop the process.

2. *Waste created by inventories.* This is the money invested in inventory and the cost of facilities where inventories are stored as well as the costs of operating and maintaining the inventory system and the cost of risks associated with inventory. These are all forms of waste and the JIT philosophy is to reduce all these wastes to a minimum.

3. *Waste of space.* Poor layout design that does not make efficient use of space is a cause of excessive travel and other types of waste. This is the waste associated with building and maintaining larger facilities. Therefore, the equipment used on the shop floor should be located in such a way that minimizes the size of the facility. In a similar way, warehouses should be designed to minimize their space and the waste associated with it.

4. *Material-handling waste*: This waste is a result of poor layout and poor management that cause excessive transportation of materials and products. The movement of materials does not add value to the customer; just the opposite. Material handling may cause defects and damages and these should be minimized.

The focus on waste and waste prevention led to distinguishing between value-added activities in the supply process and activities that do not add value. The idea is that ideally all activities in the supply process should add value to the customer. It is necessary to examine the activities that do not add value, and if possible, to eliminate them or at least to reduce such activities to the necessary minimum. For example, if material is transported between machines during the production process, an effort to improve the layout to save space and to reduce transportation distances to a minimum is required. In a similar way, inventory does not add value and it is important to minimize inventories or eliminate them if possible.

The JIT scheduling rests on the following principles:

1. *Reduction of the setup time and setup cost of machines and equipment.* As previously illustrated by the economic order quantity (EOQ) model, reduced setup time leads to smaller batch sizes, and therefore, to lower inventories. Furthermore, if a job moves between two machines, the smaller the number of units in the job batch size, the less completed units have to wait to be transferred. If, for example, the job consists of a single product unit, as soon as one machine finishes processing it, the next machine can receive it and perform the next operation. However, if the batch size is more than one unit, the first completed unit might have to wait as in-process inventory until the first machine finishes processing all the other units and the entire batch moves to the next machine for processing. This situation generates a WIP inventory and waste. To shorten the waiting period, the transfer batch can be smaller than the processing batch. In other words, as soon as a machine completes a unit, it moves to the next machine for processing, without waiting for the rest of the units in the batch to be completed. The key to smaller processing batches is setup time (and setup cost) reduction. The Toyota production system is based on redesigning machines and developing better ways to perform setups in the shortest possible time. The effort to reduce setup time is based on the principles of the method called single minute exchange of die (SMED).

2. *Reduction of in-process inventory.* To minimize in-process inventories between successive operations, the JIT scheduling system limits such inventories using kanban cards or by restricting the physical space available for in-process inventories. The kanban system is a simple information system based on cards attached to each job. Each card contains information about a specific job and has several purposes. The card functions as a production order, purchase order, and for inventory control (Ohno, 1988; Monden, 2012). A job is not created without a proper kanban card, and therefore, the number of jobs is

limited by the number of cards in the system. By reducing the physical space or the number of kanban cards, the in-process inventory is reduced as well.

3. *Reduction of material-handling activities.* Material handling includes the transportation of jobs between machines, between storage rooms, etc. It is a source of waste, as there is a cost associated with the material-handling equipment and its operators. Any handling can cause defects, and during the period the material is being handled, inventory cost is incurred. Better layout, smaller transfer batches, and good scheduling can reduce material handling and the associated waste to a minimum.

7.6.1 Illustrating the Kanban Card System

To understand the kanban concept, consider the example of two adjacent work centers A and C, in a flow shop where jobs are transferred from A to C via an intermediate inventory B located between A and C. Since the two work centers are adjacent to each other, in-process inventory is controlled by limiting the space used for such inventory between the work centers. When this space is full, the operator in A does not have space for moving the current unit he or she is processing and when it is finished he or she is forced to wait until the operator of B removes a unit from the in-process inventory and frees up some space. If the two work centers are not adjacent to each other, the kanban system can be used. Each job must have a kanban card attached to it, and therefore, the number of jobs in the system will never exceed the number of kanban cards.

There are two types of kanban cards: production kanban cards and pull kanban cards. Kanban cards moving between A and B (into the buffer) are production kanban cards, and kanban cards moving between B and C (out of the buffer) are pull kanban cards.

The operator in each work center monitors the number of cards. When the number of pull kanban cards at C reaches a predetermined level, the operator transfers the cards to B, removes the production kanban cards from the jobs in the WIP and attaches the pull kanban cards to the jobs. Jobs processed at A are transferred to the in-process inventory with the production kanban cards attached to them. The operator at A stops production when there are no more free production kanban cards; and he or she waits for the operator at C, to pull jobs and free up production kanban cards (Figure 79)

In this simple example, there are two workstations. However, the same logic can be used with any number of workstations. Each kanban card is identified by the product type and process performed on the product attached to it, as well as its identification as a pull kanban or a production kanban.

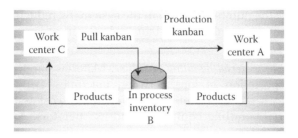

FIGURE 7.9
Scheduling using kanban cards.

7.7 Theory of Constraints and the Drum Buffer Rope Approach to Scheduling

The TOC is a philosophy introduced by Eliyahu M. Goldratt in his 1983 book titled *The Goal*. The philosophy is applied to production systems based on the assumption that the ability of a production system to maximize production (or profit) is limited by a few resources known as bottlenecks or constraints. We touched upon this subject briefly in Section 7.3.2.

Scheduling a production system based on TOC comprises the following four steps:

1. Identify the constraints by comparing the load on each resource in terms of working hours required to complete all the jobs it has to process with the capacity of the resource or the number of hours available for processing on that resource. The resources where the load is greater than the available capacity are bottlenecks. If no such resources exist, the resources whose load represents the highest proportion of their capacity are the bottlenecks.

2. Schedule the bottleneck to maximize the goals of the organization; assuming that the goal is to maximize profit, the bottleneck is scheduled to process jobs that have maximum contribution to profit.

3. Schedule all the resources that precede the bottleneck to ensure that the bottleneck will not be idle and will have all the jobs scheduled for processing when needed.

4. Schedule the resources that succeed the bottleneck so that the jobs processed by the bottleneck will be finished and available for delivery as soon as possible.

The book *The Goal* illustrates this theory using a group of boys on a field trip. The analogy to the bottleneck is a boy who is relatively slow, and therefore, he slows down the boys that follow him. The first solution is to put this boy at the head of the group but this is not possible in the production analogy as the routing or the sequence of processing on the different machines is predetermined. Another solution is to give the slow boy a drum and by having him beat on it, he will dictate the speed of the whole group. The analogy is to use the bottleneck machine as a drum in the production system. To protect the bottleneck machine, a buffer inventory is kept in front of it so in case a preceding machine fails, the bottleneck is not idle. It can work on the buffer. In the production analogy, the bottleneck machine uses a rope to open the gate and release new raw material based on its pace. Hence, the name drum buffer rope for this scheduling technique.

7.7.1 Illustrating the Drum Buffer Rope System

To illustrate the DBR system, we use the three-machine three-job example again (Table 7.12).

In this example M2 is the bottleneck, as the total processing time on this machine is 10, while the due date of job B is 7 and the due date of jobs A and B is 8. When we tried to minimize the number of jobs finished by their due date, the best schedule that we found for this problem is shown in Figure 7.10.

In this schedule, jobs B and C are completed on or before their due date. This schedule, however, creates idle time for the bottleneck M2 between time 3 and time 4.

Based on the DBR principle, M2, the bottleneck, is the drum and the goal is to schedule the three jobs on M2 to minimize this machine's idle time. This is achieved using several priority rules that we examined earlier. The FIFO, LIFO, EDD, and SPT rules, respectively, are illustrated in Figures 7.11 through 7.14.

From this example, it is clear that the selection of a scheduling model depends on the system's goal. In the resulting schedule, only one of the three jobs is finished by its due date. We also saw that it is possible to finish two of the jobs by their due date but idle time is generated on the bottleneck.

TABLE 7.12

Summary Data for the Three-machine Job Shop Example

Machine Job	Routing (Sequence of Processing)	Processing Time on Machine 1	Processing Time on Machine 2	Processing Time on Machine 3	Due Date of Job
A	M1, M2, M3	2	3	2	8
B	M2, M1, M3	3	3	1	7
C	M1, M3, M2	1	4	3	8
Total time per machine		6	10	6	

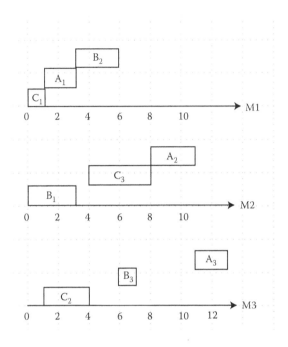

FIGURE 7.10
Three Gantt chart example with two on-time jobs.

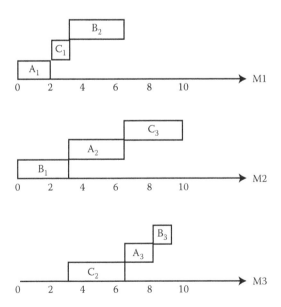

FIGURE 7.11
Three Gantt charts of the FIFO three-machine example.

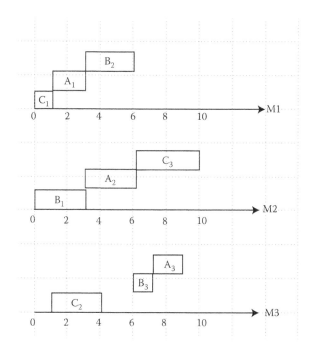

FIGURE 7.12
Three Gantt charts of the LIFO three-machine example.

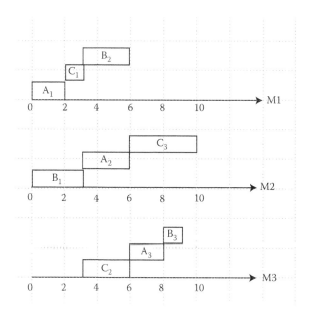

FIGURE 7.13
Three Gantt charts of the EDD three-machine example.

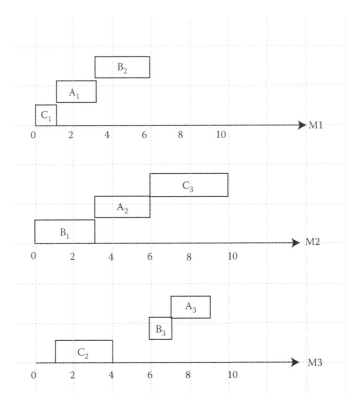

FIGURE 7.14
Three Gantt charts of the SPT three-machine example.

7.8 Summary

This chapter presents a discussion on the scheduling of day-to-day operations in an industrial organization. The focus is on the single production
organization—a typical link in the supply chain and its internal operations.
Scheduling these operations, along with the flow of material, the flow of
information, and the decision-making processes transforms the input to the
link (raw materials, parts, subassemblies, or even finished goods from suppliers) into the link's output, that is, the goods and services that the organiza-
tion/link provide to its customers.

General concepts are presented along with specific scheduling tools and
models for production scheduling. Specific models for the single machine
scheduling, the job shop, and the flow shop are presented in detail. The
just-in-time (JIT) scheduling approach and the theory of constraints are discussed as general scheduling philosophies that can be used in any type of
organization or link of the supply chain.

References

Goldratt, E.M. 1983. *The Goal*. Great Barrington, MA: North River Press.

Goldratt, E.M. 1999. *Theory of Constraints*. Great Barrington, MA: North River Press.

Goldratt, E.M. and Cox, J. 2012. *The Goal: A Process of Ongoing Improvement*. 25th anniversary edn. Great Barrington, MA: North River Press.

Monden, Y. 2012. *Toyota Production System: An Integrated Approach to Just-In-Time*. Boca Raton, FL: CRC Press.

Ohno, T. 1988. *Toyota Production System: Beyond Large-Scale Production*. Cambridge, MA: Productivity Press.

Pinedo, M. 2012. *Scheduling Theory, Algorithms and Systems*. 4th edn. New York: Springer.

Womack, D., Jones, J., and Roos, D. 2007. *The Machine That Changed the World: The Story of Lean Production—Toyota's Secret Weapon in the Global Car Wars That is Now Revolutionizing World Industry*. New York: Simon and Schuster.

8

Streamlining the Transformation Process: Material Requirements Planning Systems

8.1 Need for Material Requirements Planning

In previous chapters, we discussed several aspects of the process that organizations apply to transform input into output (the "transformation" process). In Chapters 5 through 7, we discussed three essential aspects of the transformation process that takes place in many organizations where these aspects are the management of the interfaces that comprise a supply chain:

1. Managing the interface with suppliers and subcontractors (the input to the transformation process).
2. Managing the internal operations of the organization that transforms input into output (the inner operation of the transformation process).
3. Managing the interface with the customers (the output of the transformation process).

Each chapter presented a short discussion, models, tools, and techniques developed to handle each aspect.

In the past, marketing, purchasing, and operations decision makers worked separately and concentrated on "their" function, assuming that if each function runs properly, the organization's entire transformation process will function well. In reality, this created many problems. Each function tends to develop its own goals and constraints, its special information systems dedicated to support the specific issues of the function. For example, marketing information systems focus on the interface with customers, purchasing information systems focus on the interface with suppliers and subcontractors, and the information systems that support the organization's internal operations (that transform input into output) focus on production planning and control. Each separate information system, collectively known as "legacy systems," contained the specific data and models needed by the relevant function. Thus, the marketing information system that focused on the customers, their orders, their past demand, etc., had no connection to other information systems. The purchasing information system was also a stand-alone system and dealt with suppliers, the goods they supply, and their past performance information such as the quality of goods supplied, on-time delivery, and cost. There was scarcely any connection between these legacy systems, and consequently, they could not provide a solution that fit the organization's entire transformation process of input to output. Each system had only part of the needed models and only part of the needed data—hindering coordination. Even worse, some data needed by two or more legacy systems were entered and stored in each system separately, causing extra work and waste. At times, different information regarding the same issue existed in different legacy systems. Each legacy system focused on parts of the problem, and the management of the interface between these parts was not supported efficiently.

To understand the situation just described, consider a manufacturing organization that assembles the product depicted in Figure 8.1 the diagram showing the components of a product is called a bill of materials (BOM). The final product A is composed of part B that is manufactured at the same site, and part C purchased from a supplier. Each unit of A comprises a unit of B and two units of C. The lead time to assemble a typical batch of A is a week, and the lead time for manufacturing a batch of B is 2 weeks, whereas the lead time for ordering and receiving C from the supplier is 3 weeks.

According to marketing (according to the master production schedule—MPS), 10 units of A are needed in week 10. Current A inventory is 1 unit, B is 2 units, and C is 3 units.

The legacy information system uses the economic order quantity (EOQ) model to manage the ordering of B. In a similar way, the EOQ model also generates orders for A and C. The EOQ model uses the approximation of

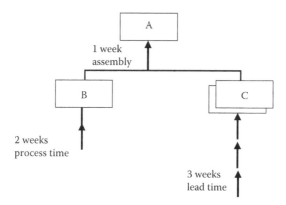

FIGURE 8.1
Bill of materials (BOM) with lead times for product A.

the constant average demand rate and, therefore, has a fixed replenishment point with a fixed order quantity. The EOQ inventory diagram is a "saw tooth" diagram. These models are not coordinated with the MPS, and therefore, the system cannot ensure that the 10 units of B and 20 units of C required to assemble 10 units of A will be available when the assembly operation is scheduled to start in week 9 (1 week of lead time before week 10). Since the EOQ model is assuming that demand for B and demand for C are independent, the probability that not enough units of B or C or both will be available is not negligible. In addition, the fact that the current inventory of a unit of A, two units of B, and three units of C will be carried in inventory until week 9, when they are needed, worsens the situation from an expense perspective. Furthermore, the marketing unit that manages the MPS does not have the data and models to ensure that the capacity needed to assemble an additional 9 units of A starting at week 9 will be available, or the capacity required to manufacture the B units will be available when needed. Availability may be limited if the resources are scheduled to work on other products/jobs.

This example demonstrates that the legacy systems that do not support the needed coordination among the purchasing, marketing, and operations provided only limited support for decision making during the transformation process.

The development of affordable and relatively powerful computers in the 1970s opened up new opportunities for managing the organization's entire transformation process, from start to end, more efficiently. The team of Orlicky (1973) leveraged the new computing power to develop MRP logic and software. MRP systems are integrated in the sense that the data (in the database) are used by all the functions involved in the transformation process. The same MRP model is used by all three functions (purchasing, marketing, and operations) to recommend when to order, and how much to order, to maximize value and minimize waste.

The MRP systems support many decisions in the transformation process including

1. Demand forecasting
2. Configuration of products for quotation
3. Configuration of products for manufacture
4. Scheduling sales orders and delivery
5. Processing of sales orders
6. Establishing agreements with suppliers
7. Processing purchase orders
8. Processing subcontracting orders
9. Planning and scheduling work orders

10. Managing the MPS
11. Managing the safety stock
12. Modeling the logic of the MRP
13. Managing the capacity requirements planning (CRP)
14. Sequencing orders on the shop floor
15. Scheduling the shop floor
16. Managing product BOM
17. Managing inventory
18. Preparing kits for production orders

To demonstrate the MRP logic, consider the previous example. There are four types of input to the process:

1. *Marketing input:* In the form of the MPS that contains information on requirements for product A, which is the output of the transformation process.
2. *Engineering input:* The information that each unit of A is assembled from 1 unit of B and 2 units of C and that the lead time for the assembly operation is 1 week.
3. *Purchasing input:* Information from suppliers of B regarding the promised lead time of B: the input for the transformation process.
4. *Inventory input:* The number of units of A, B, and C currently in inventory.

The MRP logic uses this input to calculate how many units of A, B, and C to order and when to order these units.

The data processing logic of the MRP system is very simple and straightforward.

Gross to net: Since 10 units of A are needed in period 10 and 1 is available in inventory, only 9 should be ordered. This is "gross-to-net" logic, the logic that subtracts available units in inventory or in the pipeline from the gross requirements to calculate the net requirements.

Time phasing: The lead time for A is 1 week and, therefore, to have 9 units of A ready at week 10, the assembly of B and two C parts for each A, should start a week earlier in week 9. This is known as "time phasing" logic. It takes into account the lead time for issuing the work order, so that units will be ready exactly when needed.

The same gross-to-net logic and time-phasing logic are used to decide how many units of B to order and when to place the order. However, since B is not an output of the transformation process (the terminology used is "end product") that is sold to the organization's customers, its gross requirements are

TABLE 8.1

Summary of the MRP Steps

	Week 1	Week 2	Week 3	Week 4	Week 5	Week 6	Week 7	Week 8	Week 9	Week 10
Demand: A										10
Inventory: A										1
Net: A										9
Assembly									9	
Demand: B									9	
Inventory: B									2	
Net: B									7	
Process: B							7			
Demand: C									18	
Inventory: C									3	
Net: C									15	
Order: C						15				

based on the work order for A that assembles B and C into A. A summary of this planning process is depicted in Table 8.1.

The assembly of 9 units of A should start at week 9 and, therefore, 9 units of B will be required (each unit of A needs 1 unit of B). There are 2 units of B in inventory and, therefore, only 7 units should be ordered. The lead time of B is 2 weeks and, therefore, the materials and parts for 7 units of B should be ordered from the supplier at week 7.

Finally, the gross-to-net logic and time-phasing logic help the managers decide how many units of C to order, and when to place the order.

Two units of C are required for each unit of A. Therefore, to start assembly of 9 units of A in week 9, 18 units of C are required. Since 3 units of C are in inventory, only 15 should be ordered and the work order should be placed for week 6 as there is a 3-week lead time for C.

This simple logic is implemented by a model called the MRP record—a record that is used throughout the MRP system.

8.2 Basic MRP Record

The MRP logic applies two basic steps: gross to net and time phasing. To facilitate clear visibility of these steps, the simple MRP record was developed and is used as the building block of the MRP system.

The first line in the MRP record is the *timeline* or *time bucket*. The length of a period in the timeline could be a day, a week, a month, or a quarter, depending on the resolution or the level of details of the planning process. Deciding on the timeline is an important decision in the design of the MRP implementation. The number of periods in the timeline is equal to the duration of the planning horizon, which is another important decision in the implementation of MRP systems.

The other entries in the MRP record are related to the timeline in the following way:

Gross requirements: Time-phased future demand (actual customer orders or forecasts, or both) during each period in the planning horizon.

Scheduled receipts: The orders for the items that are due in at the beginning of the period. These are a result of past decisions to issue work or purchase orders.

Projected available balance: The projected balance after scheduled orders have been received and gross requirements have been satisfied at the end of each period.

Planned order releases: Recommendations for issuing orders at the beginning of each period based on the projected available balance. When the projected available balance for an item is negative, it must be corrected by issuing an order that will satisfy the net requirements.

Additional information that may appear follows:

Lead time: The expected time from the issuance of a work order or purchase order until the items are finished and available for further stages in the transformation process.

Lot sizing: The batch size or lot size for this item (based on some criterion, such as EOQ, number of items in a container, etc.). If there is no recommended lot size, a lot for lot (LFL) policy is assumed; that is, the order size is equal to the net requirements for the item (Table 8.2).

The same record is used for all items (see Tables 8.3 through 8.5).

The gross-to-net logic and time-phasing logic along with the MRP record integrate the input, output, and processing of the transformation process in a simple yet effective way.

TABLE 8.2

The MRP Record

Period	1	2	3	4	5	6	7	8	9	10
Gross requirements										
Scheduled receipts										
Projected available balance										
Planned order releases										
Lead time =										
Lot sizing =										

TABLE 8.3

MRP Record for End Product A

Period	1	2	3	4	5	6	7	8	9	10
Gross requirements										10
Scheduled receipts										
Projected available balance	1	1	1	1	1	1	1	1	1	0
Planned order releases								9		
Lead time = 2										
Lot sizing =										

TABLE 8.4

MRP Record for Part B

Period	1	2	3	4	5	6	7	8	9	10
Gross requirements								9		
Scheduled receipts										
Projected available balance	2	2	2	2	2	2	2			
Planned order releases							7			
Lead time =										
Lot sizing =										

TABLE 8.5

MRP Record for Part C

Period	1	2	3	4	5	6	7	8	9	10
Gross requirements								18		
Scheduled receipts										
Projected available balance	3	3	3	3	3	3	3			
Planned order releases					15					
Lead time =										
Lot sizing =										

8.3 Input Data and Data Quality Issues

The MRP system gets its data from three major sources: (1) MPS, (2) BOM, and (3) inventory records. MRP itself is the source for purchase orders and work orders (see Figure 8.2).

The quality of data used as input for MRP systems is very important poor quality of input data will result in poor MRP plans. Thus, for example, if the inventory data are not correct and, in reality, there is less inventory than stated in the records of the information system, shortages will be generated. In the opposite situation when the inventory data are not correct and, in reality, there is more inventory than the records in the information system state, excess inventory will be generated.

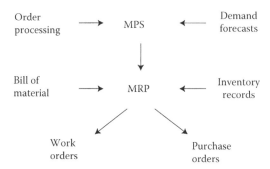

FIGURE 8.2
MRP and its main input and output.

BOM: This file contains the information (including quantity and lead time) on the required raw materials parts, subassemblies, and assemblies needed for the assembly of the finished product.

In the previous example, finished product A comprises 1 unit of B and 2 units of C. The indent BOM presents this case thus:

> A
>> B
>> 2C

Another way is to present it graphically:

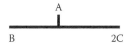

In many products, the BOM has several levels. For example, suppose that C in the above example comprises 1 unit of D and 3 units of E. The single-level indent BOM illustrates this as follows:

- BOM for A

> A
>> B
>> 2C

- BOM for C

> C
>> D
>> 3E

The single level BOM represents a finished product (also known as the father) and the components directly under it (also known as the sons).

The quality of information in the BOM is extremely important. If it does not represent accurately, the components of the finished product, the MRP logic will result in ordering the wrong parts and components, and it will not be possible to assemble the finished product. Ensuring the accuracy of a BOM presents a challenge in the face of frequent changes in engineering design, which occur for many reasons. For example, safety may necessitate design changes, if a fault in the existing design emerges. Achieving higher quality may force design changes if new technology can improve existing quality. Economics may dictate design changes if the latter enable cost savings, etc. Updating, also known as maintenance of the BOM, is an important task of the engineering function in many organizations.

In some companies, a finished product may have many optional features such as the features offered in cars, for example, manual transmission or automatic transmission, two-door or four-door models, etc. In these cases, it is advantageous to use a single BOM, known as a modular bill, to represent the different models. A modular bill is a common tree with parts (branches) relating to the specific product model of the product it represents. The maintenance of the single BOM is easier and less expensive to perform than maintaining several similar trees.

Inventory files: Information about inventories is required for performing gross-to-net analysis. The quantities of current and future inventories are an important input to the MRP record, and the more accurate the data on inventories in storage, in process, and in the pipelines, the better. *In process inventory* refers to orders already issued that will be realized in the future. *Pipeline inventory* refers to orders issued in the past that are scheduled to be realized in the future. The quantities and expected dates of arrival are used in the gross-to-net process of the MRP logic. Errors from many possible sources can affect this data negatively. For example, quality issues may result in production of defective units as part of a work order. In this case, the inventory file must be updated as soon as the number of good units is known. If quality is a recurring problem, efforts to fix it are required, along with systematic collection of information on the "yield" of the production process, that is, the percentage of good units in each production order. Yield information can serve as a basis for quality improvements efforts. It also helps the organization meet future orders by providing information it needs to order enough units so that the expected number of good units will suffice. In a similar way, a supplier, not having all the units ordered by the organization, will ship the order in stages. The actual size of each shipment must appear in the inventory system to ensure that pipeline inventory is correct.

Lead time information: This information is important for the time-phasing process of the MRP. Supplier's lead times are based on the supplier's commitment but it is important to track a supplier's actual lead time and record the deviation of actual delivery of orders from the promised due date. This information can be used to help the supplier solve scheduling problems (if they exist), as a basis for a search for better suppliers, or at least to ensure meeting future orders by placing orders earlier, taking into account past delays in delivery.

The lead time for work orders can be estimated based on the actual lead time of past work orders. The problem is that the actual lead time of work orders fluctuates when the load on the shop floor varies, or the mix of work orders changes. In some advanced systems, the estimated lead time is based on analysis of the actual load on the shop floor.

Master Production Schedule: The MPS is the source of information on time-phased future demand (actual customer orders, forecasts, or both) during each period in the planning horizon. The master schedule is constantly changing as customers place new orders, orders are supplied, and existing orders change. Time-phased future demand is the driver of the gross-to-net and time-phasing processes, and, therefore, continuous updating of the MPS is necessary, as the MRP output is very sensitive to inaccuracies in the MPS.

A major problem with the MPS is its feasibility. The MPS may present future demand that cannot be met due to limited capacity of existing resources. This was a major problem with the early MRP systems, also known as MRP1 systems. To solve the problem, industrial engineering researchers developed a new generation of MRP systems, known as MRP2 systems.

8.4 Capacity Considerations: The Evolution of MRP2 Systems

The gross-to-net logic and time-phasing logic used in MRP1 systems integrated the data from different legacy systems and generated work orders and purchasing orders designed to provide what is needed, when it is needed, and in the quantities needed. Implementation of these systems revealed that they are very sensitive to the quality of data. Poor data quality led to ordering the wrong raw materials, components, and parts (a typical result of errors in the BOM) in the wrong quantities (a typical result of errors in the inventory files), and at the wrong time (a typical result of poor lead time estimates). Worst of all, some MRP work orders could not be executed due to lack of capacity. To overcome this capacity problem, information about required capacity and available capacity became an integral part of the new generation of MRP systems known as MRP2 systems. An obvious effort to increase data reliability in MRP2 systems led to better connection of MRP2 to its sources of information.

Two models helped add capacity information to the MRP2 system:

1. *Rough cut capacity planning (RCCP):* This model uses MPS information to check whether the required capacity in each time period is available.

2. *Capacity requirement planning (CRP):* This model uses MRP output to check whether the required capacity is available at each time period.

Sections 8.4.1 and 8.4.2 explain the two models.

8.4.1 Rough-Cut Capacity Planning

The RCCP, using MPS information, is performed prior to gross-to-net and time-phasing calculations. Information regarding the processing time per unit on each machine or work center is also required.

The following example of four machines and four products illustrates the logic used for RCCP.

8.4.1.1 Example of RCCP

The unit processing time (in minutes) of each of four products A, B, C, and D on the four machines is summarized in Table 8.6. The last column shows the MPS information for the production planned for the next period. The second-to-last row in the table is a simple estimate of the capacity required for each machine, excluding setup time. The estimate is the sum of the processing time of each product on the machine multiplied by the planned quantity of the same product, according to the MPS. For example, the estimate for machine 1 is:

$$6 \times 60 + 4 \times 80 + 2 \times 200 + 3 \times 70 = 1290$$

Assuming that each machine is available 5 days a week, 8 h a day, the available capacity of each machine is 2400 min ($8 \times 5 \times 60$). Machine 2's load exceeds its capacity. Machine 4's load is very close to its capacity; only 70 min (2400–2330) will be available for setups, and since each setup takes 30 min, only two setups can be performed. The conclusion is that this MPS is not feasible because machine 2 does not have capacity to process the MPS. The gap between the required capacity and the available capacity may be even larger as machine breakdown is likely to reduce the available capacity of

TABLE 8.6

Example of Processing Times Using RCCP

	Machine 1 (min)	Machine 2 (min)	Machine 3 (min)	Machine 4 (min)	MPS Quantity
Product A	3	5	2	7	70
Product B	2	7	4	6	200
Product C	4	4	3	5	80
Product D	6	9	6	4	60
Required capacity	1290	2610	1540	2330	
Setup time	30	25	30	30	

some resources. In some RCCP models, past information on machine break-down is used to forecast future capacity loss.

The RCCP is an important tool added by MRP2 for the management of the MPS. In the early MRP1 systems, the decision to accept a new order from a customer was based primarily on intuition and experience of production planners. RCCP, in contrast, creates a sound basis on which to make the decision to accept new orders. Furthermore, it can help determine due dates for such orders. From the discussion on the theory of constraints and bottle-necks, we know that bottleneck identification and commensurate planning is crucial for optimizing the throughput. While a loss of production time on a bottleneck resource is a loss to the overall throughput, the loss of production time of a resource with excess capacity could be gained back by accelerating production on this resource.

To summarize, it is very important to check the feasibility of the MPS and to closely monitor the resources that are heavily loaded (possible bottlenecks) to avoid unnecessary idle time of these resources.

8.4.2 Capacity Requirement Planning

The RCCP enables managers to estimate the load on each resource in the MPS *before* executing the MRP (Drexl and Kimms 2010). It enables making rough-cut corrections either to the actual capacity or to the MPS, before executing the MRP. However, using the MPS as the basis for capacity planning could lead to significant errors. For example, the MPS does not take into account existing inventories of components, WIP, and raw materials, and the capacity already invested in them. By using existing inventories, the load on the shop floor machines may be reduced. Furthermore, the actual load on resources in a given time period is not generated by the future demand presented by the MPS for that period in the planning horizon. This is due to time-phasing logic.

A better estimate of the load on resources is developed using the work orders generated by the MRP logic. In other words, the size and timing of work orders produced by the MRP system combined with information on the time it takes to process each job on the machines, and the sequence of processing, produces a good resource load estimate. This logic is known as CRP.

The data on the processing required by each job is stored in the product routing file.

The product routing file: This file contains information on the sequence of processing each item on the shop machines, and the time it takes to process each item on these machines. In many applications, each job has a copy of the relevant routing information attached to it, which helps shop workers correctly move the job among the job shop machines.

To summarize, the evolution of MRP systems from MRP1 to MRP2 added the dimension of capacity at two different stages: (1) before executing the MRP—this is RCCP, which is used to revise the MPS, and (2) after executing the MRP—this is CRP, which is used to fine-tune the MPS (Figure 8.3).

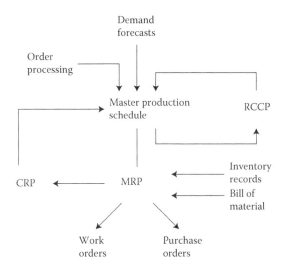

FIGURE 8.3
Input and output of MRP2 with its capacity planning modules.

8.5 Using the MPS for Available to Promise Analysis

Thus far, we considered a very limited version of the MPS assuming that the only information it contains is on time–phased future demand (actual customer orders or forecasts, or both) during each period in the planning horizon. In its simplest form, the forecast of future demand and actual orders are summed up for each future period. In some applications, the MPS is constructed based on forecasted demand and lot sizing considerations. Such applications distinguish between *forecast* of future demand and *actual orders* for each future period. This method helps marketing people estimate how many units of future gross requirements have not been assigned to a specific customer order and, therefore, is available to promise (ATP). To explain the ATP logic, consider the following MPS example.

The planning horizon is 9 weeks. There are confirmed orders for the next three periods: 10 for period 1, 10 for period 2, and 7 for period 3. Forecasted sales for the next 5 weeks are 15 units a week, and for the rest of the planning horizon, it is 20 units a week. We assume that there is an initial inventory of 10 units and the batch size (EOQ) is 35 units (Table 8.7)

In this version of the MPS, the first step is to decide on when to issue work orders. Orders are issued based on the forecasted inventory for each period. This is calculated by subtracting the maximum between the forecast for the period, and the firm orders for the period (from the calculated inventory of the previous period). If the calculated inventory for a period becomes negative, an order for 35 units is placed in this period.

TABLE 8.7

Data for the MPS Example

	Week 1	Week 2	Week 3	Week 4	Week 5	Week 6	Week 7	Week 8	Week 9
Forecast	15	15	15	15	15	20	20	20	20
Start inventory	10								
Confirmed orders	10	10	7						

In period 1, the calculated inventory is the inventory at time 0, which is 10 minus the maximum between the forecasted demand of 15 and the firm orders of 10 for period 1.

$$10 - \max(15, 10) = 10 - 15 = -5$$

Since the result is negative, an order is placed for 35 units in period 1. As a result, the calculated inventory at the end of period 1 is $35 - 5 = 30$.

In period 2, the calculated inventory is 15:

$$30 - \max(15, 10) = 30 - 15 = 15$$

Since the result is not negative, no order is placed and the calculated inventory at the end of period 2 is 15.

In period 3, the calculated inventory is 15:

$$15 - \max(15, 7) = 15 - 15 = 0$$

Since the result is not negative, no order is placed and the calculated inventory at the end of period 3 is 0.

In period 4, the calculated inventory is negative:

$$0 - \max(15, 0) = 0 - 15 = -15$$

Since the result is negative, an order is placed and the calculated inventory at the end of period 4 is $35 - 15 = 20$.

This calculation is performed for all nine periods in the planning horizon. The result is summarized in Table 8.8.

When the MPS calculations are completed, the number of ATP units is calculated as follows.

For each period in which there is a requirement in the MPS, that is, "the calculated period," the ATP is equal to the forecasted inventory at the end of the previous period plus the order for this item in the calculated period minus the confirmed orders from the calculated period until the period before the next MPS order.

In the above example, the first MPS requirement is in period 1; therefore, the ATP in period 1 is equal to the forecasted inventory in period zero (10)

TABLE 8.8

Summary of the MPS Example

Week	0	1	2	3	4	5	6	7	8	9
Forecast		15	15	15	15	15	20	20	20	20
Starting inventory	10									
Confirmed orders		10	10	7						
Forecasted inventory		30	15	0	20	5	20	0	15	20
MPS		35			35		35		35	35

TABLE 8.9

ATP Calculations for the Planning Horizon

Week	0	1	2	3	4	5	6	7	8	9
Forecast		15	15	15	15	15	20	20	20	20
Starting inventory	10									
Confirmed orders		10	10	7						
Forecasted inventory		30	15	0	20	5	20	0	15	20
MPS		35			35		35		35	35
ATP		18			35		40		35	50

plus the MPS order (35) minus the confirmed orders in periods 1, 2, and 3 (given that there is an MPS order in period 4).

$$10 + 35 - (10 + 10 + 7) = 18$$

Thus, in period 1, there are 18 units ATP to customers.

The next MPS requirement is in period 4; the inventory in period 3 is 0 and there are no confirmed orders for periods 4, 5, and 6. Therefore, the ATP is equal to 35. Table 8.9 summarizes the ATP calculations for the planning horizon.

The MPS can be linked to the RCCP logic and support marketing decisions regarding whether to accept new customer orders or to respond to customer requests to change the date or quantities of existing orders. ATP information along with RCCP estimates of loads on the shop floor improves the ability of the marketing division to manage the MPS and to ensure that commitments to customers are feasible.

8.6 Lot Sizing Considerations

The stated goal of MRP systems is to order exactly what is needed, when it is needed, in the quantities needed. In the last example, EOQ helped determine the order size (35). However, the MRP logic that we discussed thus far could have achieved this goal by applying a lot-sizing concept called LFL. The LFL concept is simple and straightforward—the planned order releases are simply the time-phased net requirements. LFL makes a lot of sense if there are no significant setup times and setup costs or fixed costs. The LFL assumption that the shop floor is a deterministic environment where all the data used as input to the MRP record are known and perfectly accurate is rarely the case as both lead time and demand are based on forecasts and subject to forecasting errors. Furthermore, the LFL logic assumes a linear production or purchasing cost function. In other words, the cost of ordering $2x$ units is exactly twice the cost of ordering x units. This is not, however, always correct; for example, many vendors offer discounts on larger orders. The cost of transportation is also not linear (e.g., in overseas transportation, the cost of shipping a container is not a function of the number, volume, or weight of the items in the container) and, therefore, economy of scale exists. Similarly, the setup cost of machines is usually fixed and not a function of batch size. Consequently, larger batches reduce the setup cost (or setup time) per unit produced and economy of scale becomes an important consideration.

MRP systems take advantage of economy of scale by using proper lot sizing. LFL is recommended only when the impact of economy of scale is negligible. In other situations, batch-sizing techniques such as EOQ should be considered. One problem with EOQ is that the order quantity may not cover an integer number of time periods, and hence may lead to a situation where inventory is carried into a time period when a new order is received. To prevent such situations, a modification of EOQ called the periodic order quantity (POQ) was developed (Ptak and Smith, 2011).

8.6.1 Example of Mixed LFL and POQ Lot Sizing

In this example, item A is composed of components X, Y, Z, and W; and W is composed of T and U (Figure 8.4).

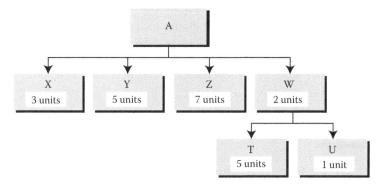

FIGURE 8.4
The BOM of item A.

The gross requirements for end product A are as follows:

Week	0	1	2	3	4	5	6	7	8	9	10
Gross requirements		2	4	3	2	1	7	4	8	7	6

Additional information:

- The lead time for production of product A is 2 weeks.
- The available inventory of A at the beginning of the planning horizon is 6 units.
- Lot sizing for A is based on LFL (Tables 8.10 through 8.15).

The MRP record of U is based on the POQ model. The batch size is calculated using the LFL method, modified so that each order covers two periods ahead. U's lead time is 1 week. The planned material requirements for the item U are summarized in Table 8.16.

TABLE 8.10

Other Data for the Example

Item	Lead Time	Inventory on Hand	Pipeline Inventory
X	1	10	5 units on week 5
Y	2	24	2 units every week (starting week 1)
Z	2	38	
W	1	10	

TABLE 8.11

MRP Record for A

Week	0	1	2	3	4	5	6	7	8	9	10
A's gross requirements		2	4	3	2	1	7	4	8	7	6
Scheduled receipts											
Projected available balance	6	4									
Planned order releases (A)		3	2	1	7	4	8	7	6		

TABLE 8.12

MRP Record for X

Week	0	1	2	3	4	5	6	7	8	9	10
X's gross requirements		9	6	3	21	12	24	21	18		
Scheduled receipts						5					
Projected available balance	10	1									
Planned order releases (X)		5	3	21	7	24	21	18			

TABLE 8.13

MRP Record for Y

Week	0	1	2	3	4	5	6	7	8	9	10
Y's gross requirements		21	14	7	49	28	56	49	42		
Scheduled receipts											
Projected available balance	38	17	3								
Planned order releases (Y)		4	49	28	56	49	42				

TABLE 8.14

MRP Record for Z

Week	0	1	2	3	4	5	6	7	8	9	10
Z's gross requirements		15	10	5	35	20	40	35	30		
Scheduled receipts			2	2	2	2	2	2	2	2	2
Projected available balance	24	11	3								
Planned order releases (Z)			33	18	38	33	28				

TABLE 8.15

MRP Record for W

Week	0	1	2	3	4	5	6	7	8	9	10
W's gross requirements		6	4	2	14	8	16	14	12		
Scheduled receipts											
Projected available balance	10	4									
Planned order releases (W)			2	14	8	16	14	12			

TABLE 8.16

MRP Record for U

Week	0	1	2	3	4	5	6	7	8	9	10
U's gross requirements			2	14	8	16	14	12			
Scheduled receipts											
Projected available balance			14		16		12				
Planned order releases (U)		16		24		26					

8.7 Uncertainty and Buffering Considerations

The simple gross-to-net logic and time-phasing logic may cause shortages when forecasting errors in demand and lead time occur. In addition, life is full of surprises and uncertainties. There are plenty of reasons for delays (e.g., machine failures, absenteeism, strikes, blackouts, and other emergencies).

TABLE 8.17

MRP Record for T

Week	0	1	2	3	4	5	6	7	8	9	10
T's gross requirements			10	70	40	80	70	60			
Scheduled receipts				10		50					
Projected available balance	15	15	60	40	30	70	60				
Planned order releases		55	40	30	70	60					

One way to protect the system from such shortages is to use buffers. MRP systems commonly use two types of buffers:

Buffer lead time: In this case the lead time used for time phasing is larger than the forecasted lead time. The time added to the forecasted lead time is called a time buffer. As a result, orders are placed earlier, and a delayed order does not cause shortages as long as the delay is shorter than the time buffer. This method protects the transformation process from uncertainty by increasing the holding time of inventory. The extra inventory carrying cost should be weighed against the cost of shortages due to late delivery of orders.

Buffer inventory (safety stock): The size of each order calculated by the gross-to-net process is increased by a buffer inventory. If the supplier ships defective items or the shipment is smaller than the order, the buffer inventory can protect the system from shortages, as long as the number of good items shipped plus the good items in the buffer inventory is equal to or larger than the actual demand.

To demonstrate the use of buffers, consider item T in the previous example. Table 8.17 shows the planned material requirements for the item T. T's lead time is 1 week but the supplier is not reliable and late deliveries of 3–5 days are common. Since T is crucial for the production of W and W is crucial for the production of A, it is important to protect the system against delays in the delivery of T.

To protect the system, a buffer time of 1 week will be added. In other words, although the nominal lead time is 1 week, for T we will use a lead time of 2 weeks.

8.8 MRP2 as a Predecessor of the ERP

The MRP2 is a natural evolution of MRP systems. MRP2 is not only concerned with manufacturing materials. Its wider scope is focused on the coordination of the entire manufacturing production, including materials, finance, and human relations. To this end, it was connected to the various legacy information systems of the factory. The goal of MRP2 is to provide consistent data to all players in the manufacturing process as the product moves through the production line (or system).

The connections between MRP2 systems and legacy information systems (such as finance, human resources, purchasing, engineering, etc.) have contributed to the consistency of data in the organization. However, it also made its users aware of the need for a single database and a single model base for the whole enterprise. Such a system, developed a decade after MRP2, is the enterprise resource planning (ERP) system, which is still the major business information system used by organizations today.

8.8.1   Enterprise Resource Planning

The ERP provides an integrated view of core business processes, often in real time, using common databases maintained by a database management system. ERP systems track business resources—cash, raw materials, production capacity—and the status of business commitments: orders, purchase orders, and payroll, including

- Accounting and financial management and costs
- Inventory management
- Production/service planning
- Product planning
- Manufacturing or service delivery
- Marketing and sales
- Shipping and payments

The ERP facilitates information flow between all business functions and uses a consistent database shared by the applications that comprise the system across the various departments (manufacturing, purchasing, sales, accounting, etc.). Most industrial ERP systems still use MRP2 logic for streamlining the production.

8.9 Summary

The MRP systems are designed to support decision making throughout the transformation process by translating gross requirements, managed by marketing, into purchase orders, managed by purchasing, and work orders, managed by the production function. The simple MRP logic is implemented by the MRP record that uses information from:

1. *Marketing:* information on firms orders and forecasts
2. *Purchasing:* information on suppliers, lead times, and pipeline inventories
3. *Engineering:* information on the BOM and routing
4. *Operations:* information on processing time and in-process inventories to issue purchase orders and work orders

Starting from master production schedule (MPS), this input is translated to phased requirements of subassemblies and parts using bill of materials (BOM) and lead time information. The basic MRP logic can be extended to include RCCP, CRP, lot-sizing models, and buffering against uncertainty. These extensions have made the MRP logic the mainstay of many organizations, and an important component of the next generation of information systems—ERP systems.

References

Drexl, A. and Kimms, A. 2010. *Beyond Manufacturing Resource Planning (MRP II): Advanced Models and Methods for Production Planning.* Berlin: Springer.
Orlicky, J.A. 1973. Net change material requirements planning. *IBM Systems Journal* 12(1), 2–29.
Ptak, C. and Smith, C. 2011. *Orlicky's Material Requirements Planning 3/E.* New York: MacGraw-Hill.

9

Enterprise Resource Planning

Educational Goals

In this chapter, the concept of a single, integrated information system for the whole organization is presented. This concept is implemented by a class of information systems called enterprise resource planning (ERP) systems. ERP systems are applied in most modern organizations big and small. The strength of ERP systems is their support of data integration in a single database and the application of a single model base for all the organizational functions (an extension of the MRP systems that focus on the organizational functions that perform the transformation process, mainly the production and inventory management activities). The idea of a single database and a single model base for the whole organization is a cornerstone in the architecture of ERP information systems along with the focus on organizational process management. This idea enables efficient management of data, easy extraction of information, consistency, and uniformity of the information technology processes. By replacing several legacy systems, ERP eliminates boundaries caused by communication difficulties between different legacy systems. ERP systems support organizational business processes by:

- Providing a definition of the process activities
- Providing a definition of the organizational function that performs each of the process activities
- Providing a definition of the information required for each of the process activities
- Providing a definition of the process flow between the organizational functions that perform each of the process activities
- Providing data, models, and information needed to perform the process activities
- Management of the process flow
- Monitoring the progress of the process

In this chapter, we show how processes are modeled and managed by the ERP system. We will also discuss the difficulties and possible failures in the implementation of ERP systems:

- The difficulty to introduce change in organizations
- The substantial costs of introducing ERP systems
- The difficulty to reengineer organizational processes to realize the theoretical benefits of ERP systems

9.1 Introduction

The history of ERP development is essential for appreciating its role and its importance. The roots of ERP systems stem from the development of the computerized materials requirement planning (MRP) systems. The early MRP systems (MRP1) concentrated on the management of materials in the transformation (main production or service) process by generating a work plan recommending when to issue work orders (WO), when to issue purchase orders (PO), and how much to order. The addition of capacity considerations in MRP2 systems was an extension of the MRP1 systems to allow better scheduling of resources used by the transformation process. The MRP system did not satisfy the need for data and models to support functions that are not directly involved in the core production or service process. These functions (such as accounting, human resources [HR], and sales) were supported by separate information systems (known as legacy systems) developed for each organizational function. Each legacy system had its own data and models. Its input came from data entry like typing the data or from other information systems in the same organization, and its output was essential for other computerized information systems of the same organization. For example, sales departments had their own software, while accounting departments used another separate software package; HR departments had their specialized software package, etc. When a business process was performed by two or more organizational units, data had to be transferred between the legacy systems used by these organizational units. This was done initially by retyping the data that existed in one system into the other. This double data entry was a waste of time and effort and a source of data entry errors that caused low data quality. Later on the transfer of data between legacy systems was done by transferring files that contained the data. In the 1970s, a group of former IBM employees formed the SAP Company, which developed an early solution to these problems by software that supports business processes performed by several organizational functions sharing data and models needed by the different organizational units.

In parallel, Oracle Corporation started its database business, which would eventually develop to become a major ERP business. This was the beginning of a new era of information systems—systems that support decision-making and business processes performed throughout the organization.

ERP systems replaced the legacy information systems that were dedicated to a single organizational function. The original MRP systems that supported most organizational functions involved in the transformation process developed into organizational information systems that stored all the organization data in a single database so that the data could support any function in the organization. The ERP concept to use a single database and a single model base that supports the entire organization helped in breaking the barriers between business units and increasing the efficiency of employees in performing their part in business processes. A well-designed and implemented ERP system provides high-quality information (accurate, precise, reliable, updated, and consistent information) to employees, which enables them to make better decisions faster. In addition, new modules were added to the main base of ERP systems. Typical modules are customer relationship management (CRM), supplier relationship management, and interactive voice routing for incoming calls. In addition, new models were added that could find various relationships by analyzing large quantities of data stored in the organizational database performing new functionalities known as analytics and data mining. These functionalities are the foundation of business intelligence (BI). Over the years, thousands of organizations in different sectors implemented ERP systems and the lessons learned from these implementations formed a collection of best practices in different areas such as supply chain management.

Some additional layers of ERP systems added to the MRP2 systems are shown in Figure 9.1.

While MRP systems were used mainly by organizations in the manufacturing sector, ERP applications are used in many sectors including service-oriented organizations and organizations in the government sector. Examples include organizations in aerospace and defense, agriculture, apparel, automotive, banking and mortgage, chemicals, construction, consumer products, food and beverage, education health care, high tech, insurance, nonprofit, oil and gas, pharmaceuticals, professional services, public sector, security, software development, telecommunications, travel, transportation, utilities, wholesale, and retail. Based on lessons learned and best practices, some ERP software companies offer solutions that meet the unique requirements of specific industries.

In Chapter 1, we introduced five basic processes that exist in many organizations:

1. *The development process:* This process starts with an idea for a new product or service and ends with the design of the new product or service and a working prototype.

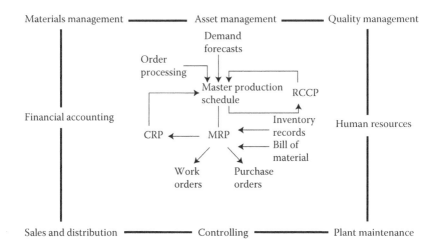

FIGURE 9.1
The typical ERP layers and main modules.

2. *Preparation of infrastructure:* This process starts with a working prototype of a new product, and ends with the successful completion and testing of the product's production facility or the service facility.

3. *Sales:* This process starts with market research and ends with an order from a customer.

4. *Supply:* This process starts with an order from a customer, and ends with delivery and receipt of payment from the customer, who received the products he or she ordered.

5. *Service:* This process starts with a customer request for service, and ends when service is provided to the satisfaction of the customer.

Modern ERP systems are designed to support these processes in different sectors and by a variety of organizations. They provide both information and communication to users of these processes. These systems are based on client–server architecture (Klaus et al., 2000) or on the Internet infrastructure and cloud computing. They are designed to transfer information between the various parts of the organization, through a computer network or the Internet. A relatively new development is to provide ERP software as a service also known as web-based ERP. In these applications, the ERP software resides on the vendors' servers and the user access it via the Internet. The main advantage is that the user's initial investment is low and the implementation can be quicker. In addition, the user pays per use only and does not need IT staff to maintain servers and other hardware. Some vendors also developed mobile interfaces to the ERP software to allow access anytime from anywhere.

9.2 Functionalities and Components of ERP Systems

Most ERP systems contain a library of application modules. These modules can be implemented as an integrated solution, but it is also possible to select only some of the modules for implementation based on the needs of the organization or the implementation plan (some organizations prefer to implement the ERP system module by module to reduce the risk of implementing all the modules simultaneously). Furthermore, as we will see later, some organizations select and integrate modules from different ERP systems in an effort to find the most appropriate combination of modules for the company. Typical modules in ERP systems are discussed in the following sections.

9.2.1 Production Management

This module is usually an extension of MRP logic; it is used to plan and control the manufacturing activities of the company. This module typically includes bills of material, routings, work centers management, master production scheduling, shop floor control, management of production orders, and management of production orders (POs).

Some ERP vendors developed specific applications for different types of industries. These industries operate in a way that is different than traditional MRP industries, and therefore require different logic for their management. A typical example is a module for continuous production or process industries like oil and gas or beverage industries. Another example is a module for discrete mass production like the manufacturing of cars and electronic appliances.

The MRP logic of gross to net and time phasing is widely used in the production modules of ERP systems. The master production schedule is typically the driver of the system based on forecasts, actual customer orders, or a combination of the two.

Capacity planning models are available in many ERP production modules. Models for rough cut capacity planning and capacity requirement planning along with models that take into account the limited capacity of resources in the scheduling process are available. Some ERP systems apply logic similar to the drum buffer rope model (known as finite capacity loading) while some ERP systems support just in time and the kanban scheduling approach and a variety of strategies like "make to stock," "make to order," and "assemble to order."

Most ERP systems generate production orders and purchasing orders and monitor and control the actual progress of these orders based on active data collection. The data may be typed in, or collected using technologies like bar-code readers, RFID scanners, speech recognition, and pattern recognition. The integrated approach of ERP makes it easy to estimate the total cost of production using cost accounting analysis. ERP enables to estimate the cost of manufacturing different products, to track man-hours and material used on each batch of each product type, and to use information from sales and distribution to update production plans. Most ERP systems introduce changes in real time so the database always reflects the current status with high accuracy. ERP supplies high-quality information on canceled or added orders, such as POs, WOs, and sales orders (SOs). Most ERP systems also provide current information on changes in suppliers' delivery times or quantities, machine breakdown, and employee absenteeism. This information is used to update plans and to identify and fix problems as early as possible.

9.2.2 Project Management

This module supports the planning, execution monitoring, and control of nonrepetitive undertakings or projects, performed by the organization. Models for scheduling, resource management, budgeting, risk management, and portfolio management are available in many ERP applications.

Projects may be part of the internal processes of a company, or they may be performed for external contractors. They may be performed by project-oriented organizations, functional organizations, and organizations that use the matrix structure. Therefore, all of these organizational structures must be supported by the ERP project module. In some ERP applications, this is done by the project organizational breakdown structure that represents the organizational units participating in the project, and the work breakdown structure that represents the division of the project scope, or work content, among the participating organizations. The integration of the project management module with other ERP modules makes it easy to track cost and man-hours invested in the project and to monitor and control the technological aspects such as changes in design and specifications.

9.2.3 Human Resources

This module supports personnel management and its administration, including recruitment, salaries, incentives, benefits and compensation management, personnel cost planning, HR funds and position management, and retirement pension plans. In some ERP applications, it supports time management including work schedules, shift planning, time recording, time evaluations, and incentive wages. An important component is the payroll module that in some ERP systems can deal with country-specific requirements for payroll processing. Support is also provided for training management including catalog of courses, dates, costs, billing, follow ups, and appraisals. Information related to hiring, such as vacant positions and job descriptions, is also managed by some ERP systems.

9.2.4 Materials Management

This module provides information on the materials moved in, within, and out of the organization and the valuation of those materials. It can also support tracking, controlling, and management of those materials including inventory management, purchasing, and invoice verification. The inventory management models support lot-sizing decisions and material consumption forecasting. The module may include models that control and track material movements, receipts, returns, and physical inventory including the actual cost of material used to calculate the actual cost of production.

Supplier's management modules may be available to help determine possible sources, to issue requests for quotes (including conditions and prices), to issue POs, and to monitor actual delivery of materials. These modules also track changes to the invoices, canceling invoices, and determining invoice variances.

In addition to the management of material, this module is extended in some ERP systems to support outsourcing; that is, the process of procuring external services from vendors, including the definition of the service specifications, writing the bids and managing the bidding process, awarding the order (signing a contract), and controlling the execution of these services. It is used to enter the services performed, accept those services, and then verify the service invoice.

9.2.5 Financial Module

This module supports the management of the general ledger, as well as the management of other accounts such as accounts receivable and accounts payable. It provides financial information on all the activities of the organization including transactions in sales, bank accounts, and expenses. It maintains customer and vendor records and supports reporting for payment history, overdue items, and currency risk. The financial module may also provide accounting support including cost center accounting showing the expenses of the organization from various views, such as cost per product type, cost per period, cost per department, cost per process, etc., as well as the planning, budgeting, controlling, and allocation of departmental expenses. In some ERP applications, product costing models focus on the creation of the standard product cost estimates, charging of overhead to production, capturing actual costs of production, calculating production variances, and settling production expenses to the appropriate profitability market segment.

9.2.6 Asset Management Module

This module supports the management of fixed assets of the organization including purchase and sale of machines, equipment, furniture, and building. The ERP maintains the records for the assets: asset description, the producer, the vendor, the payments, maintenance instructions, and maintenance contact details. Some ERP systems support country-specific requirements of depreciation calculations.

9.2.7 Plant Maintenance

This module supports the management of scheduled preventative maintenance, as well as unscheduled breakdown maintenance of machinery and equipment that needs service. It tracks the actual time and cost to perform routine and nonroutine maintenance operations and supports the scheduling monitoring and control of maintenance activities.

9.2.8 Quality Management

This module supports the management of quality planning, quality assurance, and quality control. Quality planning applies tools such as quality function deployment, design of experiments, and failure mode and effects analysis to identify opportunities and mitigate risks. Quality assurance supports the management of measuring equipment, inspection of inbound goods, and vendor rating. For preventing quality problems, quality assurance applies tools such as root-cause analysis, fishbone diagrams, and analysis of variance. Some ERP systems contain models that analyze, document, and file quality data from the manufacturing processes of inbound parts to guarantee product quality and to minimize the risk involved in product liability. The result is improved quality of products through defect prevention. The quality control models analyze inspection data using predefined statistical models (such as control charts) and report incidents of processes that are out of control and, in many cases, predict when processes are about to get out of control.

9.2.9 Sales and Distribution

This module helps the sales function in selling, shipping, and billing products and services. This starts with presales support including inquiry processing, quotation processing, and continues with SO processing, delivery processing, and billing. By integrating this module with the material management and production planning modules, the salesperson can get information on "available to promise" goods—these are quantities of products that could be promised to a customer for a required date. ERP also enables checking the feasibility of changing promised due dates and promised quantities. Figure 9.2 shows a schematic structure of the sales information module. It shows that its main components are sales and sales support, shipping, and billing with close connection to financial accounting and material management.

9.3 The Database and the Model Base

Important components of the ERP system are its model base and database. A single database that handles all the enterprise data is a cornerstone in the ERP

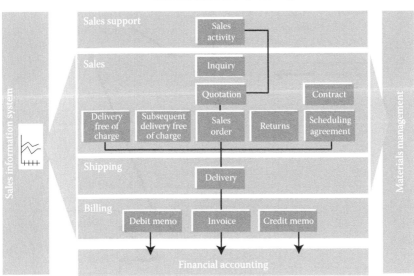

FIGURE 9.2
Schematic structure of sales information system.

concept. To appreciate the importance of a single database, think of multiple files holding the same data. For example, an employee's data may be stored for salary payments in one file and for business correspondence in another file. Consistency of data requires that updating the employee details must be done in both files at the same time in an identical manner. Any failure in simultaneous duplication of updates would cause inconsistency. A single database eliminates these duplications and inconsistencies. ERP database includes all the data that support all the processes performed by the organization, and it can be retrieved by authorized employees from any functional unit and processed by proper models from the model base. By having each data item stored only once in one location in the integrated database, the quality of data is easier to maintain and everyone is using the same data so collaboration between different functions in the organization is easier. In addition to the model base that contains models similar to those discussed in earlier chapters, several functionalities are built into ERP systems (such as workflow management and business intelligence) to support the organizational processes.

9.3.1 Workflow Management

Workflow management is a capability that many ERP systems possess. This feature enables supporting processes, automatically advancing and controlling them without regard to interorganizational boundaries. Figures 9.3 and 9.4 describe examples of such processes.

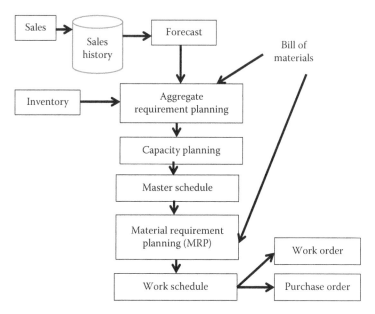

FIGURE 9.3
Workflow of planning production and acquisition.

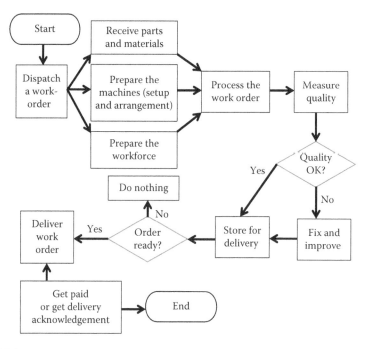

FIGURE 9.4
An example of a WO process.

Business processes are a collection of activities performed by one or more employees from one or more organizational functions. In many organizations, the same process is performed by employees from different units of the same organization. This used to be a major problem before the ERP era as each organizational unit had its own (legacy) information system along with its own data. When the process crossed the boundary (also known as the "wall" between organizational units), the flow of information from one legacy system to the other was not always smooth and in some cases information had to be typed in when the process crossed this boundary. The single database and the single model base of ERP systems eliminate this problem.

Typically, the person performing process activities gets input from the person preceding him or her in the process and generates output used by the person succeeding him or her in the process. The activities may be performed in a simple linear sequence (i.e., each activity follows one predecessor and has a single successor and there are no loops in the sequence), but in some cases nonlinear sequences where several activities precede one or more activities or several activities succeed one or more activities also exist. Furthermore, in some processes, there are loops as the process returns to a person who already took part in it. More complicated processes may have a variable sequence of activities that varies depending on some predetermined conditions. For example, consider a relatively simple process—a purchasing process that starts with a person filling out a form with information on the goods to be purchased, the required quantities, and the time the goods are needed. The succeeding activity may be a simple check if that person has the authority and the budget needed. In case the answer is positive, the process goes on to the purchasing agent who contacts a supplier and issues a PO; however, if the answer to one or both of the above questions is negative, the process returns to the person who initiated it with information about the problem found and the actions required to fix the problem. Figure 9.5 illustrates two simple workflow examples.

Many processes have many different variations to each of the subprocesses. For example, purchasing processes include the process of purchasing office supplies, which is very different from purchasing low-cost bolts, screws, and nails for the production process, which in turn is a different process from purchasing raw materials, which is different from ordering parts for maintenance, which is different from ordering furniture, etc.

In many companies, processes are broken down into many subprocesses. For example, the assembly of cars would be broken into the following main processes: engine assembly, gear assembly, power-train assembly, chassis assembly, white body assembly, painting, trimming, and final assembly. Each of the main processes is broken down to its subprocesses, and there are process variations as mentioned above. The management of all of these processes to ensure that they are performed correctly and each

(a)

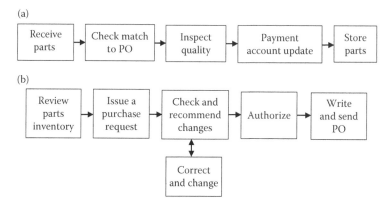

(b)

FIGURE 9.5
Two simple workflow examples. (a) A linear workflow example and (b) a workflow example with a correction loop.

person involved has all the information needed to perform his or her task is becoming increasingly complex. Workflow management modules of ERP are the main tools that can control and manage this immense complexity. When many processes are performed simultaneously, the ERP workflow management system ensures that each process is performed without delays or errors due to reasons like lack of information, a sick employee, misunderstanding, etc.

Workflow management systems are designed to support the definition, administration, coordination, monitoring, and control of processes. Process definition is done by defining the activities required to complete the process, the precedence relations among these activities, the person performing the activity, the data needed, the models needed, and the time allotted to perform each activity in the process. Software tools are used to support the process planning and definition. Many tools are object oriented, and the icons of the process building blocks (such as activities or precedence relations) are simply dragged and connected to design the process. User-friendly graphic user interface translates the flow diagram created by dragging, dropping, and connecting the icons of process components into a code that is used by the process management system.

To continue with the purchasing example, here is how ERP may be used in this process: after the employee enters the information on the items to be purchased, required quantities, and dates, in a specifically designed form, the finance department will automatically receive this information along with an approval form. If the request is approved, the system exports all necessary data to the purchasing department. When the PO is issued, the employee who triggered the process is automatically informed. If the request is not approved, the form is returned to the employee with instructions on what to do next.

Workflow management systems are used for planning, monitoring, and control of processes especially when employees have to work together and share information. In the purchasing example, the person that started the process can monitor its progress by getting information from the workflow management system on the current status of the process he or she initiated. Control can be implemented by setting a maximum number of days for each predefined activity, and whenever an activity does not finish within the allotted time a warning is issued.

9.4 Business Intelligence

The ERP systems store large amounts of data in their databases. Some data are used on a regular basis with specific models to support processes and routine decisions. BI is used to perform nonroutine, ad hoc analysis based on large amounts of unstructured raw data that reside in different locations in the database. This data can be accessed and analyzed in order to extract new insights that can help ad hoc decisions (Leon, 2014). A typical example is customer profiling based on all past transactions of a customer, his or her payment history, and other pieces of data. Such transactions may take place over a number of years and consist of purchases, request for service, complaints, and even customer inquiries. BI systems are designed to help management in retrieving the data and transform it into meaningful information that can be used to support ad hoc decision making and sometimes even strategic decisions.

9.4.1 Data Mining

The availability of ERP system single database presents an opportunity to search for seemingly unrelated data items and to see whether there are patterns from which useful information can be extracted to support decision-making processes. The analysis of large amounts of data and its presentation to decision makers using proper visualization can lead to new insights about the organization, its customers, its processes, and its competition. The goal of data mining is the extraction of patterns and knowledge from the data in the ERP database. In many ERP systems, data mining is performed by automatic or semiautomatic analysis in search of unknown interesting patterns. These patterns are used in further analysis, for example, by forecasting models to predict the results of possible decisions, to cluster (or to group) similar records that might be advantageous to be treated simultaneously, or to test hypotheses regarding cause and effect relationships between different variables. Data-mining success depends on the availability of large amounts of seemingly unrelated data. The single database of ERP systems, where all the data of the organization reside, is an excellent candidate for data-mining activities.

9.5 Process Design and Reengineering

Some ERP packages come with built-in templates of business processes that represent a competitive business model or best practice. While some organizations find that these business processes are good enough for their needs and adopt them, other organizations find that these processes do not fit their specific environment and therefore they must be modified or cannot be used at all. Careful examination of such templates is important as the adoption of a process template that does not fit the strategy, environment, and culture of the organization may lead to a failure (Grabot et al., 2008).

The types of best practices and process models in each area of ERP applications depend on the vendor of the ERP system. Different vendors developed different best practices, and models based on their experience with customers who implemented their ERP systems successfully. As a result, different vendors excel in different areas of ERP applications. As an example, PeopleSoft ERP is considered strong in its HR module, while Oracle is considered strong in its financial module and SAP is very strong in production planning and materials management. However, many organizations look for ERP systems that have a strong HR module as well as a strong financial module.

To overcome this problem, some organizations integrate different modules from different ERP systems, an approach known as "best of breed," while other organizations are investing resources in the redesign of their business processes prior to implementing ERP systems, an undertaking known as business process reengineering (BPR). The goal of BPR is to achieve closer alignment of process models with the business processes of the organization. An important tool type in BPR is diagram. Industrial engineers use diagrams to construct, present, and analyze processes. Diagrams can help in understanding the process flow, and the relationship between the participants in the process. It can be used to detect logical errors, bottlenecks, and opportunities for process improvement. A commonly used diagram type is the flowchart, which helps in understanding a process and its features. Flowcharts use a variety of symbols such as rectangular boxes to represent activities, diamond-shaped boxes to represent decisions in the process, and arrows to represent the flow of information and material. Some flowcharts use swim lanes—a lane for each function or organizational unit that participates in the process. The activities and decisions performed by an organizational function are within its lane so that the interfaces between functions can be easily seen and the responsibilities to perform activities and take decisions are clear. Figure 9.6 shows an example of a simple flowchart for the process of making tags, starting with receiving the order and ending with finalizing the shipment.

A technique commonly used in the reengineering of business processes is to compare the current process to best practices. Comparison of key

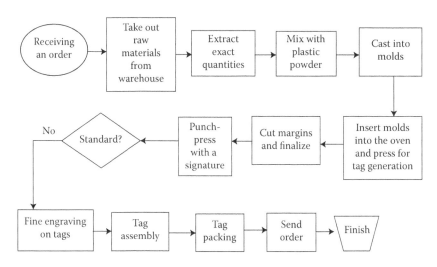

FIGURE 9.6
Example of a flowchart for a process that follows an order of tags.

performance measures in the areas of competition like quality, time, and cost can help identify potential areas for improvement and use successful process designs from the best practices. The focus is on what can be improved and how it can be improved.

Based on the results of benchmarking the best practices, most organizations can develop plans on what to improve and how to improve processes.

9.6 ERP Implementation Projects

The need to reengineer business processes and to integrate production and inventory management with other functions in the organization calls for the participation or even leading of industrial engineers in ERP implementation projects. The training of industrial engineers in project management, information systems, process design and reengineering, and production and inventory management systems provides the knowledge needed to select the most appropriate ERP system, to plan its implementation project, and to manage its execution while monitoring and controlling it. Such projects are risky and difficult, and the stakes are high as they touch every aspect of the organization.

Well-known problematic ERP implementation projects were surveyed by Wailgum (2009).

Some of the reported projects are as follows:

ERP implementation at Hershey Foods: Following an ERP implementation project, Hershey had a problem delivering $100 million worth of Kisses for Halloween during the Halloween season in 1999 and caused the stock to drop 8%.

ERP implementation at Nike: Nike suffered $100 million in lost sales, a 20% stock dip, and a collection of class-action lawsuits in 2000, following the implementation of supply chain and CRM modules into the ERP. These projects were aimed at upgrading the Nike systems.

ERP implementation at HP: HP's centralization of its disparate North American ERP systems onto one ERP system in 2004 cost HP $160 million in order backlogs and lost revenue—more than five times the project's estimated cost.

ERP implementation at the University of Massachusetts: More than 27,000 students at the University of Massachusetts as well as Stanford and Indiana University were unable to find their classes and, even worse, were unable to collect their financial aid checks.

ERP implementation at Waste Management: Garbage-disposal giant Waste Management got into a $100 million legal battle with its ERP provider over an 18-month installation of its ERP software. The initial deal began in 2005. In March 2008, Waste Management filed suit and claimed its ERP provider fault resulted in the massive failure. Similar stories are quite frequent in other failures of ERP projects.

ERP implementation at FoxMeyer Drugs: FoxMeyer Drugs, a $5 billion company, implemented a $100-million ERP system and went completely bankrupt not long after. They launched the project in 1993, and implementation began between 1994 and 1995. By 1996, FoxMeyer had been driven into bankruptcy, and by 1997, the pharmaceuticals company was suing the ERP project vendor as well as Andersen Consulting, responsible for integrating the system for $500 million apiece.

ERP implementations by the U.S. navy: The U.S. Navy spent about $1 billion on four different ERP pilot projects since 1998, and all four have failed. All four installations turned out to be incompatible and redundant, ultimately failing to meet the requirements of the Navy.

While ERP implementation failure continues to be a major risk, the success rate of these projects has climbed significantly since the turn of the century. Success factors of ERP implementation projects were studied intensively (Loh and Koh, 2004)

We summarize below the main success factors we found in most studies.

9.6.1 Critical Success Factors in ERP Implementation Projects

9.6.1.1 Listening to the Stakeholders (Do the Right Project)

ERP projects are undertaken to satisfy the needs and expectations of stakeholders. The identification of the important stakeholders and the understanding of their needs and expectations are crucial to the project success. It is very important to translate these needs and expectations into a clear definition of project goals and performance measures that can be used to assess success. Lack of consensus on the desired outcome can lead to problems later on. Without clear and agreed consensus on goals, it is impossible to develop an agreeable project plan.

9.6.1.2 Developing a Good and Agreed on Plan (Do the Project Right)

A good master plan is the first step in the planning process. The plan must be realistic, acceptable by the stakeholders, and describe the way the goals will be achieved. A good plan starts from the requirements and defines the people responsible to achieve them. The master plan is developed further into a level of detail where a knowledgeable person is assigned to each part of the project called a work package. This person, the work package manager, is the expert who provides estimates of the duration it will take to perform each of the project tasks, the resources required, the estimated cost, and the risk associated with each task. The project manager should integrate the information provided by the managers of the work packages and introduce the logical sequence of tasks and integrate all the information into a good plan that shows what needs to be done to achieve the goals when the project will end and how much it will cost.

9.6.1.3 Selecting the Project Manager

A dedicated, experienced project manager is a key success factor. The project manager is an integrator who must see the whole picture and lead his or her team to the predefined goal. He or she must have project management knowledge and experience as well as knowledge and experience in information technology and process reengineering. Although it is possible to train project managers in some of the required skills, there are some skills that are the gift of God (known as charisma). These are mainly soft skills or people skills like the ability to lead with limited formal authority, and the eloquence combined with emotional intelligence to communicate effectively. A dedicated and capable full-time project manager with proper training and experience is essential for project success.

9.6.1.4 Risk Management

The project plan is based on forecasts or estimates of the tasks that will be performed, their duration and cost, and the resources required to perform these tasks. Any estimate is subject to estimation error and a source of risk. Risk management should start by identifying the major sources of risk (e.g., by learning from ERP projects that failed). Examples of risk sources are insecure funding, inappropriate software selection and evaluation, inexperienced project leaders, noncommitted project partners, poor project planning, ineffective communication, unclear role and responsibility, unskilled personnel, and poor teamwork.

Analysis of these risk drivers is needed to decide which of these risks should be mitigated (i.e., protect the project from this risk). Mitigation can take many forms such as time buffering, safety stock, reserve funds, insurance policy, or actions that reduce the probability of risk or the severity of damage. In terms of time, mitigating risks includes a time buffer or setting dates that leave some slack in case the project is late. In terms of capacity risks, mitigating risks includes resource buffers or assigning excess resources to the project and planning for absenteeism of key workers, etc. The project budget should have a management reserve or a budget buffer that is used when a risk is materialized. In addition, monitoring throughout the project execution should focus on early detection of risks that materialize, and the implementation of corrective actions when needed.

9.6.1.5 Process Reengineering

ERP implementation or maintenance provides an opportunity to improve business processes that are managed and supported by the ERP system. It is essential to design the processes supported by the ERP system, because keeping the existing processes may lead to marginal improvements or, even worse, the existing processes may not fit the ERP system. Reengineering of processes starts with the analysis of requirements like time, cost, quality, and flexibility of the process. Next, benchmarking can be used to identify good solutions implemented by other organizations, including the best practices supplied by the ERP vendor. It is important to design processes that the ERP system and its workflow engine can support.

9.6.1.6 Customization

A major source of risk is customization. Very few companies implement ERP systems with little or no customization. The reason is that customization is part of the needs and expectations of the stake holders. Tight control should be applied because scope creep in the form of uncontrolled changes resulting in a continuous growth in the project's scope could lead to a failure. Planning early and deciding on the required customization, along with tight change control, could help solve this problem.

9.6.1.7 User Training

Adequate training of users is very important, as this determines the impression and attitude of the user in relation to the system. ERP systems are used by many different users with a large variety of previous knowledge and skills, doing different tasks that are part of a large variety of processes. Thus, it is important to train each user according to his or her future use of the system. Training all the users to perform the right tasks in the right way is a difficult undertaking that should be planned as part of the project plan, and adequate resources and budget should be allocated to it.

9.6.1.8 Testing

The purpose of testing in an ERP project is to make sure that the system meets the project goals and fulfills the stakeholders' needs and expectations, managing processes, supporting decision making, and producing the output needed. Testing is part of risk mitigation as it identifies early the potential of ERP failures, such as missing important functions or producing an erroneous output or error message. Treatment of such findings prevents a situation in which the ERP is not well accepted by end users.

9.6.1.9 The Role of Industrial Engineers in ERP Projects

The training of industrial engineers in process design and reengineering, in information systems and information technology, in project management, and in planning and executing the transformation processes makes them ideal members in ERP implementation projects. With the proper experience, they are the right leaders of such projects. In previous chapters of this book, we discussed all these areas of knowledge and the way they are taught in industrial engineering programs. In the following chapters, we will discuss the integration of ERP systems of different organizations within the supply chain.

9.7 Summary

Today, an efficient ERP system is a competitive must for all organizations. ERP systems evolved from MRP2 systems into an organizational software platform that encompass most of the significant functions of any organization. This includes the database, and the model base to support all the major activities in the firm. Thus, ERP in most organizations supports the finance, communications, the marketing and sales, the human resource management and payroll management, the procurement, the quality management, the operations and production, the project management, the service management, and risk management.

ERP systems have a single efficient database that prevents the duplicity of data and inconsistencies. Due to the sheer size and the wide scope of ERP systems, their installation and implementation are always large problematic projects. Therefore, replacing an old ERP system by a new ERP system is always a major effort that is done in stages.

Since industrial engineers intensively deal with organizational processes and since most (if not all) computerized organizational processes are part of the ERP, most industrial engineers deal with the ERP system on a daily basis and must know how to operate and manipulate it. In some cases the industrial engineer must reengineer the business process, so that it would run appropriately and could be computerized.

In recent decades, the attention of the computerized world focused on the supply chain and the communications along the supply chain (including computerized communication). As industrial engineers manage supply chains, they are often involved in moving information along the chain and bridging between different ERP systems.

References

Grabot, B., Mayere, A., and Bazet, I. 2008. *ERP Systems and Organizational Change: A Socio Technical Insight*. London, UK: Springer-Verlag.

Klaus, H., Rosemann, M., and Gable, G.G. 2000. What is ERP? *Information Systems Frontiers* 2(2), 141–162.

Leon, A. 2014. *ERP Demystified*. 2nd edn. New-Delhi, India: McGraw-Hill.

Loh, T.C. and Koh, S.C.L. 2004. Critical elements for a successful enterprise resource planning implementation in small- and medium-sized enterprises, International. *Journal of Production Research* 42(17), 3433–3455.

Wailgum, T. 2009. 10 famous ERP disasters, dustups and disappointments. *CIO Magazine* (10), 1–5. http://www.cio.com/article/2429865/enterprise-resource-planning/10-famous-erp-disasters--dustups-and-disappointments.html?page=2

10

Human Factor

Educational Objectives

There is no organization without people, and there is no management without people. People have needs, feelings, limitations, and characteristics that overall are summarized under the term the *human factor*. This chapter discusses the human factor in the work of industrial engineers. The unit exposes students to ergonomics used in the context of workspace design and to key factors in human resource management, including occupational design, occupational evaluation, and organizational occupational evaluation systems.

The unit introduces the main terms and various methods used by industrial engineers in dealing with the human factor. These terms and methods allow the students to understand how dealing with the human factor is integrated into the work of industrial engineers.

10.1 Introduction

We are all humans, and as such, we all have values, perceptions, feelings, limitations, and social interactions. Moreover, our body is susceptible to fatigue and exhaustion, to backaches, headaches, and numerous other symptoms that may stem from problematic work habits. Industrial and service organizations are composed of humans. In Chapter 2, we considered organizational structures and discussed the advantages and disadvantages of the functional structure, the project-based structure, and the matrix-based structure. In order for groups of individuals to work in coordination and to achieve common goals, it is important to design the structures in detail. In the context of these structures, the members of the organization will carry out their various tasks within the organizational system: the employees' roles, the manner in which they perform their duties, and the employees' work environment.

10.1.1 Employee's Work Environment

10.1.1.1 Physical Aspect

- Where will the employee work?
- What is the most comfortable body position for accomplishing the employee's work?
- What tools and appliances are at the employee's disposable and how will the employee use them?

The answers to these questions involve the science of anthropometry and ergonomics.

10.1.1.2 Functional Aspect

- How will the work be carried out?
- What is the input?
- What is the output?
- What method will the employee use to transform the input into output?

The answers to these questions involve industrial engineering, work design, and work measurement.

10.1.1.3 Organizational Aspect

- What is the employee's position in the organization?
- To whom is the employee subordinate?
- Whom does the employee supervise?
- To whom does the employee report, and who reports to the employee?
- What is the definition of the employee's position in the organization?
- What comes under the employee's authorization?
- What responsibility is assigned to the employee?

The answers to these questions require specialty in organizational structures, work analysis, and method design.

10.1.1.4 Compensation Aspects

- What is the salary paid for a certain position?
- What are the other benefits (social benefits, for instance)?

- What compensation, aside from financial compensation, will the employee receive?

The answers to these questions require specialty in remuneration and incentive systems.

The industrial engineer has to give the answers to the above questions, and therefore must be knowledgeable about:

- Anthropometry and ergonomics
- Work design and work measurement
- Organizational structures
- Work analysis and methods design
- Remuneration and incentive systems

Since these physiological, psychological, and sociological aspects need to be integrated while dealing with the human factor, it is important to be able to synthesize information from various sources and to integrate different types of tools—some of which are quantitative (hard tools) and some of which are qualitative (soft tools).

10.2 History

Interest in the human factor in the workplace arose at a relatively late stage because of the industrial revolution. One of the reasons for this interest was the workers' distress because of the speedy industrialization and urbanization processes that accompanied the industrial revolution. These troubles became a management problem when workers began to respond to their distress both as individuals and as an organized group.

Taylor (1911) developed a systematic approach for handling the state of the human resource in a factory. His work was continued by a group of engineers in the United States. This approach is known as *scientific management*. Taylor argued that management is obligated to plan the work process based on meticulous analysis of the required tasks, tools, and skills, pick the most suitable workers for each phase in the work, supply each worker with the most suitable work tools, and instruct and supervise the quality of their production.

Taylor also researched motivation in workers. He assumed that the workers' primary motivation is financial. In terms of the Maslow Pyramid (Hjelle and Ziegler, 1992), discussed in Chapter 2, Taylor assumed that the workers in his era aspired to meet the needs at the lowest levels of the Maslow scale. As we will see further, even if this assumption were true at the time, the changes that have occurred since, require today's industrial engineers to examine compensation methods that also address needs that are higher up in the Maslow scale, among them social needs, personal fulfillment, and so forth. Taylor's assumption that human beings are rational creatures and therefore would be willing to sell their labor for the highest possible price and for as little effort as possible may also be true today in some cases and in some countries; however, it is certainly not true for every organization and for every role. Moreover, it is rarely the case that the amount of effort exerted is linearly related to the amount of compensation. Therefore, industrial engineers must understand the factors that will motivate workers for whom they are responsible and design a suitable compensation system.

Since Taylor assumed that people would be willing to work in monotonous jobs and would not be interested in the planning and analysis process of the work, but only in the pay they receive for their labor, he suggested minimizing the need for worker discretion and dividing work into simple tasks, to closely supervise workers, and to ensure proper physical conditions and tools.

The ancient Greeks were probably the first to use ergonomics (from Greek: *ergo* = work, *nomos* = natural law). In modern history, ergonomics developed to counter the occupational diseases that developed due to the industrial revolution, and the repetitive jobs derived using the Taylor approach for maximizing work repetitiveness. Ergonomics soon developed to capture other human factor–related issues to be discussed later in this chapter. Today, these subjects are studied as the basic courses in industrial engineering programs.

Parallel to Taylor's theory and to the methods he developed, other approaches evolved originating from the sociological–psychological theory of humans as complex creatures with social needs (in other words, needs that are higher up on the Maslow scale (Hjelle and Ziegler, 1992), as discussed in Chapter 2) that need to be met in the context of the role in the organization. One of these central approaches was developed by Herzberg et al. (1959), who suggested dividing the environmental factors in an organization into two groups:

1. Factors that prevent negative attitudes toward the system are called *hygienic factors*. These include a regular salary, financial security, work and environmental conditions, and responsibility suitable to the authority and hygienic conditions.

2. Factors that encourage excelling in performance and achievement called the *motivating factors*. These include work challenges, opportunities to be creative, interest in the work content, financial incentives, or identification with the purpose of the work. For example, enriching a position is considered a motivating factor.

Other approaches with a psychological–sociological orientation are known as approaches based on human relations analysis. These approaches have led to the understanding that the work environment needs to be planned while taking the workers' social needs into consideration. This led to the development of an industrial engineering specialization focusing on organization and methods (O&M) as well as a management specialization focusing on human resources management.

It is customary to note the Hawthorne studies (Mayo, 1949) carried out by Blau and Scott, as the beginning of the growth of human resources management approaches at work and as the basis for developing a new and scientific management technique for motivating workers.

The Hawthorne studies took place at the Hawthorne factory, owned by a subsidiary of Bell Telephone Company. The factory supplied telephony equipment to the parent company. At the time of the experiment, the Hawthorne factory employed approximately 29,000 workers in various positions related to manufacturing communications equipment. At its outset, the study examined how various physical factors such as temperature, lighting, humidity, and break times were related to workers' productivity.

Five female workers participated in the first phase of the Hawthorne studies. For the purpose of the study, they were separated from the other workers in their department. From April 1927 and through June 1929 changes were made to various factors in their working conditions. To everyone's surprise, it turned out that regardless of the changes, productivity kept increasing, as did the workers' morale. This was reflected by fewer absences, satisfaction with work and with colleagues, willingness to cooperate, and willingness to participate in after-work social encounters. These outcomes were called the *Hawthorne Effect*. In other words, this was positive behavior resulting from meeting social needs such as appreciation, importance, and prestige. The researchers attributed the results to the supervision patterns and to the relationship with superiors, who, for the purpose of the experiment, were open-minded, sharing, and considerate.

As previously mentioned, the Hawthorne studies were the foundation for new managerial philosophies and for the field of human resources management. What began in the first years of the twentieth century as departments that recruited employees and paid salaries, from the 1940s onward developed into manpower departments and later on into human resource departments that deal with screening, training, and planning, as well as development of management skills for implementation of social and managerial philosophies, managing employee welfare programs, and so forth.

The Hawthorne studies also laid down the scientific foundations for a new scientific discipline—organizational behavior. Organizational behavior courses constitute part of industrial engineering programs and provide the required knowledge in O&M and in handling the psychological–sociological aspects of the human resource.

Parallel to developments in the Western world, the Toyota factories in Japan developed a managerial philosophy, some of which, such as time-wasting prevention and just-in-time delivery, have been discussed in previous chapters. An important component in the managerial approach developed by Toyota was the handling of the human resource, a component that includes several principles:

1. Providing employment security to loyal and ardent workers—minimizing layoffs.
2. Respecting human beings regardless of their position and status in the company or in society.
3. Minimizing the effect of fluctuations of workload on the workers.
4. Teamwork and emphasis on partnership of the team members in effort and responsibility.
5. Decision making resulting from agreement and consensus.

By emphasizing the human factor, industrial engineers aspire to find the right integration of ergonomics and organizational behavior in order to increase organizational effectiveness. Ergonomics optimizes the physiological and cognitive aspects of work, for example, by improving the man–machine interface. Organizational behavior optimizes the incentives and social aspects, by improving the man-to-man interface and by using the knowledge of organizational behavior and human behavior in the organization. In recent years, the boundaries between these disciplines have become somewhat blurred. Thus, for example, ergonomics deals with the design of man–computer interfaces while utilizing psychological tools.

The human factor in the workplace may be positioned along a continuum that ranges from the "hard" engineering field to the "soft" organizational behavior approach. The engineering field includes ergonomics, which in extreme cases in the past used to consider human beings as machines with characteristics, limitations, and abilities that are to be analyzed, characterized, and taken into consideration when designing or planning the work. Anthropometry is only one example of this "hard" approach. Another perspective in ergonomics was to analyze the physiology of people and plan the work environment accordingly. This approach emphasized optimizing the workload, optimizing required efforts of gestures and body movements, and the avoidance of occupational diseases. Organizational behavior considers the human factor in the workplace as a complex human creature whose performance is affected by psychological and sociological considerations and by interaction with the other people in the workplace.

In recent years, the human factor in the workplace has developed considerably as a field of study, and today it is customary to assume that a combination of the various aspects of the human factor is essential for efficient human performance in the workplace. This combination is described in Figure 10.1.

In Figure 10.1, ergonomics is located on the right side. Ergonomics places emphasis on the study of worker performance, which is also influenced by the "machine" (the object) with which workers interact. In this case, the term

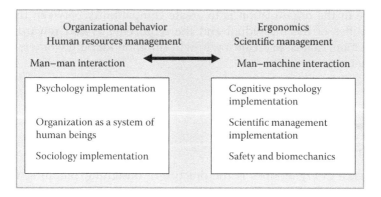

FIGURE 10.1
Approaches for handling the human factor in organizations.

machine describes the physical work environment, for example, a cashier's workstation in the supermarket, a computer programmer's workstation, a dentist's workstation, a lathe, a milling machine or any other manufacturing system operated by a worker. Ergonomics uses information regarding the body measurements, abilities, and limitations of the people who are supposed to use the machine, as well as information about safety rules and regulations, processes of perception, analysis, and problem solving, and so forth. This represents the implementation of scientific management and of cognitive psychology.

Fields marked by greater interpersonal differences are located on the left side of Figure 10.1. These include, for example, behaviors that originate from and are impacted by other people in the organization, or in other words, the implementation of social psychology and the study of organizations as a human system, as well as the implementation of sociology. The terms *organization* and *the study of the organizational behavior* originate mostly from sociology, which emphasizes the organization's structure, its human systems, and the interaction between the systems.

Organizational behavior brings together two terms from the social sciences: *behavior* and *organization*. This field aspires to examine the human behavior in organizational frameworks on three levels:

- The individual
- The group or unit within the organization
- The organization as a whole

Additional fields such as anthropology and economics also influence the design of the workplace, as discussed further. The implementation of organizational behavior knowledge is carried out through human resources management in the organization. The purpose of managing the human resource in the organization is to create compatibility between the needs and abilities of the individual and the requirements and rewards of the organization, that is, to create a win-win situation for both workers and the organization.

10.3 Ergonomics

The science of ergonomics is the practice of designing/planning workers' work environment while considering the relevant abilities and limitations of working in such an environment. Ergonomics started out as a scientific-biomechanical perception of the worker, and therefore initially addressed anatomical and physiological aspects. Today it also addresses psychological aspects.

The field of ergonomics and human factors engineering serves as a basis for improving man–environment interfaces and making them more efficient. The purpose of ergonomic design is threefold: first, to improve worker performance (including supervision and decision making); second, to increase worker output, productivity, and satisfaction; and third, to protect and improve worker health. These objectives are realized by designing man–machine interfaces and man–computer interfaces, by designing the workplace and designing products. For this purpose, ergonomics utilizes tools for the analysis of occupational tasks and work environment requirements, including those related to worker safety.

Safety is the first and foremost aspect to be taken into consideration in work environment design. It is a professional responsibility of industrial engineers to make sure that the work environment is safe for workers and is designed in accordance with safety rules and regulations. Therefore, it is incumbent upon industrial engineers to be extremely knowledgeable in this field, which is studied in the context of industrial engineering programs.

It is important to note that safety has important financial implications. For example, a 2004 study published in Spain based on 2002 data indicates that the cost of accidents in Spanish industry reached 12 billion euros, constituting 1.72% of the GDP (Josep Espluga, 2004). A total of 21,597,604 workdays were lost in 2002 due to workplace accidents (http://www.eurofound.europa.eu/eiro/2004/03/feature/es0403211f.htm).

An example of a workstation is illustrated in Figure 10.2 and its safety regulations in Figure 10.3.

Anatomical factors are second in importance in comparison to safety, together with physiological and psychological factors. The anatomical factors focus on the human body's physical structure and include measurements, motion ability, and the geometry of various body parts. It is therefore customary to divide *anatomy* in the context of ergonomics into two:

FIGURE 10.2
An example of a workstation.

- *Anthropometry:* A discipline that studies the human body's measurements and matches tools and work environments to these measurements.
- *Biomechanics:* A discipline that studies the exertion of forces by the human body.

FIGURE 10.3
Work safety regulations with earmuffs and gloves.

The *physiological* factors are also divided into two:

- *The physiology of effort:* a discipline that studies the human body's ability to produce energy and to use it for the performance of work.
- *Industrial physiology:* A discipline that studies the influence of the work environment on the human body.

The *psychological* factors in ergonomics focus on the cognitive interface between man and work environment.

A comparison can be made between the human being as an information processor and the organizational information system. The human being can be described as an information processor through the terms studied in Chapter 4, which discusses information and its use. The four basic functions of human beings as an information system are:

1. *Information input:* Receiving information from outside the organization and feedback from within the organization. Receiving the information is dependent on the form and intensity of the stimulation. Communication and perception mechanisms are involved in this process.
2. *Processing information:* Analyzing the information with an appropriate decision-making model.
3. *Storing information:* Memory plays a central role in this process, which also includes coding the information (in various forms: auditory, visual, and so forth).
4. *Decision making and action:* A decision is made and acted on based on the analysis. This decision is implemented while receiving feedback and control.

These functions are performed by human beings that handle the information. It is the job of industrial engineers to understand how they are performed (psychologically) and to support them, for example, by designing decision-supporting systems and man–computer interfaces.

It is incumbent upon industrial engineers to design a work environment in which workers can receive all the required information at the right time, handle the information, make decisions, and carry them out optimally (in other words, making the best decisions at the greatest speed and with the least amount of effort) in two types of tasks: motoric and cognitive.

10.3.1 Motoric Tasks (Primarily Manual Labor)

Industrial engineers must design workstations that are based on ergonomic principles in order to make the work environment more safe, comfortable, and efficient. The engineer does so by taking into consideration the physiological aspects of the human body while planning the task and the workstation environment. Ergonomic principles minimize the attention and effort exerted by the operator, thus supplying more time and enabling workers to be more attentive to handling disturbances and deviances that require making nonroutine decisions.

10.3.2 Cognitive Tasks (Primarily Office Work)

Industrial engineers must ensure efficient handling of components such as information presentation, information processing, situation evaluation, design and planning, solving routine or one-time problems, and decision making. The cognitive component is significant in many tasks, and sometimes it is more important for the completion of the task than the physical component. These tasks are called *cognitive tasks* and include such tasks as designing, planning, computer programming, medical diagnoses, process control, flight control, and diagnosis of defects and malfunctions in technological systems. Industrial engineers must make sure that the work environment is appropriately built based on the types of tasks workers must perform. For designing the work environment, industrial engineers must know and implement motoric and cognitive scientific principles as discussed next.

10.4 Motoric Tasks

The human body is a system of muscles and tissue controlled by the nervous system and supported by the skeleton. The bones are attached to one another at the joints, and the skeletal muscles are attached to the bones by tendons. Each component in this complex system requires proper use to prevent vulnerability. Dedicated care is also required for the system to function. In other words, "preventative maintenance" is in order.

Specialization in motoric tasks requires developing work expertise through practice, repetition, and improvement of the work method. This type of specialization is a key factor in the efficient performance of motoric duties. Motoric specialization may be cultivated by training and practice drills. Drilling leads to building reflexive responses and to sharpening skills by creating patterns.

Using our muscles requires expending energy that originates from the food we eat. Chemical energy is transformed into mechanical energy in the metabolism process. Once the body embarks on a physical task, its demand for oxygen increases, and the heart rate and respiration also increase in order to provide this oxygen to the organs. The muscles are burdened because of effort. After the body has experienced a burden, it becomes physically tired: the ability of the muscles to contract declines, and fatigue develops when a muscle has been working for a while and waste residue (of materials that had stored the energy before it was exerted) accumulates in the body (Lerman et al., 2012).

The repetition of motoric tasks that characterized the industrial revolution caused serious bodily injuries. One of the known chronic diseases caused by motion repetitiveness is *cumulative trauma disorder* (CTD). This disorder is caused by repetitive motions (regardless of its intensity or speed). For example, even daily intensive typing on a computer keyboard can cause CTD. External factors such as heat, humidity, vibrations, and noise also impact motoric abilities and the work environment must be designed so that the levels of such factors are reasonable.

Ergonomics deals with designing workstations, planning work methods, and designing work tools while taking the abilities and limitations of the human body into consideration. These abilities and limitations derive from the systems that comprise our body and from the processes that take place in our body.

In the following, we describe three major motoric fields in ergonomics— body posture, access and space, and use of force.

10.5 Body Posture

People spend time in various body postures in the context of performing their jobs. For example, a car mechanic must lean over the engine to do his or her job, or use a hydraulic crane to raise the vehicle over his or her head. Even typing (or driving) every day for many hours is carried out in a sitting posture, which is risky in and of itself, as described next. We choose the body posture that enables us to invest the required force for performing an action (theoretically, anyone can move the world with a fulcrum, according to Archimedes). Some body postures cause pinching of the nerves between two bones, causing a numb, tingling, and ticklish feeling in the organ. Some body postures completely block the blood flow to certain organs, thus cutting off the nourishment and oxygen required for their functioning. A common posture such as resting the palm of the hand on the edge of the table can cause damage to the nerve that enervates the external part of the palm (in the ring finger and little finger where the ulnar nerve is located). Multiple repetition of this posture may cause complete paralysis of those two fingers (many people who work on computers have experienced this). Raising the arm up and backward (to reach an item on a high back shelf, for instance) causes pinching of a nerve track that runs through the shoulder, leading to damage in the nerves running through the forearm and the palm of the hand. Remaining in this posture for several minutes will only feel uncomfortable and may at most cause temporary damage. However, being in this posture for long periods will cause cumulative and even irreversible damage.

Even standing and sitting postures can be dangerous for the muscles and the skeleton. Extended standing exerts pressure on the knee joints and on the foot bones. Many muscles are activated to keep the body stable in this unnatural posture (from an evolutionary perspective this is not a natural posture). After an extended period, the burden on the muscles and skeleton can cause bloating in the leg blood vessels, ulcers in the feet, and of course pain and discomfort.

The sitting posture is also a posture that when kept for extended periods of time can cause damage (kneeling, bending, and crouching are more natural postures for humans). Sitting creates an unnatural bend in the spine—when a person sits, the pelvis is tilted forward and pulls on the spine. This causes hardening of the lumbar part of the spine—kyphosis. As a result, uneven pressure is exerted on the intervertebral discs. These discs are unevenly "squished" and in time lose their original shape. If the burden continues, they also lose their agility and become susceptible to herniation. The intervertebral discs, which were intended to serve as suspension cushions and to provide the agility required when walking as well as the needed separation between the vertebrae (for nerves to connect between the spine and the body's periphery), lose their ability to perform all these roles. The result is back pain (caused by pinching of the nerves between the vertebral protuberances) and even a herniated disc (deterioration of the muscular ring that surrounds the disc and protects its internal material; Figure 10.4) (Kroemer et al., 2010).

FIGURE 10.4
Example of a sitting work posture in soldering tiny electric circuits.

All body postures, aside from a supine position, place a burden on certain organs and muscles. Findings (Punnett et al., 2005) indicate that each year at least 15% of all adults around the world have back pain. Worldwide, 37% of low back pain was deemed attributable to occupational risk factors. However, in the United States, the numbers are higher. The U.S. government website (http://www.niams.nih.gov/health_info/back_pain/) states: "In a 3-month period, about one-fourth of U.S. adults experience at least 1 day of back pain." It is one of our society's most common medical problems. Eighty percent of workers experience at least one episode of back pain in their lifetime. Approximately 35% of all work accidents are caused due to lower back injuries. According to Punnett et al. (2005), work accidents in the United States are caused by overexertion in lifting, disposing, folding, carrying, pushing, or pulling of items. The total lost time due to work accidents was 818,000 disability-adjusted life years lost annually.

10.6 Access and Space Design

Aside from focusing on work postures, ergonomists who design workstations must consider clearance and reach design.

Designing the workspace includes preventing problems caused by lack of room for the knees under the table, the absence of a place on which to place the wrists, no room to walk between or around machines, and so forth. These types of problems can lead to improper and even dangerous body postures.

Issues of space are dealt with by designs geared to meet the needs of the tallest or widest worker, for example, according to the relevant body measurements of a man in the 95th percentile (in other words, a man who is larger than 95% of the relevant male population).

Design for easy access includes handling problems that originate from placing controllers in areas that are too far, or too high, or that do not allow convenient access to control and command panels. In designing range of reach (e.g., convenient operation reach for a controller), the design should be according to the measurements of the shortest person—the height measurement of a person in the 5th percentile in the body measurement table, in other words, a size that would suit 95% of the relevant worker population.

10.7 Anthropometry

Ergonomic design of tools, workstations, equipment, and control panels is based on knowledge of the human body's abilities and limitations. Industrial engineers, who practice work environment design, examine matters such as at what angle of the elbow are the biceps at maximum strength? What is the optimal distance between the feet for maximum force when lifting a load? What is easier—pulling a load that is at a 120-degree angle from a person or pushing the load? What is the maximal weight a person can lift/pull/push? How frequently can workers perform an action throughout an 8-h workday without hurting their health?

Industrial engineers use *anthropometric* data (anthropometric = human body measurement) to design workstations. The measurements of individuals vary greatly. This variance originates from gender differences (women usually have smaller measurements than men do), interracial differences (Japanese people have smaller measurements than Westerners do), age differences (older people are usually shorter yet heavier than younger people are), and from all of us being different people with different DNA. There are anthropometric tables that contain data on various human body measurements, such as the distance of the elbow from the floor when sitting and when standing, the width of the pelvis, the distance of the eyes from the floor, the width of the head, as well as a variety of other data. These tables detail measurements for men and women and are usually divided by percentiles. Three important percentiles are the 5th percentile, the 50th percentile, and the 95th percentile. The engineer decides which table to use (men or women) as well as what percentile. Thus, for example, the distance between two machines in a factory that employs men and women needs to be designed according to the measurements of the widest man (the 95th percentile). The same is the case with respect to the height of a table (short people can use a stool). On the other hand, a critical operation button should be designed according to the shortest woman (the 5th percentile).

It is very important to design workstations that are suitable for the great variance between workers. Each item in a workstation, and certainly each tool used over an extended period, must be carefully planned, considering how it should be designed to make it convenient, healthy, and effective for the entire worker population using it.

Example: The following table provides partial anthropometric data taken from an anthropometric table built according to the data of British women aged 19–65 (Grandjean, 1988; Kroemer and Grandjean, 1997; Kroemer, 2008). The measurements in the table are in millimeters and correspond to the measurements marked in Figure 10.5.

Measurement	95th Percentile	50th Percentile	5th Percentile
$a + c$	610	555	505
b	445	400	355
c	280	235	185

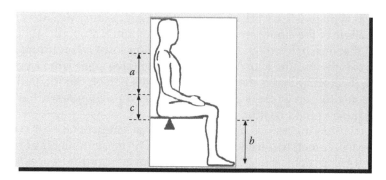

FIGURE 10.5
Anthropometric data and measurements for the example.

In designing a chair to be used at a British woman's workstation, the chair measurements should be determined in accordance with anthropometric data as shown in Figure 10.6.

Measurement 1: Studies by Grandjean (Kroemer and Grandjean, 1997) revealed that workers prefer chairs with a high backrest. A chair with a high backrest is more effective in supporting the body's weight than a chair with a low backrest. This chair's measurements correspond with the $a + c$ measurements in the anthropometric table. In order for the chair to fit the body measurements of all British women, we calculate the average between the largest measurement and the smallest one: $(610 + 505)/2 = 557$ mm). Since leaning back when sitting on the chair generates an angle, approximately 5 cm need to be subtracted from the average value. In other words, measurement 1 is 507 mm.

Measurement 2: This measurement corresponds with the b measurement in the anthropometric table. The chair in the illustration is

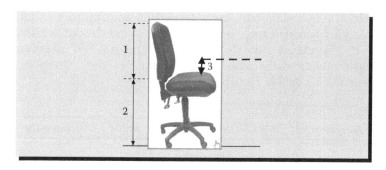

FIGURE 10.6
An example of chair design.

an adjustable one. In other words, it can be raised and lowered. Therefore, its lowest position needs to be adjusted to the measurements of the tallest woman. A short woman can use a footrest to elevate her feet. Therefore, the value of this measurement is 445 mm, the measurement size corresponding to the 95th percentile.

Measurement 3: The height of the table platform is determined according to the measurements of tall workers; otherwise, they would not be able to sit in front of it. Shorter people can raise their chair and use a footrest. To prevent fatigue in the arm muscles and forearms, the table platform must be approximately elbow high. For this measurement, we take the size from the 95th percentile corresponding with measurement C in the table, in other words 280 mm.

10.8 Workload Lifting and Exertion in the Workplace

Fatigue in the workplace is caused by excessive workload and not only is counterproductive but also a source of safety risk as it decreases the worker awareness and alertness (Lerman et al., 2012). Disruption in normal sleep hours is a major contributor to fatigue. For example, working consecutive shifts is known to increase the fatigue level. Therefore, regulations for work and rest hours in most places limit the amount of continuous work that a worker can do. This is crucial for aviation, railroad, and transportation workers—as their function and responses can risk the lives of many people.

To ovoid fatigue and body injuries, the American Occupational Safety and Health Administration (OSHA) and the National Institute for Occupational Safety and Health (NIOSH) have issued several guidelines. Most notable ergonomic guidelines are for maximal workload and lifting in the workplace (Waters et al., 1994). These are known as NIOSH lifting guidelines considered for standard standing posture and standard environment as:

1. Recommended weight limit (RWL) equation
2. Lifting index (LI)

RWL is a multiplication of the following factors:

$$RWL = LC \times HM \times VM \times DM \times FM \times AM \times CM$$

where
   LC = load constant—related to weight
   HM = horizontal multiplier factor—distance of hands from the midpoint between ankles

VM = vertical multiplier factor—starting height from ground
DM = distance multiplier factor—vertical distance of lifting
FM = frequency multiplier factor—time between lifts
AM = asymmetric multiplier factor—angle of load in relation to the body
CM = coupling multiplier factor—quality of grasp

The lifting index reflects the exertion level LI and is computed as follows:

$$LI = \frac{\text{Load weight}}{\text{RWL}}$$

Professionally using these formulas is part of ergonomic courses and is beyond the scope of this book.

For maximal force exerted in push and pool activities, there are also guidelines.

Maximum horizontal push and pull workloads:

Standing (whole body involved): Limit = 50 lb or 23 kg

Standing (only shoulders and arms involved): Limit = 24 lb or 11 kg

Kneeling: Limit = 42 lb or 19 kg

Seated: Limit = 29 lb or 13 kg

Maximum vertical Push and Pull workloads:

Pull down above head height: Limit = 120 lb or 55 kg

Pull down shoulder level: Limit = 45 lb or 20 kg

Pull up 25 cm above floor: Limit = 70 lb or 32 kg

Pull up elbow height: Limit = 33 lb or 15 kg

Pull up shoulder height: Limit = 17 lb or 7.5 kg

Pushup shoulder height: Limit = 45 lb or 20 kg

Pushup elbow height: Limit = 64 lb or 29 kg

(*Source:* Eastman Kodak Company, *Ergonomic Design for People at Work*, vol. 2, Van Nostrand Reinhold, 1986.)

10.9 Workplace Environmental Factors

As human beings, we are sensitive to our environment: noise, lighting, temperature, strong odors, and ventilation affect our well-being, our performance, and our ability to concentrate. Therefore, it is quite reasonable to expect that a workplace would be designed to be quiet, lighted, ventilated with comfortable temperatures, and without odors. This is easier said than done, and we shall now expand a little on each of these factors.

10.9.1 Noise

Occupational hearing loss is one of the most common work-related illnesses in the United States and in most other industrialized countries. According to NIOSH website, 22 million U.S. workers are exposed to hazardous noise levels at work and an estimated $242 million is spent annually on hearing loss.

Noise is measured in decibels. The decibel A filter is widely written as dBA or dB(A). dBA roughly corresponds to the inverse of the 40 dB (at 1 kHz) equal-loudness curve for the human ear.

Using the dBA filter, the sound level meter is less sensitive to very high and very low frequencies. Measurements made with this scale are expressed as dB(A). Workplaces should be designed to minimize the dBA. Different environments strive to meet different dBA levels: while offices may have around 10 dBA, manufacturing facilities may be in the range of 25–50 dBA. It is recommended that beyond a certain average threshold, the worker should wear protective earphones (the specific threshold should be less than the maximal noise exposure to be described next).

10.9.1.1 Noise Exposure Limits

According to NIOSH, noise exposure should be controlled so that the exposure is less than the combination of exposure level L and duration t. The maximum time (hours) per day of exposure can be calculated as:

$$t = \frac{480}{2^{(L-85)/3}} \tag{10.1}$$

where
  t = maximum exposure duration (seconds)
  L = exposure level (dBA)
  3 = exchange rate (dB)
  85 = Recommended exposure limit–REL (in dBA)

Under 80 dBA, there is no evidence for hearing losses. A maximal dBA in an 8-h shift is 85 dBA. Maximal dBA for 1 full hour is 95, and for a quarter of an hour, we can tolerate 100 dBA. Under no circumstances shall a worker be exposed to 140 dBA or more.

Sites that are prone to high dBA require the use of protective equipment such as earplugs or earmuffs. The sign that requires such a protective measure is shown in Figure 10.7.

FIGURE 10.7
A common sign requiring the use of protective noise equipment.

10.9.2 Lighting

Lighting is measured by Lux or the number of lumens per square meter. Lux measures the luminous flux per unit area. One Lux is equal to one lumen per square meter.

Some examples of lighting levels are:

- Office lighting: 320–500 Lux
- Sunrise or sunset on a clear day: 400 Lux
- Overcast day; typical TV studio lighting: 1000 Lux
- Full daylight (not direct sun): 10,000–25,000 Lux

The New Zeeland Labor guideline standards (1680.1:2006) for lighting indicate the following minimal lighting levels:

Movement and orientation only	40 Lux
Rough Intermittent	80 Lux
Simple tasks	160 Lux
Ordinary tasks	240 Lux
Moderately difficult tasks	320–400 Lux
Difficult	600 Lux
Very difficult	800 Lux
Extremely difficult	1200 Lux
Exceptionally difficult	1600 Lux

In addition, to promote task efficiency and visual comfort, many more factors must be taken into account. Some of the important ones include

- The lamp selected
- The avoidance of glare and reflections

- The evenness of the illumination
- Modeling (revelation of the three-dimensional aspects of objects)
- Control systems
- Maintenance and cleaning

Proper illumination is essential for the optimization of both productivity and comfort in the workplace. Workplace lighting affects quality of perception, mood, and performance of employees. Different activities require different levels of light. In general, the more detailed the task, the greater the light requirement. A process control room should be lit at a luminance of 400 Lux, a corridor or walkway may only require 150 Lux, while studying an engineering drawing may require 850 Lux.

Some works, such as operating rooms, quality control, and soldering, require special lighting. In these cases, there are guidelines as to the optimal lighting intensity and other lighting features of each case.

10.9.3 Temperatures and Humidity

The combination of temperatures and humidity causes us to feel cold, comfortable, or hot. Most of the world uses Celsius (°C) for temperature and the United States uses Fahrenheit (°F) degrees. Humidex is a Celsius-based measure that gauges how hot we feel at a given combination of temperature and humidity. It is an equivalent scale intended for the public to express the combined effects of warm temperatures and humidity, just like windshield tells us how cold we feel with certain combinations of temperatures and wind conditions.

The following is a general guideline for workplace humidex (based on °C):

0–14: Cold
15–18: Cool (slight discomfort)
19–24: Comfortable
25–29: Warm (slight discomfort)
30–39: Discomfort
40–50: Extreme discomfort

The Humidex values for combinations of relative humidity and temperatures are shown in Figure 10.8.

10.9.4 Ventilation

Fresh air is needed for respiration, to dilute and remove impurities and odors, and to dissipate excess heat. Replacement air should be as free of

°C	Relative Humidity (in percent)																		
	100	95	90	85	80	75	70	65	60	55	50	45	40	35	30	25	20	15	10
49																			50
48																			49
47																		50	47
46																		49	46
45																	50	47	45
44																	49	46	43
43																49	47	45	42
42															50	48	46	43	41
41															48	46	44	42	40
40														49	47	45	43	41	39
39													49	47	45	43	41	39	37
38												49	47	45	43	42	40	38	36
37											49	47	45	44	42	40	38	37	35
36									50	49	47	45	44	42	40	39	37	35	34
35								50	48	47	45	43	42	40	39	37	36	34	33
34							49	48	46	45	43	42	40	39	37	36	34	33	31
33					50	48	47	46	44	43	41	40	39	37	36	34	33	32	30
32			50	49	48	46	45	44	42	41	40	38	37	36	34	33	32	30	29
31	50	49	48	47	45	44	43	42	40	39	38	37	35	34	33	32	30	29	28
30	48	47	46	44	43	42	41	40	39	37	36	35	34	33	31	30	29	28	27
29	46	45	43	42	41	40	39	38	37	36	35	33	32	31	30	29	28	27	26
28	43	42	41	40	39	38	37	36	35	34	33	32	31	30	29	28	27	26	25
27	41	40	39	38	37	36	35	34	33	32	31	30	29	28	27	26	25		
26	39	38	37	36	35	34	33	33	32	31	30	29	28	27	26	25			
25	37	36	35	34	33	33	32	31	30	29	28	27	26	26	25				
24	35	34	33	33	32	31	30	29	28	28	27	26	25						
23	33	32	31	31	30	29	28	28	27	26	25								
22	31	30	30	29	28	27	27	26	25	25									
21	29	29	28	27	26	26	25												

FIGURE 10.8
Humidex table.

impurities as possible. Air inlets should be sited where they can draw fresh air; they should not be sited near source of fumes or other impurities. Legislation requires that every enclosed workplace have effective and suitable ventilation, which provides a sufficient quantity of fresh or purified air. In many cases, windows or other openings will provide sufficient ventilation. If they do not, mechanical ventilation systems must be used. Recirculated air (e.g., in air-conditioning systems) should be adequately filtered to remove impurities and the purified air should have some fresh air added.

10.10 Specialization and the Development of Workers' Medical Problems

Work that is based on specialization, mainly in a system of high-paced mass production, can find expression in performing the same action hundreds and thousands of times during an 8-h workday. Sometimes the work is also performed during overtime hours, every day, 5 days a week.

The result is that the individual's whole body or some part of it (e.g., a certain position of the palm of the hand) remains in a certain position for an extended period of time. Very frequent repetition of movements is the main cause of CTD. Disorders of this type develop gradually over time. Very frequent repetitive activity causes tissue and joint fatigue in certain parts of the body. The trauma stems from mechanical load, which leads to physiological damage of the overburdened organs. The human body needs sufficiently long periods of rest between periods of cyclical work so that the body can repair itself (recovery times for preventative maintenance of the human body). If the recovery time is not long enough and the application of forces or the problematic posture occurs too frequently, workers will be at high risk for CTD (Figure 10.9).

A simple way to avoid this type of cumulative damage is to frequently change postures and take breaks so that oxygenized blood can flow to the overburdened organs and muscles and the body can eliminate waste residue (primarily lactic acid that causes muscle pains).

A production line workstation in a mass production factory—for example, in a manual assembly line—usually does not allow workers to change postures or take breaks when they want to. Consider, for example, a worker who sits on a chair by a conveyor belt and needs to place a washer on a tube that moves along the conveyor belt. The designer of the production line dictates the pace at which the tubes appear on the belt. (From a timing perspective, this offers a great advantage that makes it easier to supervise the work and avoids the need for timing and synchronization between the workers on

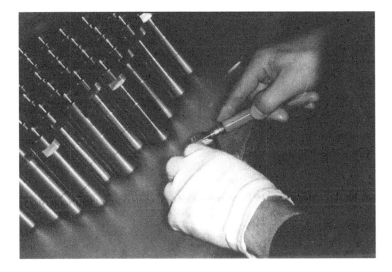

FIGURE 10.9
Repetitive work could be dangerous even when using gloves.

the line.) Usually a worker does not work at a pace that is comfortable (or healthy) for him or her, but rather at the production line pace (the speed of the belt) determined by the factory's efficiency and profitability factors. If the pace is too high, the worker does not have time to lean back and stretch his or her arms upward for example, and certainly does not have time to stand up and walk around to ensure improved blood flow. The harm caused to the worker in these types of tasks may accumulate quickly and cause pain and fatigue, expediting the path to developing medical problems in the back, the hands, and the neck, as well as to absences and severance pay.

Therefore, while designing the work environment to maximize utilization of the human resource may result in profits in the short term, it also has the potential to cause serious expenses in the long term. Such design is unethical. It represents an erroneous approach to management and planning, as the end product quality will be impaired as well, because the worker will suffer from pain and will not be focused.

Industrial engineers are responsible for maximal utilization of the production floor area. In designing the work environment, they must take into consideration sufficient and safe spaces for the workers. For example, they must ensure that even a person with a wide body can safely pass between two machines and that there will be room to move under equipment that is placed on high shelves, sufficient legroom, and so forth.

10.11 Design Flexibility

It is important that the work environment could accommodate different workers. For example, in a production line that is run for three shifts in a 24-h day, the workstations are manned continuously. In these cases, several workers are assigned to the same station (they sit on the same chair and use the same computer). The work environment needs to be designed so that each worker can easily and simply adjust the workstation to his or her body measurements. Workers tend to avoid making complex adjustments. Poor quality design will lead to workers who carry out their work, sitting on chairs that are either too high or too low. Chairs too high cause legs swinging in the air, and pressure on the lower back. Chairs low cause folded legs. Platforms that are too high cause raising the shoulder blades and holding the shoulders in the air will place a load on the deltoid and trapezius muscles in the shoulders and the back. Tables that are too low cause bending of the back and the neck. Auxiliary equipment such as cushioned support for the palm of the hand, support for the forearms, and a footrest for the legs sometimes makes it difficult for each worker to adjust the workstation. Industrial engineers must take these factors into consideration and design workstations that can be easily and efficiently adjusted for each worker.

The arrangement of equipment in the workstation must consider human posture. Sometimes workers need to perform the same action hundreds and even thousands of times a day. For example, when the design is flawed and the worker must frequently reach backward with his or her arm, the tissues in the shoulder joint will wear down, and a major load will be placed on the muscles that assist in this motion. Work efficiency in the speed of manufacturing as well as compensation based on production may cause the worker to relinquish pauses and breaks and make things worse. Therefore, it is important to make sure that the design is flexible and enables simple adjustment for each worker, as well as to establish mandatory breaks that will enable the worker's body to recover.

10.12 Cognitive Tasks

Cognitive tasks include mental processes such as perception, information processing, thinking, and concentration. Human mental effort includes attentiveness, recognition and identification, information representation, memory, understanding, learning (Cunha and Louro, 2000), language processing, problem solving, and deduction. Cognitive tasks are all those tasks in which we use these processes, and in order to perform them, we need the human brain and its resources, known as attentiveness resources.

The cognitive tasks in the world of work can be classified according to several characteristics, among them the task's level of complexity, the amount of attentiveness resources invested in the task, the variety of actions performed simultaneously, and the timing of the task, that is, whether the actions are performed sequentially or in parallel.

Many cognitive models handle processes such as information processing, situation evaluation, decision making, and planning. These models are manifested in areas such as technical design, manufacturing management and planning, computer programming, medical diagnoses, process control, and so forth. These models refer to the characteristics of cognitive tasks. Next we discuss three models of the human cognitive system relevant to the fields of technology and management.

The first model is Wickens' *human information-processing model* (Marmaras and Kontogiannis, 2001). This model focuses on the way in which humans process information.

This model is based on a computerized information system analogy.

According to this analogy, information is received by sensors and is passed into short-term memory. In the next phase, the individual becomes aware of the received information, and as a result, the information is processed by means of prior knowledge (which is extracted from long-term memory) and allocation of attentiveness and processing resources. Finally, a response is given to the received information (Figure 10.10).

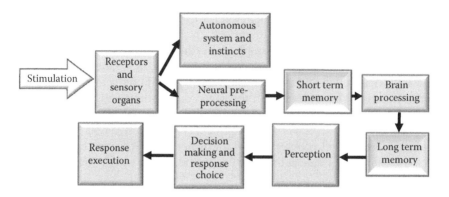

FIGURE 10.10
Person as an information processor.

This process includes several important characteristics and processes, among which are the period of information storage (the duration of time in which the information is available), the potential range of storage (the quantity of information that can be remembered), the availability of attentiveness resources and the amount of existing resources (the extent of the ability to relate to the available information and to process it efficiently and properly), and the individual's (or group's) prior knowledge and experience.

Attentiveness is the central control mechanism that determines when the transition takes place, from perception to cognitive processing, or to long-term memory extraction. Therefore, the amount of attentiveness resources plays a central role. When an individual becomes skilled at certain actions, these actions become automated and less controlled, and therefore, the individual does not need to invest as many attentiveness resources in performing them. The individual can perform these actions faster and usually more accurately (Marmaras and Kontogiannis, 2001).

The second model is Norman's *action-cycle model* (Marmaras and Kontogiannis, 2001). This model examines how people reach their goals while using external world feedback. According to this model, there are two major cognitive processes—a goal and intention evaluation process and an execution process—through which people achieve their goals. The point of origin in this process is setting the goal. In the execution process, the goal is translated into intentions, and later into specific actions to be implemented, and these can be performed in more than one way. The performance action includes implementation of the evaluation process in which the individual perceives what is happening in the world, understands what is happening in terms of needs and intentions (provides interpretation), and finally compares what occurred in practice with the individual's initial goals.

This model helps explain gaps between goals and intentions on the one hand and actual performance on the other, as well as gaps between the viewed situation and the interpretation given to it. The model identifies two central gaps: one is a performance gap, which reflects the difference between intentions and actions—the more the environment enables the individual to realize his or her intentions easily, directly, and with no additional mental effort, the smaller the performance gap. An evaluation gap reflects the extent of effort that needs to be exerted by the user in order to understand the environmental situation and to determine the extent of correspondence between the situation and the intentions. The gap will be small when the system provides easily understandable information about the environmental situation that corresponds with the user's way of thinking about the specific system.

The third model is Rasmussen's *skill, rule, and knowledge-based model* (Marmaras and Kontogiannis, 2001). This model assumes that control and performance of human actions are a function of a hierarchic cognitive control system that operates on three levels: the skill level, which performs automatic control; the rule level, which enables conditional control; and the knowledge level, which enables compensatory control.

The skill level is the lowest level and includes human behavior that is controlled by planned patterns of actions. This type of behavior is characteristic of routine situations in which workers are highly skilled. A skill develops in response to certain conditions, for example, when a familiar situation is recognized, and a response is activated and implemented, without the need to analyze the situation and examine solution alternatives.

In the rule level, the human behavior is controlled by conditioned rules: if (situation X), then perform (action Y). Behavior on this level requires awareness that includes identifying the need to act, extracting familiar actions from memory, as well as methods for the creation of new rules. This level is slower and requires more cognitive resources in comparison to the skill level.

The highest level is the knowledge-based level. Performance is controlled by in-depth analysis of the situation and systematic comparison of various response alternatives. An attempt is made to maximize efficiency and to minimize risk as much as possible. This level is common to unfamiliar situations. The division between the levels in this model is not absolute. For example, a person can transfer from a rule level action to a knowledge level action in accordance with his familiarity with the situation and with the extent of its new aspects.

The three models are significant in work environment design. The first model—the human information-processing model—is mostly relevant for interaction between humans and a permanent and stable work environment and the manner in which the work environment is designed (the machines or work tools).

The second model—the action-cycle model—is relevant for the work environment design when some uncertainty is present. According to this model, in a system's design process it is important to emphasize a clear definition of its goals, to detail the goals in accordance with intentions and specific performance methods, to properly plan the physical aspects of performing that task, and to integrate a system that will provide information on the performance method at each and every stage, and will facilitate problem evaluation and identification.

The first two models provide a framework for designing the work environment for maximal utilization of the operator's abilities. These models focus on perception, information processing and response, as well as on the means of designing the machine, and are not suitable for behavior in the context of tasks that require knowledge and use of strategy.

The third model—the rule and knowledge-based model—adds a distinction between three cognitive levels. The first two levels—the skill level and the rule level—are relevant for implementation in the interaction between humans and work environment, with which the first two models also deal. The industrial engineer should try to simplify the work and make it suitable for the skill level. In particular, skill level is suitable for the operation of a machine. Proper design of the work environment, as suggested by the first two models, will allow reaching an automatic skill level with minimum errors. In the rule level, integrating a system that provides situational alerts in the operation of the machine, as suggested by the second model, facilitates action according to rules. For example, when X happens (e.g., a red light turns on), Y must be performed (e.g., activating a system for lowering the temperature). Being skilled at cognitive actions at the rule level can facilitate basic deduction based on information from the machine, and the creation of new basic rules.

The knowledge level in the third model adds a cognitive level that is more complex than the skill level or the rule level. Operating on the knowledge level is relevant in nonroutine decision making, in planning, in problem solving, and in situations that require in-depth situational analysis and the consideration of various alternatives. These actions usually require greater mental effort (cognitive burden) compared to cognitive actions on the skill and rule levels. It is difficult to transition between levels. Even a skilled operator may find it difficult to transition from rule-level thinking patterns to knowledge-level thinking patterns. Therefore, as on the first two models, according to the third model correspondence between the work environment design and the required task level is also important.

In summary, the design of the task and the work environment should be adjusted to all cognitive levels. For example, the machine should include a direct action feature (skill), consistent mapping of limitations, and subsequent alerts (operation on the rule level), and representation of the field of work in a way that supports problem solving (knowledge). Transition from a rule pattern to a knowledge-level operation occurs following an unusual event, such as a plane that has to land due to a malfunction, weather conditions, and so forth. In such situations, the operator must transition from control in accordance with rules to decision making based on knowledge and experience (where should he or she land the plane, what is the proper order of the landing procedure, and so forth). The operator must consider a large number of parameters in order to make the right decision.

10.12.1 Complex Cognitive Systems: Diagnosis and Decision Making

The knowledge level in the third model discussed previously can be divided into various sublevels, ranging from simple decisions to complex decisions, such as the in-depth analysis of a situation that includes uncertainty and decision making based on partial information. Cognitive actions that involve analysis and decision making are usually important actions with future implications, and require considerable time and investment. The mental burden involved in such actions is significantly higher than the mental burden involved in cognitive activities based on skills or rules (Marmaras and Kontogiannis, 2001; Wickens et al., 2004).

Analysis is defined as a process in which a person tries to identify the factors causing an event, frequently an unwanted event or situation, such as a technical malfunction in a system or the diagnosis of a disease. This process involves perceiving stimulations or symptoms that point to the existence of a problem and searching for additional external knowledge about the problem. At the same time, it involves extracting information from long-term memory, based on learning and prior experience (Cunha and Louro, 2000), developing assumptions about the problem's causes and examining assumptions.

The analysis process is affected by the number of components in the system, the relationships between components, the available information, the level of uncertainty, the extent of operation freedom of the system, and the number of problems that may arise simultaneously. The time pressure and the risk level in the diagnosis also constitute significant factors in the analysis process.

Analysis that is dedicated to characterizing a system's status and finding the root cause for it is called diagnosis. Diagnosis is particularly difficult when it is necessary to make sure that the system continues to operate while the problem is being located, diagnosed, and solved. In this process, an individual skilled at making diagnoses uses decision rules, shortcuts, or heuristics (approximate solution methods) that make it easier to move between different cognitive levels and shorten the duration of the diagnosis.

Examples of heuristics used in the context of diagnosis include using what is perceived as the most probable assumption although probability is subjective; the choice of an assumption according to availability (availability heuristic), anchoring bias (adhering to a certain point of origin), and confirmation bias (searching and filtering information according to an assumption and not according to the objective situation) (Marmaras and Kontogiannis, 2001).

In the emergency plane-landing example, it is necessary to make a quick decision while processing a great deal of information (a cognitive burden that sometimes reaches maximum capacity). In such a case, the decision-making process will be based on conspicuous hints such as glaring lights, noise, and pilot stress (which do not necessarily correspond with the real level of urgency). These hints serve as shortcuts and receive attention that is disproportional to the information they provide (Wickens et al., 2004).

Decision making is a situation in which a person has to decide on a course of action. The action may be as simple as purchasing the best alternative, choosing the best candidate, or choosing the best proposal, or may be as difficult as choosing an emergency landing place or optimizing the use of resources for maximal profit. Decision making also includes many criteria for choosing between alternatives. Optimization of one criterion may have a negative impact on another (such as a decision to shorten production time, which may compromise quality). The decision-making process is affected by information quantity, information uncertainty, the dynamic nature of the work environment, the complexity of the criteria, and time and risk limitations.

This process uses heuristics, which leads to selectivity in the use of information. It leads to disregarding information and reducing the amount of information that must be dealt with and examining only a limited number of alternatives. These factors often lead to suboptimal decisions.

In Chapter 4, which discusses information systems, we noted that there are systems that can support cognitive actions such as diagnosis and decision making. As discussed, it is important to distinguish between routine decisions and nonroutine decisions. The former are based on repetitive data types that are processed in the same way each time, such as how much and when to pay suppliers. It is easy to integrate routine decisions into a computerized system, whereas most nonroutine decisions cannot be fully computerized. Most nonroutine decisions, such as if and what discount should be given to a loyal customer on a one-time basis or whether to change the product design, cannot be fully computerized. The frequency of the decision or the diagnosis is also important. Frequent decisions based on a predefined and quantifiable model justify computerization, whereas rare decisions and diagnosis of situations that are based on the decision-maker's intuition and experience are difficult to computerize.

Computerized support provides a response for two needs. First, it frees up time from simple processes, enabling management to confront the more complex problems and decisions that cannot be programmed and cannot be computerized. This includes the computerization of routine and frequent decisions, computerized training of personnel, and automating long and time-consuming calculations and procedures. Second, computerized support can make it easier to diagnose complex situations and to make decisions at senior management levels. Computerized systems can provide a response to various decisions by transactions processing systems that document transactions and processes in the organization, by information systems that provide raw information, summaries and reports about what is occurring, as well as by systems that combine data and analytical models in support of decision making. These systems save time and make better use of complex heuristics that bring the products of the diagnosis and the decision-making process closer to optimal solutions.

An interesting attempt to support decision making led to the development of software tools known as expert systems. Expert systems carry out the entire diagnostic or decision-making process after they are fed the relevant data (manually or automatically). Expert systems cannot reliably confront new and unfamiliar situations. In order to support diagnosis and decision-making activities that require high cognitive levels, expert systems must provide consultation. In other words, they must supply a summary of all the relevant information (about the specific case and about the system as a whole), as well as information from prior experience (to assist in retrieval from long-term memory). Subsequently, the systems present this information appropriately and according to order of importance, showcase various alternatives, and provide methods of evaluating the alternatives, but they do not make the decision, which is ultimately made by a human being.

10.13 Key Elements in Work Environment Design

The roles performed by workers in organizations are very diverse, and therefore it is difficult to develop a unified design method based on ergonomic principles.

Several organizations have developed rules and even standards for unique implementations that require the organization to employ in-house design engineers or engage engineers that design for the organization. Thus, for example, the U.S. Department of Defense in collaboration with the armed forces has developed a military standard for work environment design, primarily in military systems.

Particularly notable are the American standards of the NIOSH (a branch of the U.S. Department of Health). These standards are very well accepted worldwide and utilize many ergonomic principles (Salvendy, 2012).

Academic institutions offering ergonomic courses propose work environment design tools and approaches, including computerized tools, such as a human body computerized simulator that enables the testing of designs through appropriate simulation.

An introductory industrial engineering course can help students understand the complexity of work environment design by means of a checklist, which will prevent disregarding or forgetting important elements.

Following is an example of a checklist that consists of a series of questions industrial engineers need to consider when designing a work environment:

1. Are there safety hazards in the work environment and, if so, how can they be handled?

2. What is the level of noise? Is hearing protection required?

3. Are there safety standards to which the design needs to adhere?

4. What are the worker's objectives?

5. What are the relevant performance measures and what are the required values of these measures? (Are there minimum requirements or threshold requirements? Does the desired solution entail maximal requirements that do not need to be exceeded?)

6. What tasks need to be performed in order to achieve the objectives?

7. What are the required actions in the context of the defined tasks?

8. Can all or some of these actions be performed mechanically? (With a suitable machine or with a suitable computer and program?)

9. If mechanization is an option, what is the machine's role, and what is the required connection between the operator and the machine (man–machine interface)?

10. What knowledge is required for performing each task?

11. What kind of information is required for performing each task?

12. What information is received directly from the work environment (e.g., by observing the conveyor belt and the items it is carrying or by listening to a colleague), and what type of information must be received from special sensors and presented on a suitable display?

13. What type of display would be most appropriate for each kind of information and why?

14. Is the displayed information sufficient? Can this information be received and processed at the required pace?

15. Is a routine decision-making process required? What types of models are used in the decision-making process?

16. Is an ad hoc decision-making process required? Is it possible to characterize models that will assist in this type of decision making?

17. Is control required? If so, what type of information is needed for the control?

18. What is the connection between humans and the information system supporting their decisions (man–computer interface)?

19. Is the space in the workstation sufficient for all workers who will work at that station?

20. Do the workers have a field of sight that covers the entire work environment?

21. Do the workers have means of communication that enable them to communicate with other workers while coordinating with their superior and their subordinates?
22. Are repetitive actions required? At what pace and how many times each shift?
23. Can performing repetitive actions harm the worker over time?
24. Are special postures required in any part of the job?
25. What is the maximal lift weight required?
26. Are there motions with load or resistance that may lead to fatigue?
27. Is special training required for performing the role?
28. Is any qualification (e.g., in safety matters) required for performing the job?
29. Is maintenance required for the work environment and its equipment? If so, is there safe and convenient access for the performance of maintenance?
30. Is maintenance training required? Do the workers themselves need to perform any maintenance?
31. What is the frequency of rest required in performing the planned tasks?
32. What is the appropriate level of illumination for performance of tasks?

The modern work environment involves a great deal of computer use in the context of information systems. Designing the interface between humans and computer (man–computer interface) becomes an important component in work design.

Anxiety and frustration are part of the daily life of the users of computerized systems. Many users find themselves "struggling" to learn how to use systems that are supposed to help them in their work. Unfriendly systems and interfaces may lead to abstaining from the use of computerized tools. Alternatively, a friendly and innovative design of a system interface will lead to shortening the system's learning period, shortening performance times, reducing the error frequency, and ultimately achieve user satisfaction (Cunha and Louro, 2000).

Each computer interface must be adjusted to the types of jobs and the users intended to work on that particular system.

10.13.1 Types of Users

Users can be divided into three main types: new users, partially skilled users, and expert users. Each of these users has different needs, and the user interface needs to accommodate each of the types of users.

New users: Usually two aspects of new users are considered. The first aspect refers to the tasks required of the users, and the second aspect refers to the interface and the system that helps users perform the tasks. The system should offer users a small number of alternatives and should keep task operation simple, while integrating online support and explanations. An interface adapted to this type of user will help the worker become gradually and fully integrated into the organization.

Partially skilled users: These users have better knowledge than new users. They are generally familiar with the tasks as well as with the computerized task performance options. For this type of user, finding features and functions is critical. Easy access and easy search for functions and options is important and they should be organized for easy retrieval. In addition, users should be able to actively use the help function.

Expert users: These users have total and complete control over the necessary tasks. They need shortcuts that facilitate quick and efficient work. Work that requires the use of elaborate and hierarchically arranged help menus is tiring for this type of user. They prefer direct access and shortcuts to both standard menu actions, and popular sequences of actions.

A good user interface supports the performance of every action through a variety of alternatives so that each option can be adapted to the user. All types of users should be able to perform the action easily, conveniently, and in a user-friendly manner, suitable to the nature of the work. Proper user interface design is the key to efficiency, safety, and productivity (Van den Bogaard and Swuste, 2006; Watson and Sanderson, 2007).

10.13.2 Nielsen's Ten Principles for Designing a Convenient System

Molich and Nielsen (1990) suggest 10 principles for designing interfaces for an interactive system. The principles are heuristically based on experiments that point at factors that increase user satisfaction.

1. *System status visibility:* The system must be developed so that the user knows what is taking place. The user should receive frequent feedback that is adapted to needs and performed actions.
2. *Compatibility between the system and the "real world":* The system should be capable of communicating in the user's language through words and expressions familiar to the user. This type of communication facilitates the appearance of information in a logical and clear manner.

3. *The user's command of the application:* Users often mistakenly choose functions other than those they intended to choose. As a result, they should have the option to exit the current location in the system. Critical and unwanted actions should be reversible.

4. *Consistency and standards:* Terms, situations, or actions that have similar or identical names should not be used.

5. *Error prevention:* The system design should lead to a low probability of potential errors.

6. *Flexibility and efficiency of use:* A variety of performance alternatives should be available suitable for expert users as well as to new users.

7. *Aesthetics and minimal design:* The dialogue boxes should contain only relevant information. Information that is not frequently used or that is not relevant should not appear in the dialogue boxes.

8. *Identifying and correcting errors:* Error messages should be displayed clearly and accurately. In addition, error correction should be clear and structured.

9. *Help and documentation:* All system options should be documented. In addition, all information should be presented in a clear, focused, and structured manner, in accordance to the user's task.

10. *Visibility:* Objects, actions, and options should be designed to be visible. Situations should be avoided in which the user needs to remember information from part of a dialogue to use in another part. The instructions for use need to be visible or, alternatively, accessible from the appropriate locations.

10.14 Human Resources Management

Human resources management is a discipline that entails:

1. Preparing the human resource for work (recruiting and screening, training, and developing employees).

2. Operating the human resource (job design and evaluation, work relations, performance evaluation).

3. Maintaining the human resource (occupational welfare and work quality of life). Industrial engineers in the field of O&M deal with various aspects of human resources management, and they must be highly involved in this field. This section focuses on three topics that are important to industrial engineers: job design, job evaluation, and introducing a job evaluation system into the organization.

10.14.1 Job Design

Job design is the process of determining the activities to be performed in the context of each job and of defining the performance method, knowledge, data, and tools required to satisfactorily perform these activities (Daft, 2000). One definition of job design is the allocation of specific goals and tasks into the framework of the job. This allocation produces the "nature" of the job. This definition can be expanded by incorporating the content of the job, the manner and location in which the various tasks are performed, and the skills and qualifications required for the worker to perform his or her job satisfactorily. This process is based on the assumption that in order to achieve and improve on satisfactory performance, the job structure needs to be designed. In addition to proper job structure design, it is important to assign the most suitable worker for each type of job in order to guarantee success (Holness et al., 2006).

The purpose of job design is to meet technological and organizational requirements as well as the workers' personal needs. Since the environment in which an organization operates changes frequently, a dynamic approach to organizational job design is a necessity. Such an approach facilitates organizational flexibility and ensures organizational achievements and performance even in a changing environment. Job design is not a static process that is defined once and preserved in a stable organizational environment, but rather a dynamic process within the organization that reflects relevant aspects for designing the work environment at a given time.

The purpose of job design is to help the organization advance toward meeting its goals by using its human resources efficiently and intelligently. The importance of job design is usually recognized during the process of infrastructure preparation. We defined this process as one of the five major processes in the organization, mainly in the first phase of building a new organization and during the development of new manufacturing processes or periods of organizational change. It is important to remember that changes at various levels can impact an organization's job design and generate the need for redesign. Examples of these changes are organizational growth and expansion, cutbacks, reorganization, and employee turnover. Job design should be viewed as a tool through which an organization can promote its goals throughout its entire lifecycle.

10.14.2 Basic Terms

- *Task:* A collection of actions performed by a worker (sometimes by using tools, equipment, or work aids) in transforming input into output.
- *Work:* A mental or physical activity with desired outcomes.

- *Job:* A collection of tasks/assignments to be performed by the worker according to his or her job description.
- *Job design:* The definition of tasks/assignments to be performed in the context of a given job.

Job design is intended to maximize compatibility between the workers and the organization. The assumption is that such compatibility between the workers' needs and goals and those of the organization will contribute to increased effectiveness and efficiency, as well as to workers' satisfaction and motivation.

10.14.3 Approaches to Job Design

A historic review reveals a variety of approaches to job design. The early mechanistic job-design approach views humans as a type of machine. Later approaches recognize humans' unique needs. Perceptual/motor and biological job-design approaches emphasize the importance of humans' physiological and cognitive needs and emphasize the compatibility between man and machine. Motivational job-design approaches emphasize emotional–social needs and the importance of the compatibility between humans' abilities, skills, and requirements, and those of the organization.

- *The mechanistic job-design approach* involves the design of tasks assigned to the worker, while emphasizing standardization and clearly and "narrowly" defining the required activities. This approach creates simple jobs involving a great deal of repetition, so that the extent of automation and specialization in these jobs grows. The advantages of this type of job design include a reduction in work errors, a decrease in worker fatigue and mental burden, and a decrease in the need to train and qualify workers. In addition, filling open positions in the organization is much easier because the required abilities and skills are low.

  Mechanistic job design leads to higher manufacturing efficiency and larger profits. However, this type of design diminishes the value of the worker as a sophisticated and thinking human creature. The potential outcome is a decline in worker motivation.

  Medsker and Campion (2001) argue that this type of design will lead to a higher rate of absenteeism, boredom, and lower satisfaction and motivation rates.

  Unlike the mechanistic job-design approach, the perceptual-motor approach, the biological approach, and the motivational approach are based on recognizing that human beings have needs that must be taken into consideration in designing jobs. The emphasis is on the compatibility between man and machine. These approaches support the need to design the equipment and the work environment as well as the job tasks.

- *The perceptual-motor approach* emphasizes the importance of perceptual approaches in the context of work. This approach focuses on manipulations of factors such as lighting, display, amount of input to be processed, memory, and so forth. The object of this manipulation is to improve perceptual processes required for performing the job tasks.

This approach is quite similar to the mechanistic approach. This similarity originates from the attempt to reduce the worker's mental burden through user "friendly" equipment and to reduce information processing requirements—a strategy of simplifying perceptual tasks. One field of specialization that evolved due to this approach is the field of gauge displays and signs. For example, Figure 10.11 shows a typical dial gauge.

The prominent advantages of the perceptual-motor approach include error reduction, mental burden reduction, reduction of training and qualification time, and making it easier to fill available positions because the needed abilities and qualifications are at a low level. Job design under this approach leads to high boredom rates, and to a decrease in satisfaction due to the disregard of human needs on the Maslow scale, such as the need for belonging, self-esteem, and personal fulfillment.

- *The biological approach* emerges from the field of work physiology and biomechanism and emphasizes that job design needs to take biological needs into consideration. The important factors in this approach are physiological work variables, such as the extent of force and stamina required to perform the work, the convenience of seating arrangements, the adjustment of the chair to workers with different body measurements, noise level, humidity and heat, and so forth. Taking care of physiological factors will help reduce

FIGURE 10.11
Example of a dial gauge display for utilization.

physical fatigue, reduce the pain common in an environment that is not compatible to worker physiology and biomechanics, as well as help reduce accidents. The actions required to adapt the workplace entail costs, and the adaptation of the work environment may lead to worker inactivity.

- *The motivational approach* recognizes the needs of the workers in the organization and emphasizes their social and emotional needs as well as the need for compatibility between these needs and the organization's requirements. The biological approach, the perceptual-motor approach, and the mechanistic approach may meet the workers' physiological needs (appropriate sensory stimulation, work temperature and humidity suitable for the human body, proper illumination, rest times, and so forth) as well as their needs for security and safety (forecasting ability that derives from simple and known work, rules, and regulations that provide a sense of stability). However, they do not meet Maslow's scale of human needs, such as the need to belong, self-esteem, and personal fulfillment.

  In this context, it is noteworthy to mention what Clarence Francis wrote in an article in *Fortune* magazine (Milkovich and Bouudreau, 1998):

  "You can buy a man's time, you can buy a man's physical presence at a certain place, you can even buy a measured number of skilled muscular motions per hour or day. But you cannot buy enthusiasm, you cannot buy initiative, you cannot buy loyalty; you cannot buy the devotion of hearts, minds, and souls. You have to earn these things."

The motivational approach seeks tools to help generate motivation and dedication among workers by matching the workers' skills and abilities to the organization's requirements. This approach strives to design jobs in the organization that will fulfill the workers' potential by challenging workers in the context of their abilities, in order to develop and promote them. Workers are perceived as the organization's added value and as contributors to the creation and preservation of a competitive advantage.

10.14.4 Job Design Strategies

There are several strategies for implementing this approach:

1. Job enrichment
2. Job enlargement
3. Job rotation

These approaches attempt to provide a response to the social and emotional needs of the workers.

Job enrichment: This job design approach includes an enrichment element that creates motivation factors higher up on the Maslow scale. Under this approach, workers are assigned tasks and duties that contribute to autonomy, responsibility, authority, and involvement in the decision-making process. This type of design generates more interest in the work and contributes to increased commitment and satisfaction.

Job design based on job enrichment leads to eliminating unnecessary levels of supervision and control, at the cost of an increase in the level of abilities and skills required from the workers. Under-skilled workers will not be sufficient, and therefore training and qualification become critical. This could lead to difficulty in recruiting personnel suitable to fill open positions. The nature of the work requires a great deal of investment on the part of the workers. The extent of responsibility imposed on the workers may contribute to their sense of self-esteem and personnel fulfillment; however, this may also lead to salary demands as compensation for the intensive involvement.

The job enrichment model is based on the assumption that the core characteristics of a job have an impact on critical psychological situations, which in turn also have an impact on motivational and emotional outcomes.

This model includes the following five factors:

1. *Skill variety:* To what extent are workers required to perform a variety of actions and use their various skills?
2. *Significance of the position:* To what extent does the job influence the lives of other people within and outside the organization?
3. *Identity of the position:* To what extent are workers required, in the context of their job, to perform a task from start to finish?
4. *Autonomy:* To what extent does the job provide independence and discretion to individual workers?
5. *Feedback:* To what extent do workers realize how well the job is performed based on the performance itself, rather than on feedback from intermediaries?

The assumption is that job variety, significance, and identity will contribute to a situation in which workers feel a sense of purpose with respect to their job. Autonomy will lead to a sense of responsibility, whereas feedback will enable workers to know about the outcome of the job and its tasks. These in turn will contribute to high motivation, high performance quality, work satisfaction, and a decrease in absences and turnover.

A general rule for job enrichment is to enrich the position vertically, in other words, not only performance responsibility (horizontal responsibility) but also managerial responsibilities such as planning, decisions, and supervision.

When the job is not central to a worker's life, when his or her skills and aspirations are outside of the workplace, job enrichment may cause undesirable outcomes: unwillingness to take responsibility or exploit new opportunities, thus leading to a decrease in performance.

Job enlargement is a job-design approach that enlarges the range of activities/tasks in the context of a given job, attempting to overcome the problem of boredom at work and the problems of an overly defined job. The job is assigned additional tasks that are characterized by authority, responsibility, and autonomy, identical to those previously performed by workers in that position (and not higher, as is the case in job enrichment). This type of design increases task variety and job challenge.

The difference between job enrichment and job enlargement is inherent in the nature of responsibility. Job enrichment expands the worker's existing tasks by adding tasks accompanied by a higher level of responsibility; job enlargement, on the other hand, is characterized by more responsibility, though it is the same type of responsibility present in the role before it was expanded.

Among the advantages of job enlargement are high levels of satisfaction, interest, and commitment. However, as in the job enrichment approach, this approach also requires a great deal of investment in training and qualification, and there is a risk that workers will require higher monetary compensation for the effort they are required to make.

Job enlargement sometimes causes workers in an organization to feel as though their autonomy and authority have been compromised. Additional tasks identical in complexity to the original tasks provide variety in the required abilities, but do not significantly expand them. Exposing the worker to a greater quantity of tasks may lead to identification with the role; however, autonomy and feedback are not reflected in the context of this strategy.

Job rotation is a variation of job enlargement. This approach attempts to diversify the work by systematic rotation of workers between jobs that involve different tasks. Workers perform simple jobs that frequently change. This approach adds satisfaction, commitment, and interest. However, it increases the need to train the workers and provide them with higher payment and compensation. In addition, it provides a response to the problem of boredom at the workplace, but it may create conflicts due to the transition of workers between jobs. The rotation approach necessitates training workers with unique skills that will enable them to alternate between jobs. The advantage of this approach is that it enables flexibility in worker placement because the worker has been trained to fill several different roles.

Job design examines the design required for each type of job (defining the work content), while recognizing the fact that different jobs require different designs. For example, job rotation and enlargement may be suitable for manufacturing floor workers who have basic skills and job enrichment may be suitable for mid-level workers with a higher skill level, whereas job design based on teamwork may be suitable for workers who engage in complex tasks.

Job design requires considering workers' characteristics (skills, abilities, motivation level, willingness, and initiative) as well as the organization's characteristics (structure, stability, and competition). The organization's characteristics and environment will establish the objectives and the limitations and constraints that impact the job design.

10.14.5 Job Evaluation

Job evaluation is a process intended to determine the relative contribution of various jobs to the organization and accordingly determine the compensation and reward system (salary, status, promotion, and so forth). The main objective of job evaluation is to maintain internal consistency in the organization so that workers with identical jobs receive similar pay and workers with different jobs receive different pay, according to the contribution of their jobs to the organization (Das and Garcia-Diaz, 2001).

The assumption is that internal consistency will yield positive results, such as improving worker satisfaction and performance. The lack of internal consistency may lead to negative results such as turnover, bitterness, and decreased motivation (Swezey and Pearlstein, 2001).

Job evaluation is important in the dynamic and competitive environment in which organizations operate today. Workers are a central asset of the organization. Their satisfaction is a condition for promoting performance because a happy worker will work faster and better. Approximately 70% of the organizations in the United States use various methods of job evaluation (Hannon et al., 2001).

1. *The ranking method:* This method describes the various jobs in an organization and ranks them according to their defined value or contribution. There are three ranking methods: simple ranking, alternation ranking, and paired comparison ranking.

   Simple ranking jobs are ordered according to their contribution to the organization, starting from the most important job and ending with the least important one.

   Alternation ranking jobs are alternately ordered, beginning with the most important job and following with the least important job. In the next phase, the second most important job is selected, followed by the next to least important one and so forth until all jobs are sorted and ranked.

   Paired comparison ranking compares all potential job pairs that are in the same field in the organization.

   Once the comparisons have been completed, a job that received the highest grade is ranked at the top of the scale, a job that received the lowest grade is ranked at the bottom of the scale, and so forth.

   The disadvantages of this method derive from the subjective opinions of evaluators, mainly when the job descriptions are not sufficiently detailed. The paired comparison-ranking method requires so many comparisons that often implementation of the method is technically problematic.

2. *The classification method:* This method creates occupational categories and classifies jobs according to these categories. Jobs in one category are similar to one another and are different from jobs in other categories. Each category is defined by a department/status description. These descriptions are detailed enough to distinguish between the jobs, but general enough to make it easy to classify the jobs.

   The job of classification can be made easier by anchoring the process with stable, well-known jobs that serve as points of reference.

   The number of suitable classifications is dependent on the variety of jobs and the promotion paths within the organization. An accepted rule of thumb is to create between 7 and 14 categories.

   The disadvantages of this system derive from the pressure applied by those holding senior positions to include a certain job in a higher category, thus potentially causing distortions.

3. *The criteria comparison method:* This method involves ranking jobs according to several predetermined criteria, dividing each job into its components and assessing the value of each component in the job.

   The first phase entails determining the criteria to be evaluated for each job. The second phase involves choosing several key jobs in the organization and evaluating them according to the chosen criteria. This is followed by a job analysis: each job is examined according to its components and recorded on an index card. The cards are ordered according to the importance of the various job criteria (e.g., for a managerial job, the first item will be professional aptitude or responsibility, followed by physical requirements, whereas on the card for a laborer's job the order will be reversed). In the third phase, the base salary of each of the key jobs is determined according to the criteria set for each job. A connection is built between the job criteria and the salary levels and this is used to build a scale for comparing jobs.

   This method is expensive and complex. For this reason, only a small percentage of organizations that use job evaluations use this method. Furthermore, it is difficult to explain this method to workers who are not satisfied with their salary.

   There are those who argue that this method is preferable to the other two methods because the job evaluation criteria are defined and agreed on in advance.

4. *The point-factor rating method:* This system is extensively used by the U.S. government. It is based on criteria, grades for each job in each criterion, and weights that reflect the relative importance of each criterion.

The major steps of the point-factor rating method are as follows:

a. Job analysis (the first step in all job evaluation methods)

b. Criteria selection: the usual tendency is to choose criteria that belong to four categories:

   i. Required skills

   ii. Required effort

   iii. Required responsibility

   iv. Work conditions

c. Determining grade scales for each criterion. The grade scales need to be clear and easy to use.

d. Determining a relative weight for each criterion (e.g., 50% professionalism, 15% effort, 25% responsibility, 10% work conditions).

e. Determining a final score for each job by adding up the grades in each of the weighted criteria.

5. *Single factor systems:* Assuming that in certain cases, a job's content or value is one-dimensional, this system attempts to determine the value of a job by comparing jobs based on a single criterion (e.g., salespeople by sales ($)).

Two popular programs use the single factor approach. The first is based on time span of discretion (TSD). In this program, the value of a job is measured according to the period of time in which the worker is not under direct supervision by his or her supervisor and in which he or she is required to use independent discretion. The second refers to the required level of decision making in each job. It is customary to refer to approximately six levels, from the simplest level through the most complex level. The value of each job is examined based on the highest decision-making level in the job. For this reason, this system is known as *decision banding*.

In using job evaluation, one will encounter biases, prejudice, and subjectivity that may influence the workers and the relationship with them. It is advisable to consider workers' perceptions and feelings and to make an effort to have the entire process viewed as fair, consistent, without prejudice, fixable, explainable, and considerate. The fairness approach is based on the assumption that people in organizations want to be treated fairly and equally. Unfairness leads to negative worker attitudes (work dissatisfaction, pay dissatisfaction, high turnover, and so forth), which could significantly impact the organization and its ability to survive in a competitive environment.

To prevent subjective biases in job evaluation, attempts have been made to make the process computerized and more objective. The trend is to develop a systematic and quantitative methodology for job evaluation programs that will serve as an aid in decision making regarding pay. The objective is to maintain a fair payment and compensation system so that workers will not feel their pay does not match their job. A classic example of a decision-supporting system is a prototype is called JESS (Mahmood et al., 1995).

The JESS system was developed to compare between jobs in the field of Information Systems (hereinafter, IS). JESS evaluates IS jobs according to key roles, such as information system manager, system analyst, and computer programmer. These roles are analyzed according to criteria—for example, skills and knowledge, extent of supervision and training required on the job, extent of job complexity, and additional criteria as well.

The JESS system requests the name of the job, after which it evaluates the job by means of questions that are related to the criteria. In the next stage, the system calculates a grade for each job and compares the result to information from a market survey. Finally, it provides information analysis about the job, such as market survey information, job evaluation, and so forth. The whole process is carried out by computer, while the program itself includes a database and information, a user interface and inference engine. The process is saved on the computer's memory and enables the user to receive detailed information about the outcome.

Use of an expert system may enhance workers' sense that the organization is paying them fairly because the job is evaluated in an automatic and therefore consistent manner. A JESS system neutralizes the use of evaluations and subjective decisions, which are required in manual job evaluation methods, and thus expedites the process. Use of a computerized system helps the organization make reliable and trustworthy changes and updates, especially during a period of frequent changes in the organization's goals and in the environment in which it operates, thus resulting in reliability in the job's evaluation and the paid salary. It is important to note that the expert system does not replace the role of human evaluators, but rather assists them, simplifies their work, and provides them time to attend to more important tasks. The system does not decide on a worker's pay structure, but rather assists the evaluator in performing the evaluation and in explaining it to the worker. The decision remains in the hands of the evaluator. The system can provide good results only if the information entered into it by its operators is up to date and accurate.

An additional decision-supporting example is based on a statistical computer program that was programmed to be objective, systematic, and performance oriented for designing a payment and compensation plan. The purpose of this program is to improve the choice of payment factors and to design the pay scale in accordance with statistical data. Since the list of original criteria may be long, it needs to be filtered. The filtering methodology is based on a sampling of key positions. A decision-support system is used to filter out undesirable criteria or to change the grade scales.

The JESS program contains additional parts, such as the ability to store new information and to retrieve old information, or the ability to interactively update information. The input includes the list of criteria and the number of levels for each of the scales, a list of key positions and their salaries, and the results of key position evaluations. Once the program has undergone a trial run, the output details the number of levels in each criterion, the description of criteria by level, average, median, and distributions, frequency of key positions in each scale, analysis of connections between criteria and pay grades, analysis of the connection between each pair of criteria, and finally analysis of the distribution curve for each criterion.

In a modern society, the use of a decision-supporting system that is free of subjective influences is very important because payment is what motivates many people to work and attracts them to the organization. When workers feel that the organization takes them into consideration and that the organizational processes are fair and reflect how they perceive reality, then their performance improves, and thus the organization and the workers both profit. In a competitive market, this fact helps the organization survive and even thrive.

Despite the importance of internal consistency, which can provide workers with the sense of fairness in the evaluation process, it is important not to disregard the external environment because we live in a competitive world. It is important to remember that the organization's environment includes competitive organizations that have similar jobs and a job market that has worker supply and demand, which affect the value of salary. In addition to job value, three additional components are taken into consideration in calculating workers' salaries:

1. *Market-based pay systems:* Some companies determine their workers' salary levels according to the salary levels customary in the market and according to their ability to pay. This kind of system puts a great deal of emphasis on the external market and does not emphasize internal consistency (in other words, the relations between the jobs within the organization). This method is common in a competitive market.

   For most companies, comparison between all the positions in the organization (sometimes there are hundreds and even thousands of positions) and those in the market is unrealistic. Therefore, sometimes this system is only used to determine the salary for unique positions in the organization.

2. *Knowledge-based pay systems:* Knowledge-based systems reward workers for what they know and not according to the specific role they perform. The two basic assumptions of such systems are: (1) the higher the level of a worker's knowledge, the more the market will compete for him or her; (2) as the level of a worker's knowledge rises, so too does his or her contribution and value to the organization.

3. *Skill-based pay systems:* Pay in skill-based pay systems is based on the number of tasks an individual is authorized to perform.

The advantages of this system are that it encourages the development of quality labor. Workers who are able to perform a large number of various tasks contribute to flexible timing and worker placement. Under this system, motivation increases to learn and attend educational seminars because rewards and performance are closely related.

The disadvantages of this system include higher salary rates, higher training costs, an administrative burden due to record maintenance, and erosion of knowledge or skills that are not used.

10.14.6 Introducing a Job Evaluation System into the Organization

An organization that chooses to introduce a job evaluation system must make three major decisions:

1. A decision about the purposes of the job evaluation—for example, to establish a pay structure that will be perceived by workers as fair and will be consistent with the organization's objectives and strategic vision, to simplify the connection between the job and the salary, and to help determine salary for new and unique jobs.
2. A choice of the job evaluation system suitable for the organization—the system must be suited to the organization's objectives. The organization must ask a series of questions, such as Is there a difference between the performed jobs? Will job evaluation create a distinction that is clear to the workers? Do we use one system or more?
3. The organization must decide who will participate in the job evaluation process. For example, a committee that represents key positions from various units in the organization, and includes the representatives of workers and the professional union, in order to increase their trust and commitment to the process and its outcomes.

Introducing a job evaluation system is a very important project, and it must be planned and performed using project management tools as discussed in Chapter 3.

Job evaluation is associated not only with the pay system, but also with the promotion system (what is the next job on the worker's promotion track), the training system (what knowledge must the worker acquire to be promoted), and more. When the organization in question is stable and competes in a stable market in which the same products and services are sold over time, there will be no significant addition of new jobs that will force the job evaluation system to be frequently updated. Therefore, the initial cost may be high, but the maintenance is foreseen to be low (Hannon et al., 2001).

In a project-oriented organization based on one-time tasks performed over time and on a defined budget and resources, job evaluation is important and significant for other reasons: projects are performed under budgetary limitations. It is important to establish the budget allocated for salary payment based on job evaluation.

Once a job evaluation system has been chosen, it needs to be implemented in the organization by planning and performing an implementation project that includes these work tasks (Huber, 1991):

1. Training the evaluators and providing them with background and information about the jobs, particularly if they are external evaluators.
2. Evaluating new jobs and making amendments and examinations in evaluating existing jobs.

The job evaluation system must be supported by senior management through open and fluent communication to make sure it is in fact compatible with the organization's objectives and must ultimately approve it so that the results are acceptable to the workers.

When the process is completed, its success must be examined and the results must be assessed. The evaluation will be based on components such as

- *Reliability:* Are the results consistent under various conditions?
- *Validity:* Does job evaluation indeed achieve the desired results? What is the percentage of correct decisions made?
- *Effectiveness:* To what extent were the goals of the program achieved? Are the results effective in terms of worker behavior (decrease in turnover, increase in satisfaction, and so forth)?
- *Extent of discrimination:* Is the job evaluation free of bias and prejudice? (e.g., with regard to women and minorities).
- *Extent of acceptability:* Are the results acceptable to workers and managers?

The organization must ask these questions and make sure the system indeed receives the highest grade on each of these components.

A good reward and compensation system obtains positive results for both the workers and the organization. It provides feedback about the value of each worker in the organization. The basis for a good reward and compensation system is the job evaluation process.

Often, job evaluation is carried out by a committee that includes mid-level managers and senior level executives, as well as representatives of the professional union. This combination reduces the level of objectivity and increases the levels of conflict because participants seek to maintain their interests.

(Although management recognizes the human resource as an asset, it operates in a competitive market and therefore is interested in maintaining salaries lower than the levels desired by the workers' representatives.) As discussed previously, various attempts have been made to integrate computerized systems into the process to increase objectivity.

10.15 Summary

This chapter discusses the industrial engineer tasks associated with human resources. Unlike other resources such as equipment, tools, and materials, human resources require special handling of matters such as ergonomics, sociology, and psychology. The topics related to human resources discussed in this chapter can be divided into three factors: physical motoric, perceptual cognitive, and occupational. Some of the subtopics that characterize these three factors are outlined next.

Examples of physical motoric subjects:

- Exertion and heart–lung activity
- Exertion and activity of the muscular, skeletal, and nervous systems
- Strenuous work: lifting, bending, and sitting
- Manual work and work postures
- Anthropometry and movement
- Environmental influences: lighting, noise, vibrations, climate
- Stress, fatigue, boredom, and repetitiveness of the biological cycle
- Expertise, training, and efficiency

Examples of perceptual-cognitive subjects:

- The human sensory systems: sight, hearing, and touch
- Cognition and cognitive connections in human performance
- Discovery, identification, distinction, and scaling (the ability to rank items on a cognitive scale)
- Decision making
- Displays and operation means
- Man–computer interface and screens
- Human error, accidents, and safety
- Human social factors
- Human transportation factors

Examples of occupational subjects:

- Job design
- Job evaluation
- Introducing a job evaluation system into the organization
- Introducing and managing an incentive and bonus system
- Occupational disease (e.g., CTD)
- Professional development, training, and educational programs
- Developing an organizational culture (in many cases, culture integrated with striving for quality and excellence)

Industrial engineers are responsible for designing the human work environment. Therefore, in the context of their training, they will take courses in ergonomics, psychology, and sociology. Since other engineers do not receive this type of training, industrial engineers must be involved in the design and management of systems that involve the human factor.

References

Cunha, P.V., and Louro, M.J. 2000. Building teams that learn. *The Academy of Management Executive* 14(1), 152–153.

Daft, R.L. 2000. *Job Design. Organizational Behavior.* Orlando, FL: Harcourt College Publishers, Chapter 16, pp. 554–583.

Das, B. and Garcia-Diaz, A. 2001. Factor selection guidelines for job evaluation: A computerized statistical procedure. *Computer and Industrial Engineering* 40(3), 259–272.

Espluga, J. 2004. Study highlights costs of workplace accidents and occupational illnesses. *European Industrial Relations Observatory On Line*, March, 2004.

Grandjean, E. 1988. *Fitting the Task to the Man.* London: Taylor & Francis.

Hannon, J.M., Newman, J.M., Milkovich, G.T., and Brakefield, J.T. 2001. Job evaluation in organizations. In: Salvendy, G. (Ed.), *Handbook of Industrial Engineering.* New York: John Wiley & Sons, pp. 900–917.

Herzberg, F., Mausner, B., and Snyderman, B. 1959. *The Motivation to Work.* New York: John Wiley & Sons.

Hjelle, L.A. and Ziegler, D.J. 1992. Abraham Maslow. In: Hjelle, L.A., and Ziegler, D.J. (Eds.), *Personality Theories.* Auckland: McGraw-Hill Inc., pp. 440–480.

Holness, K., Drury, C., and Batta, R.A. 2006. Systems view of personnel assignment problems. *Human Factors and Ergonomics in Manufacturing* 16, 285–307.

Huber, V.L. 1991. Comparison of supervisor—incumbent and female—male multidimensional job evaluation ratings. *Journal of Applied Psychology* 1, 115–121.

Kroemer, K. 2008. *Fitting the Human.* 6th edn. Boca Raton, FL: Taylor and Francis.

Kroemer, K. and Grandjean, E. 1997. *Fitting the Task to the Human.* New York: Taylor and Francis.

Kroemer, K., Kroemer, H.J., and Kroemer-Elbert, K.E. 2010. *Engineering Physiology: Bases of Human Factors Engineering/Ergonomics*. Boca Raton, FL: Springer.

Lerman, S.E., Eskin, E., Flower, D.J., George, E.C., Gerson, B., Hartenbaum, N., Hursh, S.R., and Moore-Ede, M. 2012. Fatigue risk management in the workplace. *Journal of Occupational and Environmental Medicine (JOEM)* 54(2), 231–258.

Mahmood, M.A., Gowan, M.A., and Wang, S.P. 1995. Developing a prototype job evaluation expert system: A compensation management application. *Information and Management* 29, 9–28.

Marmaras, N. and Kontogiannis, T. 2001. Cognitive tasks. In: Salvendy, G. (Ed.), *Handbook of Industrial Engineering: Technology and Operations Management*. 3rd edn. New York: John Wiley & Sons, pp. 1013–1040.

Mayo, E.G. 1949. *Hawthorne and the Western Electric Company, The Social Problems of an Industrial Civilization*. London: Routledge.

Medsker, G.J. and Campion, M.A. 2001. Job & team design. In: Salvendy, G. (Ed.), *Handbook of Industrial Engineering: Technology and Operations Management*. 3rd edn. New York: Wiley, pp. 869–894.

Milkovich, G. and Bouudreau, J. 1998. *Human Resource Management*. 8th edn. Irwin, pp. 139–167.

Molich, R. and Nielsen, J. 1990. Improving a human-computer dialogue. *Communications of the ACM* 33(3), 338–348.

Punnett, L., Prüss-Ustün, A., Imel Nelson, D., Fingerhut, M., Leigh, J., Tak, S., and Phillips, S. 2005. Estimating the global burden of low back pain attributable to combined occupational exposures. *American Journal of Industrial Medicine* 48(6), 459–469.

Salvendy, G. 2012. *Handbook of Human Factors and Ergonomics*. 4th edn. Hoboken, NJ: Wiley.

Swezey, R.W. and Pearlstein, R.B. 2001. Job evaluation in organizations. In: Salvendy, G. (Ed.), *Handbook of Industrial Engineering: Technology and Operations Management*. 3rd edn. New York: John Wiley & Sons.

Taylor, F.W. 1911. *The Principles of Scientific Management*. New York: Harper & Brothers.

Van den Bogaard, A. and Swuste, P. 2006. Safety: Technology and behavior. In: Verbeek, P.-P., and Slob, A. (Eds.), *User Behavior and Technology Development: Shaping Sustainable Relations Between Consumers and Technology*. Dordrecht, The Netherlands: Springer, pp. 21–31.

Waters, T., Puts-Anderson, V., and Garg, A. 1994. *Application Manual for the Revised NIOSH Lifting Equation*. Cincinnaty, OH: NIOSH Division of Biomedical and Behavioral Science, U.S. Department of Health and Human Services.

Watson, M. and Sanderson, P. 2007. Designing for attention with sound: Challenges and extensions to ecological interface design. *Human Factors* 49(2), 331–346.

Wickens, C.D., Gordon, S.E., and Liu, Y. 2004. *An Introduction to Human Factors Engineering*. Upper Saddle River, NJ: Pearson Prentice Hall.

URL

http://www.accel-team.com/work_design/
http://www.swcollege.com/management/champoux/powerpoint/ch09,ppt
http://www.wfu.edu/~randelae/powerpoint/jobdesignstudentversion.pp
http://www.eurofound.europa.eu/eiro/2004/03/feature/es0403211f.htm

11

Introduction to Supply Chain Management

Educational Goals

In this chapter, we discuss the term supply chain and its implementation. In this context, we study the historic development of supply chain management (SCM); the characteristics of supply chains; types of contracts; the value of information in supply chains; the design of supply chains; and their management, monitoring, and control.

11.1 Introduction

Until this chapter, we focused on organizations that provide products or services. Many of these organizations purchase materials, subassemblies, and assemblies of various products from other organizations. Therefore, we discussed procurement and its management. However, procurement may have several forms: procuring standard material such as sugar, wheat, or iron (with uniform standard quality) has great competition and could be done on the basis of price alone. However, a heavy consumer of a commodity would be willing to pay for having a reliable supplier that would ensure a steady supply even in times of shortages. The less standard are the components and materials, the more important is the selection of the supplier and the connection with the supplier. Procurement of large projects and continuing services often leads to strategic partnerships. The Japanese philosophies of just in time (JIT) and total quality management (TQM) acknowledge the importance of supplier–customer relationships and foster a long-term relationship between an organization and its suppliers—this brought about a culture of strategic alliances. Any organization in such a relationship can be viewed as a link in a chain wherein it purchases from their suppliers and sell to their customers. The chain typically begins with the raw materials and ends with a finished product sold to private or institutional customers. The customary term for a system of organizations that sell and purchase from one another is a *supply chain*. The links (organizations) in the supply chain are interdependent, and the market success of the entire chain is the objective of all the links in the chain. Competition in the market is between the supply chains. In other words, every organization in the chain is an individual player participating in a competition between teams. (The performance of all the players on the team contributes to winning or losing games—and similarly, success or failure in selling the products supplied by the supply chain depends on the performance of all the organizations that are part of the chain rather than the performance of an individual organization as one link in the chain.) SCM aims at providing the right product, at the right time, in the right place, and in the right quantity in order to maximize the profits of the organizations in the supply chain.

To effectively plan good integration between suppliers and their customers along the supply chain, it is important to become familiar with the supply chain characteristics and the factors that affect the chain. Optimal management of the supply chain is a major issue in many industries because of the importance of creating an efficient and economical flow of products and services in each of the links of the supply chain. Becoming part of a supply chain enables companies to create a competitive edge for themselves by being a part of a victorious team with a successful product. The day-to-day management of the supply chain enables companies to minimize uncertainty and risk, to reduce costs, to increase flexibility, and to improve service and availability for the end consumer.

In recent decades, many organizations have understood that taking the entire chain into consideration ultimately contributes to each link/organization in the chain. Many industries focus on optimal SCM because they understand that better management can contribute to significant cost reduction and performance improvement, and profits increase accordingly.

SCM entails management of the risks and uncertainty that exist in the supply process. These risks and uncertainty stem from the fact that demand is usually not deterministic, and is frequently dynamic and dependent on future data (which are unknown at the time of production and acquisition planning). The production capacity of the links in the chain is also affected by the uncertainty stemming from machine malfunctions, transportation hitches, and so forth.

Managing the supply chain requires knowledge of many factors such as demand forecasting, actual demand, actual inventory, production capacity, uncertainty factors, and logistic system data such as transport times. Proper use of as much information as possible improves system performance and contributes to better outcomes. The discussion in this chapter focuses on the integration of the various organizations that constitute the supply chain. The discussion is based on the material studied in preceding chapters of this book, which focuses on the individual organization—the single link in the chain.

11.2 Background: Terms, Definitions, and Historic Overview

11.2.1 What Is a Supply Chain?

11.2.1.1 Intuitive Description of the Term Supply Chain

To properly understand the formal definitions of a supply chain, it is advisable to first gain some intuitive background. The following discussion provides this background.

Many products that reach the end consumer go through multiple phases of processing, production, and assembly before they become a finished product. The transition from raw material to the finished product that reaches the end consumer can be described as a chain of actions performed by various organizations along the chain. This chain is called the supply chain.

The following is a simple example of such a chain. A farmer in the United States grows wheat, harvests it, and sells the seeds, which are then shipped in big trucks to regional wholesalers who store the wheat in granaries. The American wholesaler sells the seeds to the foreign wheat importer and must transfer the wheat from the granaries to the ship, which usually belongs to an ocean shipping company. The importer receives the wheat from the ship, stores it, and sends measured quantities to gristmills. The gristmills sell and transport the flour to bakeries in large sealed trucks. The bakeries prepare bread from the flour, sell it, and transport it to supermarkets and grocery stores, where it is sold to the end consumers.

Many links in supply chains provide services rather than products. For example, distribution networks do not produce but rather provide retailing services, transport companies provide transport services, and wholesalers usually provide storage services.

Clearly the supply chain of other products such as computers or cars is much more complex and complicated. However, the principle of the supply chain remains identical: all the phases, from the raw materials to the finished product, constitute the complete supply chain of the finished product.

The phases in the supply chain of a complex product are schematically described in Figure 11.1. Each arrow in the illustration includes transport and storage processes on both ends. A multitude of storage and transport during the production significantly increases the price of the product.

Product production often requires the integration of several supply chains. For example, the supply chain of a simple product such as a chocolate bar consists of the cocoa supply chain, the milk supply chain, the sugar supply chain, and the wrapper supply chain. The sale phase of the chocolate bar (one before last in Figure 11.1) also includes several stages: transport from the wholesaler, central retailer storage, distribution to local branches, retailer shelf storage (supermarkets and grocery stores), and purchase by the end consumer.

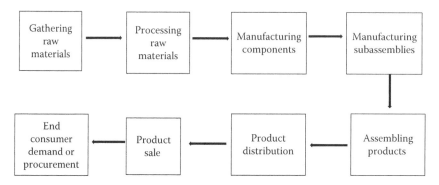

FIGURE 11.1
Schematic diagram of the phases in a product supply chain.

In many cases, treatment of the supply chain does not necessarily refer to the complete chain. For example, many organizations that are part of a supply chain that produces an intermediate product consider the purchasers of the intermediate product as end consumers. For example, the milk supply chain manufactures an intermediate product (milk) used in a variety of other products: fresh drinking milk, UHT milk, soft cheeses, yogurts, powdered milk, butter, and chocolate. In such a case, the organizations in the milk supply chain do not take the time to monitor the chain of each of the aforementioned products; for them, the manufacturers of the finished products (e.g., chocolate) constitute part of the end consumers. On the other hand, milk is one of the main components for chocolate manufacturers. For them, a regular supply of milk at a reasonable price is critical. Therefore, they monitor what happens in this chain (in other words, they monitor changes in milk prices and availability, as well as crises such as epidemics or infections).

Sometimes reference is made to a defined section of the complete chain, which usually consists of suppliers, manufactures, distributors, sellers, and customers. Each organization is a link in the supply chain, and each organization receives materials or components from suppliers as input and supplies its products as output. As mentioned, multiple phases in which inventory is stored and transported along the supply chain may significantly increase the price of the finished product.

The inventory levels of a link in the supply chain (in this case, a manufacturing organization) for deterministic and fixed demand are illustrated in Figure 11.2.

The inventories and transport in Figure 11.2 characterize practically every link in the supply chain. Thus, it indicates that inventory and transport management are major components in SCM. Since the management and transport of inventory are computerized in most organizations, computers and computer communications play a key role in SCM.

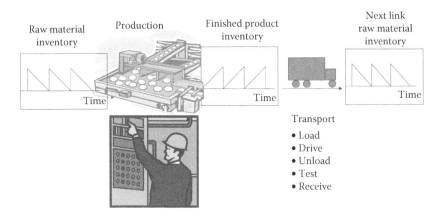

FIGURE 11.2
Inventory levels of a single link (manufacturing organization) in the supply chain for deterministic and fixed demand.

Supply Chain Management Software (SCMS) programs contain tools or modules used for transaction execution within the supply chain, for supplier relations management, and for control of relevant business processes. SCMS programs are part of an organization's information system, which is operated, maintained, and sometimes even developed by the industrial engineers.

Finally, note that in many cases, the supply chain includes a collection of the products for recycling after the consumer has used them, as well as recalling, replacing, or removing products from service.

11.2.2 Terms and Definitions

11.2.2.1 Supply Chain

The American Production and Inventory Control Society (APICS) defines a supply chain as follows:

A supply chain is a system of organizations, people, activities, information, and resources that are involved in the transfer of a product or service from a supplier to a customer. The actions across the supply chain transform natural resources, raw materials, and components into finished products that are supplied to the end consumers.

Investopedia defines a supply chain as follows:

The network created among different companies producing, handling, and/or distributing a specific product. Specifically, the supply chain encompasses the steps it takes to get a good or a service from the supplier to the customer.

The Business Dictionary (www.businessdictionary.com) defines a supply chain as follows:

Entire network of entities, directly or indirectly interlinked and interdependent in serving the same consumer or customer. It comprises vendors that supply raw material, producers who convert the material into products, warehouses that store, distribution centers that deliver to the retailers, and retailers who bring the product to the ultimate user. Supply chains underlie value chains because, without them, no producer has the ability to give customers what they want, when and where they want, at the price they want. Producers compete with each other only through their supply chains, and no degree of improvement at the producer's end can make up for the deficiencies in a supply chain which reduce the producer's ability to compete.

11.2.2.2 Supply Chain Management

The APICS defines SCM as follows:

The design, planning, execution, control, and monitoring of supply chain activities with the objective of creating net value, building a competitive infrastructure, leveraging worldwide logistics, synchronizing supply with demand, and measuring performance globally.

The Council of Supply Chain Management Professionals (CSCMP) (a worldwide association of supply chain managers from various fields and industries) defines SCM in the following manner:

SCM encompasses the planning and management of all activities involved in sourcing and procurement, conversion, and all logistics management activities. Importantly, it also includes coordination and collaboration with channel partners, which can be suppliers, intermediaries, third-party service providers, and customers. In essence, SCM integrates supply and demand management within and across companies.

Hines (2004, p. 76) provided a definition that focuses on the customer:

> Supply chain strategies require a total systems view of the links in the chain that work together efficiently to create customer satisfaction at the end point of delivery to the consumer. As a consequence, costs must be lowered throughout the chain by driving out unnecessary expenses, movements, and handling. The main focus is turned to efficiency and added value, or the end-user's perception of value. Efficiency must be increased, and bottlenecks removed. The measurement of performance focuses on total system efficiency and the equitable monetary reward distribution to those within the supply chain. The supply chain system must be responsive to customer requirements.

Lambert et al. (1998, p. 1) defined the supply chain as follows:

> The management of the supply chain is the integration of key business processes from the end customer to the initial suppliers who supply

products, services, and information that adds value to customers and to stakeholders.

Lambert (2008, p. 2) provides additional definitions of SCM:

- Flow management (back and forth, in other words includes return of merchandize) of materials, sub-assemblies, assemblies of finished products and of related information with suppliers, companies, distributors and end consumers.
- Systematic, strategic and tactical coordination of the functions of a specific company and the functions of various companies that supply to one another, for the purpose of improving long-term performance of the individual companies and of the supply chain as a whole.
- Combination of important business processes in the supply chain to add value to customers and to shareholders.

In this chapter, we will use the following definition of SCM.

A supply chain is a group of organizations directly connected in at least one back-and-forth flow of products, or services, or funds or information from a source to a customer. SCM entails setting goals and priorities, supervising the transition from the design of the supply chain to the supply chain operation and deciding how to use various resources to obtain the goals of such a chain.

Next, we define additional terms in the field of supply chains.

Supply network: In many cases, the supply chain is referred to as a supply network. In such cases, the link in the chain has many suppliers but also many different customers. For example, a manufacturer of car engine cooling pumps can manufacture the same pumps for different vehicle types and can supply its pumps to different finished products.

Lean management: An approach that primarily fosters efficient operations without spending a significant amount of money. Lean management is implemented in many organizations and minimizes delays of materials across the supply chain, as well as cutting costs and improving quality. As a result, this reduces the inventory in the various systems where it is implemented.

Supply chain events management (SCEM): SCEM considers all the events and factors that may disrupt the supply chain. It can be used to create probable scenarios and develop and implement appropriate solutions. It facilitates locating future discrepancies between planning and execution, and analyzing the reasons for such discrepancies. Big software companies (e.g., SAP) have developed software components for SCEM. SCEM programs execute and handle five major management processes: (1) monitoring the inventories and the activities in the supply chain (monitoring), (2) automatic notification and message delivery about supply chain events and situations (notification),

(3) supply chain simulation (simulation), (4) control of supply chain processes and inventory levels (control), and (5) measurement of important parameters in the supply chain (measurement).

Extended enterprise: In 1990, Chrysler suggested the term *extended enterprise* to describe its partner suppliers and companies. Chrysler realized the importance of sharing information with these companies and developed the extranet to share confidential information with the partners in the extended enterprise. Essentially, Chrysler's extended enterprise included a section of a supply chain.

Most strategic business partnership: A chain of loosely connected independent businesses that collaborate to provide a product and a service is also called an extended enterprise. An extended enterprise to some extent also operates as a team, and the extent of cooperation and information sharing affects the performance of all its members.

Thus far, we have defined the following terms: SC—supply chain; SCM—supply chain management; and supply network, Lean management; SCEM—supply chain events management; and extended enterprise.

Later in this chapter, we use terms that were defined in the preceding chapters: enterprise resource planning (ERP), electronic data interchange (EDI), and radio frequency identification (RFID).

11.2.3 Historical Development of the Term *Supply Chain Management*

11.2.3.1 Vertical Integration

Before the term supply chain was coined, it was customary to use the term *integration*. Vertical integration is defined as a situation in which an organization or corporation performs activities along the product supply chain. In other words, the production and processing of raw materials or components are carried out in a number of phases by the same organization to supply the finished product of the organization or corporation.

At the beginning of the twentieth century this concept, which preceded the notion of supply chain, was very important to developments that occurred following the industrial revolution and particularly to the invention of the assembly line. Henry Ford, for example, aspired to maintain full vertical integration of his car production operations. That is, he strived to control the entire supply chain of his car factories. The first factory he built was close to the supply centers for many of the raw materials (steel, for instance) that were used to manufacture the first cars. To gain control of supply from remote areas, Ford purchased large plots of land in the Brazilian Amazon and planted rubber trees to produce rubber for the automobile tires. Ford also purchased steel and coal mines, rubber orchards, and factories for processing the raw materials so that he could control most of the supply chain components required for assembling the cars in his factories.

The daily newspaper industry provides another example of vertical integration. The big newspapers, such as *USA Today*, integrate manufacturing and recycling of paper within their manufacturing processes. They monitor the wood prices (a raw material) and order the raw materials according to prices and demand. Newspapers print and distribute the newspapers to a complex network of regional and local distributors and collect unsold newspapers (while maintaining control of the chain).

Horizontal integration, on the other hand, refers to a situation in which an organization focuses on one process or area that may serve several supply chains—the organization or corporation specializes in a single field and becomes a major supplier in this field, as, for example, a printing house that prints different newspapers for different organizations. The content, paper, and ink are provided by the newspapers (the customers), which also do the distribution, sales, and marketing.

11.2.3.2 Creation of the Area of Supply Chains

In the beginning of the 1980s, the success of Japanese competition in the electronics and the automotive market brought to the West the adoption of Japanese JIT philosophies. The Japanese approach was to reduce inventories and foster long-term supply contracts and set partnerships between suppliers and customers. At that time also the modules of inventory management for an individual organization matured, and interest arose in management of inventory and transportation across the supply chain. There are several versions to how the term supply chain came into existence. Several sources quote the consultant Keith Oliver from Booz Allen Hamilton as the person who first coined the term SCM. The popular source of this assumption is Oliver and Webber (1982). The supply chain constitutes a chain of organizations in which supplier and customer are interdependent in the production of finished products. Historically, some of the characteristics of supply chains were mentioned in the beer game developed by MIT's Jay Forrester in the 1960s (Forrester, 1961). The beer game is still used to illustrate supply chain characteristics. However, the game was developed in the context of studying system behavior, and nearly two decades went by until a connection was made to the modern term of SCM.

When SCM was introduced, the importance of sharing information between distant sectors of the chain became apparent. In addition, factors causing a major waste of resources were discovered. The whip phenomenon (to be expanded on later in this chapter) is one of them. Evidence of these factors can be found in books that date back to that period—for example, Stevens (1989).

11.2.3.3 Information Integration Era (Present Era)

Before the 1990s, organizations did not share inventory information with other organizations. The information integration phase is the phase in which information begins to flow across the chain and various nonadjacent links in the chain are coordinated. This phase is characterized by integration of information between various organizations along the same chain.

An initial technical basis for such a transition emerged in the 1960s with the development of Electronic Data Interchange (EDI) systems. An additional basis for reliable, available, and consistent information emerged on the development of enterprise resource planning (ERP) systems. Information integration developed throughout the 1990s, mainly in the transport and collaboration categories. This topic progressed from standard distribution of information (approximately from 1998 to 2003) to bridging protocols that allow different computer systems to "talk" with each other and get information on demand (around 2003 through 2006), supplying the information on the Internet using servers and service distribution of information was the next step (roughly until 2012, there was no better technology) and ultimately developed into the provision of data, information, and even software as part of cloud services. This era has continued to develop throughout the twenty-first century together with the expansion of collaborative systems that are based on the Internet. This era in the development of the information integration is characterized by the fact that the challenges of transferring the data and extracting information from it are less technical and are more related to trust between remote links in the chain, and the will to spend time and effort on monitoring the supply chain at each link.

11.2.3.4 Supply Chain Maturity

The maturity of the supply chain can be ranked. We distinguish between three levels: level 1, level 2, and level 3.

1. Level 1 supply chains include manufacturing, storage, distribution, and material control systems that are not interconnected and that operate independently. These kinds of systems are common in large-scope operations in which the factory, the warehouse, and the distributors all operate separately with a limited interface. As time progresses, these systems are expected to become less common.

2. Level 2 supply chains include manufacturing, storage, distribution, and material control systems that are integrated under one program that is connected to the ERP systems of various organizations along the chain.

3. Level 3 supply chains are fully integrated with information from suppliers and customers. Tesco, the world's second largest merchandise retailer, is an example of such a supply chain. The company has over 6000 stores worldwide. The information systems in these stores are directly connected to the various suppliers, who get real-time updates about the in-store products and thus are prepared to manufacture and supply products accordingly.

11.2.3.5 Globalization Era (Continuing to the Present Day)

The globalization era is the next phase in the development of SCM. It followed the onset of the information integration era, though developed simultaneously. One of the globalization era's characteristics is the attention paid to global supplier relation systems and to the expansion of the supply chains beyond national borders and into other continents. While having a nearby supplier is an advantage, the best supplier may be located overseas. The success of the supply chain is related to the selection of the best team of suppliers, and this sometimes means choosing a remote supplier. Even though global sources were used by organizations' supply chains several decades prior to the globalization era (e.g., in the oil industry), it was only in the late 1980s that a significant number of organizations began to integrate global sources in their core business. This intensified in the following decades and was manifested by the heightened importance of international standards such as the ISO-9001. This era is characterized by the globalization of SCM in organizations that aim to increase their advantages over the competition through global outsourcing.

11.2.3.6 Specialization in Core Fields and SCM Outsourcing

Some organizations preferred to handle supply chain matters themselves. In the 1990s, however, organizations began to transfer part of the supply chains' operations (including inventory, transport, and information) to expert organizations. These expert organizations handled inventory, transport, and information. This trend can be divided into two levels.

1. Level 1 Specialization: Outsourcing Transportation and Distribution

In the 1990s, organizations began to specialize and focus on *core competencies*. They abandoned lateral integration, sold activities that were outside the scope of their core business (such as distribution and transport), and outsourced these functions to other companies. For example, packages could be shipped via companies like FedEx and UPS, while larger shipments could use less than truckload services of companies like Viking or Schneider. These maneuvers changed the management requirements by expanding the supply chain beyond its original organizations, and by spreading the management of interface and logistics across partnerships that specialize in supply chains. Furthermore, this also refocused the essential perspectives of each organization.

This specialization creates transportation and distribution networks that serve many individual supply chains dedicated to manufacturers, suppliers, and customers. Each supply chain operates jointly to design, manufacture, distribute, market, sell, and provide services for a product. Such a partnership alliance between manufacturer and carrier may change in accordance with the market, the region, or the channel, leading to multiple trade partner environments, each with its own unique characteristics and requirements.

2. Level 2 Outsourcing: SCM as a Service

Specialization in providing logistical services began in the early twentieth century, together with the emergence of transportation agencies and companies (e.g., UPS) that were in the business of warehouse management and transportation services. Yet the major developments in this field occurred once the importance of the supply chain was recognized. The 1980s and 1990s witnessed the development of organizations (Manugistics, i2, Optum, Escalate, E3, Logility, etc.) that offered detailed coordination and efficiency in the provision of logistic services, including supply planning, collaboration, chain operation, and maintenance as well as performance monitoring.

Service in planning, establishing, and operating supply chains developed as a result of market forces that sometimes required quick changes from suppliers, logistics suppliers, sites, or customers in the supply chain.

Similar to the outcome of purchase and distribution outsourcing, experts who provide consultancy in the field of supply chains enable companies to improve their overall capabilities. This type of consultancy enables companies to focus on their core competencies and to assemble a network of specific partners who excel in their field and who will increase the chain's overall value, and contribute to performance and efficiency in general. The ability to obtain, and quickly use, the expertise of a particular supply chain, without developing and maintaining a unique and complex capability in the organization itself is the primary reason that supply chain expert consultancy services are gaining momentum.

11.2.3.7 SCM in Shared Platforms and Portals

The next phase entailed the creation of collaborative platforms that connect buyers and suppliers across the supply chain. Sometimes these platforms are also connected to financial institutions to enable organizations along the chain to carry out automatic financial transactions. An example of one of these platforms is TradeCard and their Nipendo product—a supply chain collaborative platform that connects many buyers and suppliers to financial institutions.

In the context of collaborative platforms, it is important to note the term *SCM 2.0*. Similar to Web 2.0, this term was coined to describe a collaborative computerized framework (such as collaborative portals) in which suppliers and their customers operate on the supply chain. Just like the Internet developed social networks and tools to service these networks, SCM 2.0 also developed processes, methods, and tools to collaboratively manage supply chains in this new "era."

A collaborative portal or platform increases information sharing, creativity, and collaboration between the portal users. It typically allows forming a subportal for participants of a certain supply chain, where their internal information is shared and procurement transactions are enabled. Sharing the information on order quantities, levels of work in process, and finished products helps the other supply chain participants to assess their future demand and predict shortages. While allowing sharing sensitive internal information, the portal also allows advertising in the general part of the portal to attract new customers. These portals assist the users in navigating the abundance of available online information in order to find the desired information. The advantage of finding the desired information is achieved by a combination of processes, methodologies, tools, and supply opportunities that guide companies in meeting their goals. At the same time, the complexity and speed required from the supply chain increase due to global competition, rapid price fluctuations, short product life span, expanded expertise, and simultaneous deliveries to close, far, and international destinations. As an example, *supply chain analysis* is one of the portals that enable this type of activity. This portal is a one-stop shop for collaborative trade, supply chain information sharing, and SCM expert consultancy. SCM 2.0's improved capabilities and facilitated forecasting changes in demand and responding faster and more efficiently than in the past.

11.3 Supply Chain Characteristics

11.3.1 Major Roles of Supply Chain Organizations

Supply chains consist of organizations (and sometimes organizational units) that constitute the links in the chain. In this section, the organizations are classified according to the typical supply chain roles. These roles have a major impact on the manner in which organizations operate and on the extent to which they influence and are influenced by the supply chain. The list of the main types of organizations in the supply chain is described next.

1. *Manufacturers of raw materials:* Raw materials such as oil, steel, or wood are usually manufactured in large quantities and at a relatively constant pace. Organizations that manufacture raw materials normally guarantee price stability by purchasing or selling options (securities for guaranteeing the sale price of raw materials or finished goods) as their work pace cannot change very easily. Many raw materials are used in dozens, hundreds, and sometimes thousands of different products as well as in many supply chains (which have different finished products). Thus, changes in the demand of individual products do not necessarily impact the manufacturers of raw materials to the same extent they would impact other links in the effected chain.

2. *Component manufacturers:* A component is an item that is meant to be part of a system but has no use or function by itself. Component manufacturers specialize in the initial phases of transitioning the raw material into becoming part of a subassembly. These manufacturers typically produce a large variety of components. For example, the producer of bulbs and LED lighting for cars usually supplies its components to a large variety of car manufacturers having a variety of models each with different lighting requirements. Another example is a manufacturer of steel screws and drill bits. This manufacturer produces a large variety of different screws and drill bits that can serve in many assembly processes and be used in many drilling machines. These components are sometimes what make the difference between different models of the same product, and naturally they differ from one another in different products. Collaboration between various supply chains (that have different finished products) also characterizes component manufacturers.

3. *Manufacturers of subassemblies and assemblies:* A subassembly is a system (typically with certain functions) made up of components or other subassemblies. Some examples of subassemblies in a car are: (1) car seat, (2) engine, (3) gear box, (4) water pump, (5) alternator, (6) starter, and (7) radio-CD unit. To be useful, a subassembly must be assembled into a product or a larger subassembly. Manufacturers of subassemblies and assemblies are closer to the finished product phases. Therefore, these organizations belong to a smaller number of supply chains. The manufacturers of subassemblies and assemblies are affected by fluctuations in demand for the finished products in which the ensembles they manufacture are installed.

4. *Intermediary suppliers:* Organizations that specialize mainly in trading and supplying components, subassemblies, and assemblies are the intermediary suppliers. The intermediary suppliers' capability to contact many suppliers and customers facilitates efficient utilization of the market and of the existing supply chains. Their mere existence reduces complexity and cost in maintaining contact with many suppliers, and pooling inventory in one place.

5. *Finished product manufacturers (final assembly):* These are usually organizations engaged in assembling and packaging the finished product. When the products are very large and complex (such as planes and cars), some significant manufacturing processes are integrated into the assembly process. For example, welding and painting processes are part of all car assembly lines. Since the finished products have to be delivered directly to the customers, manufacturers usually make sure to carry out extensive and meticulous quality assurance testing of the products released for distribution. Product

distribution activities are usually completely separate from the manufacturing process and sometimes include several stages.

6. *Distributors:* Typically organizations that are spread out across widespread areas and transfer large quantities of products from manufacturers to local agencies, wholesalers, and sometimes retailers are called distributors. The distributors in the supply chain are in constant contact with specific manufacturers of finished products, and their cooperation and information sharing is extremely important for maintaining low costs. The only feedback manufacturers get on the demand levels, trends, and its mix is through the distributors. At the same time, contact with sales agencies, wholesalers, and retailers is an integral part of the distributors' operational management.

7. *Sale agencies:* Local organizations that constitute part of the distribution network of certain products are called sale agencies. They distribute the supply chain's products to wholesalers or retailers and sometimes even to customers in the region in which they operate.

8. *Wholesalers:* Large quantities of products, mainly to retailers but sometimes also to particularly large customers, are sold by wholesalers.

9. *Retailers:* Organizations that sell the products to the end user only are known as retailers. Their specialty is selling products and making them accessible to end users.

10. *End users:* In the context of consumer products, the end users are private individuals. For other types of products, they can be organizations. For example, a company that produces metal processing machines such as milling machines, laths, and drills would have organizations that process steel parts as their customers.

   The types of organizations in the supply chain of a complex product are schematically described in Figure 11.3.

   Every stage that entails a transition between the links in the supply chain (each arrow in Figure 11.3) involves

   a. Inventory of finished products for transport
   b. Transport
   c. Product reception (registration, quality assurance)
   d. Post-reception storage of raw materials

Manufacturing chains are typically characterized by an increase in batch sizes as the chain gets farther from the end user and closer to raw material manufacture. This is expected since transport costs (fixed costs of order) are higher for links that are far from the customer (e.g., transporting raw materials by sea or air is usually more expensive than transporting a finished

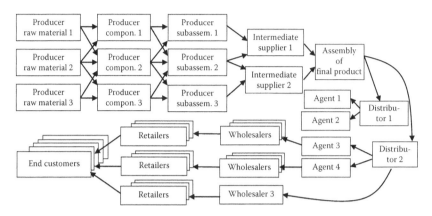

FIGURE 11.3
Schematic description of types of organizations in the supply chain of a complex product.

product from the warehouse to the point of sales). In Chapter 6, we discussed the economic order quantity formula, which demonstrates that when the fixed costs of an order increases so does the optimal batch size. Figure 11.4 illustrates this by describing the inventory levels at a local distribution center compared to those at a national warehouse.

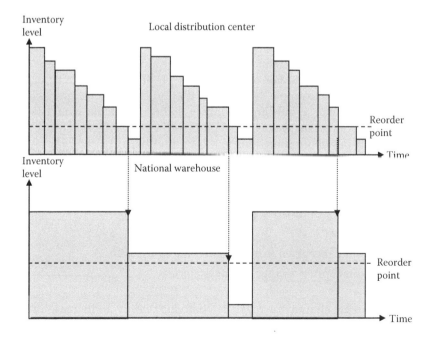

FIGURE 11.4
Illustration of differences in unit sizes between a link that is closer to the customer (local distribution center) and a link that is farther away from the customer (national warehouse).

A mistake in forecasting the increase in batch size causes more damage to links farther away from the customer, as these links order larger batches even when not required. This characteristic also contributes to the bullwhip effect described in the next section.

Intermediary suppliers, component manufacturers, and manufacturers of subassemblies and assemblies can sometimes be part of several supply chains. Suppliers that are part of various supply chains supply components for assembling many products for various companies. The market structure of component manufacturers, and manufacturers of subassemblies and assemblies, and intermediary suppliers does not necessarily mandate an increase in batch sizes in the transition from one stage to another, though such increases are quite common. Raw materials, on the other hand, are typically manufactured in large batch sizes and sometimes even continuously.

Since the end users are a major source of uncertainty, it is likely that the biggest inventory fluctuations will be close to the end users. Yet, in practice, this is not the case, as explained in the following section.

11.3.2 Bullwhip Effect

The bullwhip effect was described in the professional literature as early as 1961 (Forrester, 1961). This term refers to the inventory levels in the supply chain through an analogy to a bullwhip's wavy shape: the waves at the end of the bullwhip are small, and they become bigger as they approach the base of the whip (the handle), as illustrated in Figure 11.5.

Empirical studies have found that inventory levels across the supply chain act similarly: small fluctuations of end users' demands are expressed as larger fluctuations that grow and become larger along the chain, as they approach the raw material manufacturers.

The inventory bullwhip effect is described in Figure 11.6.

FIGURE 11.5
Illustration of wave-spreading pattern of a bullwhip.

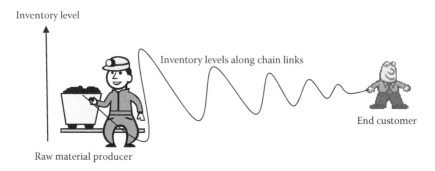

Inventory level

Inventory levels along chain links

End customer

Raw material producer

FIGURE 11.6
The bullwhip effect in the supply chain.

The end user is the source of uncertainty. Therefore, it is not intuitive that inventory fluctuations become larger as the chain links are farther from the source of uncertainty (the end users), and it was only in the 1960s that this behavior was recognized as typical of supply chains.

There are real-life examples of the bullwhip effect in supply chains in a variety of fields, among them are the automotive industry (Thun and Hoenig, 2011; Hasan et al., 2013) and laptop computers (Zhang, 2004). In fact, the bull-whip effect is common in products that undergo many different processing stages performed by various organizations (or products that consist of many different components). On the other hand, the bullwhip effect may also be present in consumer products (e.g., in the fashion and food industries) that tend to be on sale and that create fluctuations in demand.

The beer distribution game developed at MIT was one of the contribu-tors to the widespread recognition of the bullwhip effect (Khalifa, 2012). Illustrations of the beer distribution game are depicted on many web sites. Dynamic examples of the game can be found on YouTube by searching the terms *beer game case study* or *beer game simulation*.

The beer distribution game was developed to encourage a systematic thought process and to cope with fluctuating demand. The game has four stages, each of which is a link in the supply chain of a particular kind of beer: (1) brewery, (2) beer distributor, (3) wholesaler, and (4) retailer.

Each player in the beer game is assigned to one of four links in the supply chain. At each link, the player needs to decide how much beer to order for the upcoming period. The rules of the game deliberately forbid the play-ers from communicating with one another in order to show what happens when information is not shared throughout the chain. The game progresses 1 week at a time. During each week, supplies are delivered by the suppli-ers, orders are made by the customers, shipments are sent from inventory to the customers, and anything that is not supplied is recorded in a list of inventory shortages. The cost of shortages is double the cost of inventory carrying, and these shortages are replenished in the future according to orders received. The objective is to achieve maximum profits. In the beer distribution game, the demand forecast for all the links in the chain is the moving average of the last 4 weeks, and there is a pushback of supply time between the four links in the chain. Figure 11.7 describes the stages in the beer distribution game.

The many delays between the links in the chain and the time it takes the forecasts to recognize a change in demand cause decision making to be based on past events. Furthermore, an accumulation of backorders causes increases in costs and pressure to overorder. All these are causes for the bull-whip effect. Let us suppose, for example, that in October the demand for beer increased (due to Oktoberfest). However, the retailers and the rest of the chain are unaware of this. The result is higher forecasts by retailers that reflect the increase in demand. Due to these forecasts, retailers order bigger shipments than required in the months following October. This causes wholesalers also to see an increase in demand forecasts and thus they order larger quantities. However, retailers ship in crates or small pallets, whereas wholesalers ship larger batches (usually full-sized trailer trucks). Therefore, the retailer's inventory fluctuations (crate quantity) are smaller than those of the wholesaler (truck quantity). This is an early indication of the bull-whip effect. Later, a distributor that sends trucks to wholesalers witnesses an increase in demand. Therefore, its forecasting increases accordingly, and it orders an even larger quantity of beer from the brewer for its warehouses

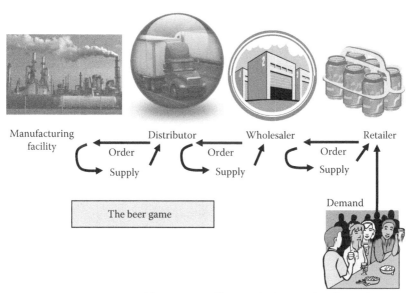

The customers: The game imitates random fluctuations

FIGURE 11.7
Major factors in the beer distribution game.

to meet the future demand for fully loaded trucks. Yet the actual demand of end users decreases in the following months (in contrast to the incorrect and inflated forecast based on October) and this causes the chain to have a large inventory surplus, even though in the weeks following October, the demand in every link of the chain is lower than the forecast. In this state of affairs, following November (after more than 4 weeks later), the forecast at all levels of the chain indicates a significant decrease in demand (since demand is met from existing inventory) and the orders are therefore postponed. However, at this point, demand levels are coming back to their regular levels and for a limited period inventory level reduces until it is completely depleted. At which point a shortage kicks in at all chain links until at a certain point the forecast for increased demand is updated. These cycles bring about a wave of demand throughout the chain, after which there will be a wave of shortage, and so forth.

Since there is a delay at every stage of the supply chain, the bullwhip effect propagates over time, as described in Figure 11.8. Therefore, for a period of time, the inventory levels in the various stages of the chain indeed are similar to the dynamic waves in the bullwhip.

Various studies have found many reasons for the bullwhip effect. The following are 10 major factors for the bullwhip phenomenon according to a literature survey on this topic (Bhattacharya and Bandyopadhyay, 2011):

1. The diverse methods of forecasting have been found to be a major cause of the bullwhip effect (Lee et al., 1997). The choice of parameters in forecasting models also affects the severity of the bullwhip effect.

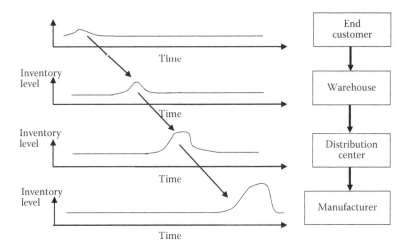

FIGURE 11.8
Schematic example: Demand fluctuation impact on the supply chain over time.

2. Order and batching policy was also found to be an important factor in the bullwhip effect. The bigger the batch size, the bigger the waves of the bullwhip effect.

3. Fluctuations in product pricing contribute and even cause the bullwhip effect (while forecasts are based on fixed selling price, in case of a discount–demand increases when price decreases, and then all forecasts and plans are updated based on a temporary demand level that completely changes when the price goes back to its normal level).

4. Demand manipulations are actions like discount sale pricing, limiting quantity per customer (to prevent shortage), and motivating customers to postpone purchases and the like. These attempts also cause erroneous purchasing policy, which feeds the bullwhip effect.

5. Supply lead time duration is also a cause of the bullwhip effect (the bullwhip effect intensifies even more when supply lead time is ignored).

6. Lack of transparency and lack of knowledge about demand between the various chain stages contribute to the bullwhip effect.

7. The greater the number of links in the chain, the greater is the bullwhip effect and the more it is problematic.

8. Local optimization for each link in the chain, instead of for the whole chain, creates solutions that become worse throughout the chain as they get farther away from the customer (Moyaux, 2007).

9. Fear of product shortage (Croson and Donohue, 2009).

10. Lack of learning and training (Yan and Katok, 2006).

11.3.3 Advantage of Integrated Chain Management: Cost Reduction and Bullwhip Effect Eradication

Integrated SCM entails full disclosure of demand and inventory levels throughout the chain. It also includes coordination in advance between the links in the chain as well as coordinating preparation for changes in demand. One of the reasons for improved performance is increased sharing of information, which is the basis for decisions in the various chain links. Integrated management contributes to eradication of the bullwhip effect.

The advantage of integrated SCM is best defined as *synergy*, a situation in which the outcome of a team of people or organizations working together for achieving the same goals is better than the outcome resulting from the combined yet separate efforts of all the entities involved. Treating the other supply chain organizations as partners contributes to better overall performance of the chain much more than separate and optimal management of each link in the chain. This is a win-win situation.

On the other hand, when each link operates separately, the master production schedule (MPS) is based on the information available to that single link. There are two plausible types of errors that occur when a link bases its production plan on forecasts:

- Shortage due to forecast lower than actual demand.
- Inventory surplus due to forecast higher than actual demand.

We now discuss the effect of integrated SCM on the 10 bullwhip factors listed above:

1. Integrated SCM based on information sharing will cause each link in the chain to base its MPS on the production, purchase, and sales plans of the subsequent links in the chain. As a result, the two forecast mistakes will be avoided because the MPS will be based on actual demand (of the next link) rather than forecasts. As previously explained, inventory fluctuations caused by forecasting errors contribute to inventory increasing and decreasing as a function of time, and thus generate the bullwhip effect (due to the resemblance to the ripples in a bullwhip). Studies have indicated that information sharing between the links in the supply chain considerably reduces the extent of the bullwhip effect, and naturally reduces inventory costs along the supply chain—a reduction that benefits all links in the chain.

2. Integration between the links in the chain usually facilitates reduction in batch sizes and, therefore, the reduction of the intensity of the bullwhip effect.

3. Integration between the links in the chain allows for better understanding of fluctuations in product price and in product components, and their impact on demand, thus reducing the intensity of the bullwhip effect.

4. Integrated SCM enables the chain's organizations to coordinate and understand demand fluctuations when demand manipulations are attempted, such as discount sales, limitation per customer quantity (to prevent shortage), motivating customers to postpone purchases, and the like. This understanding prevents overordering and underordering.

5. Integration between the organizations facilitates inventory supply synchronization throughout the chain (supply time is not ignored) and sometimes even allows the shortening of supply times.

6. Integration contributes to better transparency and to knowledge about demands between the various stages of the chain, thus reducing the intensity of the bullwhip effect.

7. Integration does not impact the number of links in the chain. However, it does contribute to better coordination between each pair of links in the chain and, therefore, the impact of the number of links on the bullwhip effect is smaller.

8. Integrated SCM makes it possible to develop a global view rather than operating according to local optimization for each link in the chain.

9. Integration means information sharing that reduces the odds of product shortage.

10. When SCM is integrated, sharing conclusions and learning from experience are more likely.

In summary, integrated SCM contributes to better coordination between the links in the chain, better understanding of changes in price and demand and, as a result, diminishment of the bullwhip effect and, ultimately, reduction of inventory and transport costs.

11.4 Major Characteristics and Considerations in Supply Chain Planning

Supply chain planning provides answers to many questions:

- *What should be the nature and characteristics of supply chain information systems?* In previous chapters, we discussed various types of information systems. However, supply chains are challenging in that they require the sharing of information between partner organizations, while at the same time ensuring this information vis-à-vis the outside world. The interorganizational interface is therefore more important in supply chain planning, and sometimes the information system is shared by several organizations. At the beginning of the chapter, we mentioned SCM 2.0 portals, and later we thoroughly discuss the challenges of SCM information systems.

- *How to determine and implement information sources, information quality, and information sharing?* Information technology has developed significantly in recent decades. For example, most one-dimensional barcode-based systems have transitioned to two-dimensional barcodes that contain a great deal of information and are much more reliable. Currently, many organizations use radio frequency identification (RFID), voice recognition, photo identification, and processing technologies. The level of information sharing is also very important in supply chain planning. The better the quality of information

available to the links in the chain (information quality was discussed in previous chapters), the better the decisions based on it will
be. However, the various links in the chain (the supply chain partner
organizations) are concerned about disclosing information that may
be leaked to competitors and hamper their ability to compete.

- *What to manufacture and what to purchase in each link of the chain?* This is
  the *make* or *buy* question described in Chapter 6 on purchasing. This
  is a major question in supply chain planning, and the decision of each
  link in the chain whether to purchase certain products and services
  constitutes the building block of the supply chain. Purchasing goods
  and services for the reasons we listed in Chapter 6 is an important
  component in the activities of many organizations that supply goods
  and services, and supply chain planning entails finding the right
  answer to the question of *what to purchase.*

- *What are the components of the chain (or what are the links in the chain)?*
  This is an elaboration of the *make* or *buy* question discussed in
  Chapter 6 on purchasing. In planning the supply chain, it is important to understand not only what goods and services each organization will purchase, but also who are the suppliers' suppliers (chosen
  by the supplier organization). The answer to this question is very
  significant from the perspective of the single link, but it is also
  important for the chain as a whole. In this context, a supplier that
  independently purchases from unreliable, unstable sources or from
  sources that have low-quality products consequently becomes an
  unreliable, unstable source itself that provides low-quality products.

- *What is the structure of engagement with suppliers?* There are a variety of engagement types, for example, one-time purchase orders as
  opposed to long-term purchase contracts (short-term engagements
  versus long-term engagements). Another example is inventory purchase as opposed to JIT purchase. A correct choice of engagement
  structure will improve the supply chain's performance.

These considerations make SCM and planning a very complex subject that
concerns many industrial engineers. Therefore, this subject is part of the
study programs in industrial engineering schools.

11.5 Types of Contracts and Engagements

Selecting the right supplier engagement (and subsequent contract) is very
important in SCM. A long-term contract with a supplier, for instance, automatically adds the supplier as a link in the product's supply chain (and

during the period of engagement, the supplier becomes part of the finished product's manufacturing team). Short-term contracts facilitate intense competition and flexibility in choosing suppliers; however, the supplier's commitment to the purchaser is not as strong as in its long-term contracts.

After making a decision to purchase goods or services, the next step is to decide with which supplier to enter into an engagement and according to what engagement structure. In an era of global competition, the abundance of suppliers in every field and the wide variety of types of engagements offer great flexibility in this choice, and it is important to understand the advantages and disadvantages of each type of engagement in order to make the right decision. Normally, for the engagement to be worthwhile for both the supplier and the customer, it must be advantageous to both parties. The literature includes numerous articles discussing these considerations in the context of game theory that involve finding effective engagement structures for a variety of situations.

In the following discussion, we examine the engagement period (short term versus long term) and the type of engagement (contract subject matter and the way it is managed during the engagement).

11.5.1 Short-Term Engagements

The simplest short-term engagement is the commercial off-the-shelf (COTS) engagement. For example, a company urgently needs standard items such as light bulbs for its offices. Therefore, the company sends a buyer to the supplier's store to purchase these items. The buyer pays for the items with a credit card or with a company check, and the supplier issues an invoice for the purchased items. The buyer receives the items at the place of purchase and the transaction is thus completed. Clearly, this type of purchase characterized by short supply time and great flexibility is appropriate for small organizations such as law or accounting firms. Note, however, that the items purchased are standard and are usually offered to the end customer. Moreover, the offices can buy a stock of bulbs sufficient for the next year or two and are not consuming the bulbs as part of their activities. Clearly, single purchase is not an effective solution for a manufacturer that purchases from the same supplier a wide variety of items in large quantities on a periodic basis. Moreover, bringing items from overseas or ordering unique items that are manufactured specially for the purchaser requires special attention, and if done on a periodic basis, should not be done as single transactions.

There are additional types of short term engagements, primarily for purchasing standard inventory items and nonunique services. Thus, for example, many companies use the Internet to electronically purchase standard COTS items (business-to-business purchases—B2B). These purchases are carried out through suppliers' web sites that advertise standard product catalogues including prices and supply terms. These can be automatic purchases through a smart purchasing system that is able to search for standard items, compare prices and supply terms, and even issue a purchase order and pay on receipt of the goods according to the agreed payment terms. In most cases, this type of purchase involves human beings and the information systems mostly provide support, but the purchase is not automated in its entirety.

A tender is issued to the suppliers in the case of a need for outsourcing a project, for purchasing nonstandard items or items that are not kept in inventory, or when the law requires. The tender is issued publicizing a request for proposals (RFP). The RFP defines the required goods, the place of purchase as well as the logistic system that will be used, for example, shipping by air or by sea. The RFP is addressed to a selected group of suppliers that submit proposals to supply the goods if they decide to participate in the tender. Alternatively, the RFP is publicized on online trade platforms, and any supplier that meets the RFP's terms can submit a proposal. After the proposals are received, they are examined according to criteria such as price, supply time, and supplier reliability. Following the selection of a supplier, a contract is signed or a purchase order is issued, which serves as the basis for supplying the goods. The contract or purchase order defines the terms of payment, as well as the guarantees each party to the transaction is to provide, if needed, for example, a bank letter of credit.

Short-term engagements are advantageous in that they facilitate competition between the participants and therefore contribute to price reduction and improved payment terms. Additionally, they allow for flexibility in purchasing the exact items that are required in the required quantity and at the right time. The biggest disadvantage of short-term engagements is that the supplier and the customer do not form a long-term relationship and there is only a one-time purchase commitment. Consequently, when certain merchandise is in low supply, an organization that operates via short-term engagements does not receive preference from suppliers who are not as committed to the customer as they would have been in the context of long-term engagements. Another disadvantage is the difficulty in guaranteeing high-quality products as well as supply reliability. Occasionally, suppliers attempt to win RFPs by lowering their profit margin to a minimum, sometimes by compromising quality and by failing to meet schedules or required quantities. To avoid such issues, it is customary to perform acceptance tests on the received goods, which makes the process slower and more expensive. In addition, it is customary to store inventory as a precaution in case of supply delays. In Chapters 6 and 8, we discussed the subsequent inventory costs.

11.5.2 Long-Term Engagements

Long-term engagements facilitate an extended relationship between supplier and customer. This is often seen as a strategic partnership. The engagement may include many details, such as the quantity to be supplied throughout the entire engagement, the agreed price, as well as the agreed terms of supply and payment. A long-term engagement may also be based on *a framework contract* to purchase a certain quantity over a long period of time, at dates and quantities to be defined by the customer according to its needs. In such a case, the customer reserves the right to change the supply dates and quantities, as needed. However, the customer enters into a long-term engagement with the supplier, and therefore, together with the supplier, can define the manufacturing system, the quality control system, and the like. This enables the supplier to be authorized to supply directly to the customer's assembly line or factory, without needing acceptance tests. The JIT system we discussed in Chapters 6 and 7 is based on long-term agreements with major suppliers that use the kanban cards of the customer's factory to manage their production. In such cases, many suppliers build factories next to the customer's assembly factory and thus are able to supply directly to the assembly line several times a day, as dictated by the kanban cards.

In many cases, long-term engagement with a supplier makes the supplier a strategic partner. The greatest advantage of long-term engagements lies in their ability to create better trust and work relationships between supplier and customer, as well the supplier's commitment to the customer. For example, an authorized supplier that supplies the required quantities at the required dates directly to the manufacturing or assembly lines without acceptance tests is a strategic asset with which the customer can share information and trust when there are difficulties and shortages. It is only reasonable that such a supplier would give preference to a long-term engagement type of customer in the case of market shortages or supply difficulties.

11.5.3 Management of Supply Contracts

There are uncertainty factors in every supply chain. In previous chapters, we discussed demand uncertainty and demand forecasting methods. There is uncertainty in manufacturing as well. Manufacturing resources malfunction or become unavailable for various reasons, such as workers' sick days, strikes, and the like. As a safeguard against uncertainty, it is important to draft a supply agreement or a contract that considers uncertainty factors, their impact on each party to the contract, and each party's ability to cope with uncertainty. For example, it is customary to use the services of commercial banks in order to guarantee that when the merchandise is supplied, the customer will not avoid payment. One alternative is for the customer to provide a letter of credit in which its bank undertakes to pay for merchandise on receipt of confirmation from the supplier, such as release from customs or the port in the customer's country. Insurance companies can also provide protection against certain types of uncertainty, such as fire, theft, and water damages in sea transport. It is very important to prepare a contract that properly represents the parties' goals and constraints, as well as their ability to absorb certain risks.

In addition to safety stock level and other safety moves decided by experience, it is important to evaluate risks in advance and to be protected against them in the context of the supply contract. Also, such a contract should include a mechanism that regulates changes during the engagement. For example, in a long-term agreement, the buyer may want flexibility in supply quantities and dates by inserting a clause that determines when changes may be made, as well as the scope of the changes.

The contract describes the agreements reached by the parties. However, due to uncertainty, it is important to include mechanisms that will allow flexibility due to changes in the business environment. Naturally the changes need to be agreed upon by both the supplier and the customer.

11.6 Information and Its Importance

11.6.1 Value of Information in Supply Chains

In Chapter 4, which discusses information and its use, we saw the great importance of information systems that include high-quality databases and appropriate model base. The SCM literature indicates that information about inventory status in the supply chain and information about the planning of inventory use is very valuable (McClellan, 2002; Olson, 2012). When such information is available, a more efficient and stable flow of inventory and product may be planned.

In addition to information about the inventory in the organization (raw materials, products in progress, finished products), monitoring inventory at other locations of the chain is also very important. This type of information can eradicate the bullwhip effect, a phenomenon that causes high inventory costs and unjustifiable shortages. Inventory accumulated at retailers, for instance, can indicate a slowdown in sales throughout the supply chain. Another example: information about an assembly malfunction can indicate that order delays and large quantities of component inventory at the suppliers constitute only a temporary problem.

Let us now examine inventory information management. Even though the classical literature (Nahmias, 2012) refers to registered inventory as actual inventory, in reality this is frequently not the case. There are gaps between the inventory registered in the information systems and the actual inventory (Lambert, 2008). These gaps derive from three main sources: inventory shrinkage (theft or total product damage), incorrect product location (due to stockroom manager mistakes or customers who mistakenly shift a product), and incorrect product scanning (due to stockroom manager mistakes on receipt by the retailer or cashier's mistake at point of sale). The outcome of such mistakes is a discrepancy between the inventory levels in the organization's information systems and the actual inventory available for sale. These mistakes make it impossible to efficiently and reliably draw up a purchase plan based solely on information systems. The potential effects of these types of mistakes are typically large inventory or shortages that harm profitability.

One method of confronting these gaps is to improve the quality of inventory data through technology. For example, if dirt is causing many errors during barcode scanning, information can be improved by replacing the traditional barcode with passive radio tags (RFID) that monitor items. A tag is attached to each item that contains a miniature structure that sends a signal to the tag and reads its response. The tag transmits the relevant information when it passes near antennae that are usually located near (or on) reading gates.

11.6.2 Information Technology in Supply Chains

We shall start the discussion with a short summary of the single-link main information technologies used by industrial engineers, which were described in Chapters 4, 8, and 9. These technologies include (1) information collection systems, (2) databases, and (3) analytical programs.

The information collection systems translate physical, financial, and economic transactions into information entered into the system (transactions management system). The transactions management system receives the transactions from various sources and stores them in the database. In the past, information collection was based mainly on manual data entry that is very costly and causes many errors and low-information quality. In recent years, information systems have undergone a technological revolution that has considerably reduced the amount of information entered manually and significantly increased the use of technology for automatic transaction collection—barcode readers and RFIDs, for example. The use of two-dimensional barcodes has also increased. The advantage of using these technological means is twofold—the cost of use is low and the quality of information is high.

Databases have also undergone significant technological development. The magnetic tape that considerably slowed down the work of databases and limited the quantity of information that can be stored in the system has been replaced by optical and electronic means that can store massive amounts of data in a relatively small volume and retrieve this data very quickly. In order to protect this information and keep it readily accessible, the use of cloud services, which store data at several server farms around the world and keep it available online, has expanded. Another feature of cloud services is that they can usually protect data by replicating it on various servers worldwide.

Analytical programs in the model base analyze data from the databases to plan, make decisions, and monitor developments. These programs include features such as forecasting, scheduling, purchase planning, and transport.

The software used in SCM has also developed considerably over the years. In Chapters 8 and 9, we saw the transition from systems designed for specific functions in the organization (legacy systems) to integrative systems such as material requirements planning (MRP) systems as well as ERP systems. In addition to information collection, transaction processing, and databases, these systems also include a model base that features models that support: routine decision making, ad hoc decisions, and decisions associated with control, supervision and workflow management. These models facilitate definition, performance, and monitoring of various processes in the organization. Recently, this field has also transitioned into cloud services, and many companies now subscribe to software services provided by cloud companies. This type of subscription-based software cloud service is called *software as a service*, and the providers of the service are called *application service providers (ASPs)*. The biggest advantage of ASPs is their expertise and competitiveness, which contribute to constant version updates and maintenance of software services. For example, a software version update is the sole responsibility of the ASP, and as such is transparent to the user.

A 2003 study (Helo and Szekely, 2005) about supply chain information technology conducted in the Canadian province of Ontario, home to numerous governmental organizations, found that:

1. Information standardization is becoming more important and even essential.
2. There is a growing trend toward sharing of large volume information among various companies in the supply chain.
3. Internet-based platforms are replacing other types of architecture.

Helo and Szekely (2005) distinguish between transaction-based programs, analytical programs, and supply chains along six dimensions, as summarized in Table 11.1.

The next section discusses the use of the information that flows through and is stored in the supply chain.

11.6.3 Use of Information

Helo and Szekely (2005) classify SCM programs according to type of use:

1. *Warehouse management systems:* These systems monitor all warehouse activities, such as item reception, storing, accessing, issuing, and shipping from the warehouse, as well as planning and monitoring the handling of items in the warehouse.

TABLE 11.1

Comparison of Transaction-Based Programs versus Analytical Programs

Dimension	Transaction-Based Programs	Analytical Programs
1. Time frame	Past and present	Future
2. Objective	Reporting, present status	Forecasting and decision making
3. The business structure	Short term	Hierarchical and long term
4. Nature of the database	Original or closely related data	Processed and summarized data
5. Query response time	Real time	Delays of several seconds are possible
6. Business repercussions	Barcode or RFID systems replace manual data entry	Improved business decisions

2. *Transport management systems (TMS):* In terms of planning, these systems plan routes to transport components and products and can even facilitate interorganizational collaboration. In terms of monitoring, advanced TMS programs provide an updated status of the transport vehicles in the fleet.

3. *ERP:* The ERP programs such as Oracle and SAP are programs that usually strive to include all organizational activities in one database.

4. *Acquisition procurement and inventory management programs:* On the one hand, these programs manage suppliers, and on the other hand, they engage in forecasting and in optimization of inventory and its transport. The following companies specialize in developing these types of programs and customizing them to the needs of various organizations: SAP APO, Oracle, Manugistics, i2, Logility, E3, Escalate, Optum, and Provia software.

5. *Enterprise application integration:* These programs facilitate connecting between the organization's information system and information systems in external organizations, or between various information systems located at different locations within the same organization.

As described in previous chapters, the main use of information in analytical programs is to support decision making. Investment in an analytical information system is justifiable when decisions made using the system are better than decisions made without the system. In other words, these decisions lower costs, improve the quality of the supply process as well as the supplied goods and services, improve flexibility, and shorten supply times.

It is important to distinguish between the use of the organization's *internal* information (a link in the supply chain) and the use of *external* information that originates from other organizations (other links) in the chain. For example, an organization usually has internal information about inventory in the raw material warehouses and in the finished product warehouses of that same organization. This internal information is available to the organization at the level on which the organization chooses to operate. In contrast, information about inventory held by the organization's suppliers or customers is external information that the organization can obtain from its suppliers or customers, but this depends on their desire to share such information.

Many studies have indicated that *external* information can significantly improve supply chain performance (Moyaux, 2007). Intuitively, it seems that an organization that is aware of the inventory held by its suppliers and is capable of evaluating their ability to supply its demand can reduce its levels of protection against uncertainty by reducing the inventory it keeps, for example. Similarly, an organization that is knowledgeable about the inventory held by its customers can reduce its finished product inventory and naturally lower the costs associated with stocking such inventory.

In addition, information about suppliers and customers' manufacturing plans, including their actual supply, can reduce the level of uncertainty with which an organization must cope. Thus, the organization can also reduce the levels of inventory required for protection against such uncertainty.

Naturally, the challenge is to encourage suppliers and customers to agree to provide information to external links in the supply chain. Such entities will be motivated to do so if the profits expected to be gained from the information are shared between the various partners in the chain–a win-win situation. One of the challenges that supply chain planners face is how to make it beneficial for the chain's participants to share information, and thus to improve the overall performance of the chain. Therefore, it is extremely important to understand and quantify the value of information.

Example: The value of customer information—A tire factory supplies tires to a car assembly factory. The uncertainty of the number of orders issued for a certain type of tires requires the tire factory to stock a safety inventory of 500 tires for each of the three most popular tire types, as well as an additional safety inventory of 300 tires for each of the next three most popular types.

The car assembly factory proposes to share information with the tire factory about its biweekly manufacturing plan, and in return asks to receive an $800 discount off the tire price every month. Is the deal worthwhile to the tire factory, assuming a cost per tire of $50 and annual interest of 10%?

Solution: The safety inventory contains 1500 of the most popular types of tires and 900 of the types next in popularity. The information allow to have no safety stock.

The cost of the safety inventory in dollars at any given time is: $(900 + 1500)(50) = 120,000$

The annual interest is: $120,000 \times 0.1 = 12,000$

The monthly dollar cost of interest is: $12,000/12 = 1000$

In other words, the savings ($1000) is bigger than the requested discount ($800), and therefore, the deal is worthwhile to the tire factory.

11.7 Designing the Supply Chain

11.7.1 Choosing the Participants

The decision regarding who will participate in the chain is usually decentralized. In most cases, each link in the chain determines the identity of its suppliers, as well as to which market segment it will deliver its products and services (the subsequent links in the supply chain). Nevertheless, ultimately the choice of participants is made in a way that maximizes the suitability of the chain to the end consumers. Since the decision is decentralized, the choice of participants is very relevant to every organization that is a link in the supply chain. These organizations must focus on the market segments for which they manufacture and make themselves attractive for the stages in the supply chain that lead to the end consumer. For example, the choice of Intel as a supplier of processors of certain PCs is made with the consumers in mind, and affects the whole supply chain. The decision about how to turn the organization into an attractive supplier also has an impact on its choice of material and component suppliers. For example, if the intended market segment is characterized by prestigious consumers, every organization throughout the chain will try to accommodate itself as much as possible to prestigious products. Accordingly, the suppliers chosen by the organization will also be suitable in quality and costs to the manufacture or supply of prestigious products. On the other hand, the suppliers of basic products, aimed at a different market segment, will be chosen throughout the supply chain according to a different set of considerations. Sometimes one of the links is dominant and influences the considerations of other links. For example, a food manufacturer that is interested in obtaining Kosher or Hallal certification will usually require that all the links in its supply chain meet the certification requirements. Similarly, a pharmaceutical company that sells its product in the United States will require that all links in the chain meet the Federal Drug Administration's official requirements. The same can be said of a company that manufactures airplanes. It will require all the suppliers in its chain to meet Federal Aviation Authority requirements.

In most cases, the considerations in choosing suppliers are those of a single organization—a link in the chain—and these include considerations of cost, quality, flexibility, and supply times guaranteed by the suppliers. There are other factors, as detailed by Newman (1988) and described in previous chapters:

1. *Process capabilities:* Can the process used by the supplier manufacture the parts at the required level of quality?
2. *Quality control:* Does the supplier's quality control system guarantee maintaining the required quality for an extended period?

3. *Supplier's financial capabilities:* What is the risk in doing business with the supplier?

   This is intended to make sure that the engagement with the supplier can be a long-term one and to ensure that the supplier will not disappear within a short time due to financial difficulties.

4. *Cost structure:* What are the actual costs of materials, work, and overhead, and what are the supplier's profits? High costs and low profits may indicate future problems. The customary approach is that in an engagement with a supplier, it is also important to make sure that the supplier is making a profit, and that the profit is within the range that is reasonable in the industry; otherwise, the engagement will not last for long.

5. *Value analysis:* This is the supplier's ability to perform value analysis, or in other words, to provide the customer with professional examination of the financial and technical value of its products. This indicates the supplier's level of understanding in the context of the customer's needs, the product features, and their relative importance.

6. *Timing:* The ability of the supplier's planning and manufacturing control system to adapt to changes and quickly provide what is required. This flexibility by the supplier improves the purchasing organization's competiveness.

7. *Meeting contractual obligations:* Delivery that is on time and of the quantity and quality determined in previous contracts serves as a basis for evaluating the supplier's performance in such contractual engagements. Information systems make it possible to monitor suppliers' performance and to examine the extent to which they met their contractual obligations.

In addition, the organization may consider the risks involved in working with each supplier, as well as its ability to protect itself against such risks. As noted, each link in the chain will typically choose the suppliers that are the best fit for its needs. Only in special cases as in those listed above will a link in the chain have an influence on the choice of suppliers made by other links, that is, those that do not supply it directly.

The marketing considerations of each link will impact the choice of participants a bit differently. For example, the decision whether to sell directly to the end customers or to work with distributors that constitute another link in the chain will usually be made by the manufacturer of each product. Similarly, the manufacturer will decide whether to build a logistics system that entails transport and storage or to use logistic services provided by other entities that specialize in these areas. The decision to build a logistics system will impact the required investment (e.g., setting up distribution warehouses and purchasing trucks) as well as the ongoing expenses (e.g., a monthly payment for leased warehouses or payment for transport by rail). Similarly, the decision whether to use expert logistics services entails expenses that need to be considered when making the decision. This decision has additional significance, such as the supply time, which is a derivative of such a decision, as well as the quality of service and the flexibility of the supply chain. In many cases, distribution and sale providers can provide their services at low cost, high quality, short supply time, and with more flexibility than manufacturers that have no expertise, experience, knowledge, and existing infrastructure for a sales and distribution system.

Some logistics systems can distribute directly from the manufacturer's factory to customers, for example, companies such as Airbus and Boeing that sell passenger airplanes directly to airlines. However, in many cases, the distribution is not direct because it is carried out by a complex network of wholesalers and retailers, as is the case with the sale of fruits and vegetables from the farmer to the wholesale market and then to the retailer shops. This market entails large food chains that sometimes purchase directly from the farmer and thus reduce the mediation gap and increase the profit of all partners in the supply chain.

Many and varied alternatives are available for choosing the participants in the chain. It is possible, for example, to choose companies that specialize in transportation, intermediary storage and transport, or any combination of these specialties. Other companies may supply maintenance and technological support services. These companies are called third-party logistics (3PL). For example, a company that uses to distribute its products by UPS and FedEx is using UPS and FedEx as 3PL in its supply chain. Similar types of suppliers can be found in storage services. One example is warehouses in which items that are subject to customs (cars) are held as a guarantee until they are sold to the customer, who pays the customs tax to release and receive the item. Similarly, some entities provide consignment storage. In other words, the manufacturer transfers goods to a warehouse that is operated by another link in the chain. Yet legally the goods still belong to the manufacturer, which only receives compensation for the goods upon the actual sale. This complex set of considerations requires an understanding of supply chains and the ability to use mathematical and statistical models that are learned in industrial engineering programs.

11.7.2 Selection of Information System

The selection of the supply chain's information system is a derivative of cost–benefit analysis. In most cases, each link in the supply chain will choose the information system best suited for its needs, according to local considerations. Occasionally one of the links in the chain will influence the choices of other links, particularly in the case of long-term engagements in which the contract also includes information system requirements. The U.S. Department of Defense, for example, requires its suppliers to meet a long list of criteria related to the accounting system, the inventory system, and payroll, primarily in cases in which the engagement is based on time and materials (T&M contract). In other words, the Department of Defense pays the supplier the cost of the materials it uses as well as the cost of the time its employees spend on the order plus an agreed-on profit.

The choice includes the information collection systems (the system that translates physical, financial, and economic transactions into information entered into the system–transactions management system). Thus, it is important to choose the technology for automatic transaction collection—barcode readers and RFIDs, for example. In addition, the choice includes the databases that store the information, the model base, as well as the system that may be a collection of systems designed for specific functions in the organization (legacy systems) or integrative systems such as MRP systems and ERP systems.

Typical functionalities associated with the information system supply chain include the following:

1. Forecasting
2. Inventory management (purchase and distribution decisions)
3. Purchase management: orders, invoices
4. Warehouse management
5. Supplier management

Coordinating between these functions is an important consideration, and many companies typically decide to use the platforms of mega companies rather than their own system. IBM COGNOS, Oracle, and SAP are examples of such platforms.

A major problem in choosing an information system stems from the difficulty in calculating the benefit that can be gained from the system. Since information systems usually do not generate income for organizations, a straightforward comparison of expenses versus income is not feasible. Therefore, in this context, it is customary to use the term *benefit* when relating to information systems. The *benefit* is equal to the level of improvement achieved as a result of decisions based on an information system rather than decisions made without the use of that particular system.

The cost structure and the need to protect information have led many organizations to upload parts of the information system (e.g., the database) to a "cloud." Thus, the information is stored with server and network providers, and these suppliers are also responsible for the security and the accessibility of the information. Many companies even use the "cloud" providers' software, as well as other computer services. Coordination between organizations in the supply chain with respect to "cloud" providers produces a great deal of uniformity and access to the information of the coordinated organizations.

11.7.3 Design Considerations and Tools

The design of the supply chain is a process of choosing between alternatives. This process examines alternatives with respect to various participants, as well as the location and role of each participant in the supply chain, the facilities that will be used, the equipment and the information system and models to be used to manage the supply chain. The choice between different alternatives is based on cost, time, quality, and flexibility considerations, as well as considerations associated with the ability to cope with uncertainty and its related risks.

The cost considerations include the fixed costs of building the system, such as building factories and warehouses; purchasing and commissioning the information system; and purchasing the equipment for manufacturing, storage, and transport. In addition, cost considerations also include variable costs, such as the manpower required for operating the system, energy costs to operate the lighting and air conditioning equipment, fuel cost, and costs associated with risk management such as various types of insurance. In the early 2000s, cost considerations caused many businesses in the electronic industry to move from the United States to countries in Eastern Asia where labor is considerably cheaper. However, the logistics of shipping products back to the consumers in the United States are expensive and complex. Another example of cost considerations is the free-trade zones that attract businesses that benefit due to considerable tax savings.

Time is a major factor in supply chain design. For example, choosing alternatives for transporting cargo is usually based on transport costs compared to the required transport time. Heavy cargo with a long-shelf life (e.g., coal or oil) is usually transported by sea, whereas relatively light cargo with a short shelf life, such as flowers and fruit, are transported by air.

Quality considerations are very important and are usually divided into three categories: (1) establishing a connection with a supplier for a long period of time requires an examination of its management of quality systems, including by means of ISO-9000 testing, (2) in the long and intermediate terms, the quality of the supply processes (supply speed, flexibility, and the ability to make changes) is very important, and (3) the quality of the products themselves is naturally critically important in all purchase processes. Thus, products that have a short shelf life, such as fruit, will be transported by air to prevent potential damage to the quality of the fruit during extended sea transport. To guarantee a high-quality supply process, sometimes participants that are more reliable, more expensive, and have a proven track record of quality on-time delivery must be chosen.

The flexibility of the supply process should also be taken into consideration while designing the supply chain. The ability of the various participants to cope with changing market requirements (such as quantities, supply times, and design of the goods and services supplied) is an important consideration in supply chain design. This is the case when the market is very dynamic and has needs that are constantly changing and are very difficult or even impossible to predict.

The uncertainty factor and its derivative risks are an important consideration in supply chain design. A structured process for identifying potential risks, reducing or preventing major risks, and preparing contingency plans for coping with other risks is a significant component in supply chain design.

Designing the supply chain of an organization must refer to the strategic level of decisions with regard to the organization's suppliers, their mode of operation, and the level of service and quality they will provide to the subsequent links in the supply chain. To cope with the many design considerations and the interactions between them, it is necessary to have appropriate tools that are studied in industrial engineering training programs. These tools are acquired in courses that teach deterministic models, stochastic models, and simulations. Many programs of study also include an integrated course that covers supply chains and their design.

11.8 Supply Chain Monitoring and Control

11.8.1 Objectives and Constraints

As previously discussed, the supply chain consists of links or independent organizations such as factories, warehouses, wholesalers, and retailers, and among these are the organizational entities that operate various means of transport by sea, land, and air. Each of these entities has its own goals and operates under its own unique set of constraints. In previous chapters, we saw that there are *objectives* and performance measurements that are common to many entities in the supply chain.

The main supply chain objectives are as follows:

1. *Reducing costs:* Since an organization's profit is the difference between its income and its expenses, this profit may be increased by increasing income or by lowering expenses. It is harder to control income because the market price and the sold quantity are affected by market forces and by the competition therein. It is easier to have an impact on expenses and particularly on costs, and therefore in most cases reducing costs is one of the prime objectives of the links in the supply chain. It is possible to lower the costs in a single organization or link and that was our focus in previous chapters. However, it is also possible to lower the chain's overall costs, for example,

by sharing information between various links in the chain, and as a result lower the inventory costs throughout the chain. In the past, organizations did not comprehend that negotiation between links in the chain only shifts costs from one link to the next and does not increase competition because eventually it is the end consumer that bears all the costs. Therefore, competition is not between organizations in the same chain, but rather between one supply chain and another.

2. *Shortening supply times:* We have seen that in time-based competition, the quantity sold, and consequently the income, can be impacted by shortening supply times. Inventory is one of the tools that can achieve this goal. To guarantee immediate supply (in other words negligible supply time), it is important to keep finished products in stock. In-process inventories as well as raw materials inventories also assist in shortening supply times. A question arises in the supply chain regarding the location (link in the chain) at which the inventory will be held. In the context of a single link, we discussed examples of models through which the order point and batch size could be established. There are similar models in the context of the entire chain. One example is the transshipments model, which operates according to the assumption of cooperation between the links in the chain. This cooperation is apparent in a situation in which links that have a shortage receive supplies from other links in the chain that have a surplus. In a study conducted at the Technion (Avrahami et al., 2013), this method was used in the distribution network of a magazine. The routine management of the supply chain included a fixed quantity of newspapers distributed to every retail point of sale, based on sales history and future sales forecasts. Newspapers that were not sold were returned to the supplier when a new edition came out, and the retailer was not charged for unsold items. In this case, the retailers' objective was to avoid shortages and receive as many newspapers as possible even if the chances of selling all of them were low because the inventory cost is negligible and unsold newspapers do not cost the retailer anything. The outcome was a big surplus of newspapers that were returned and destroyed, while at the same time some retailers had shortages, and as a result potential sales were lost.

The solution that was tested in the study had two components—a technological component that entailed inserting RFID tags in each newspaper, and a management component that dictated supplying to the retailers in two stages. In the first stage, a quantity of newspapers was distributed to the retailers that in most cases met the demand forecast, but not in all cases (in other words, there was some probability that retailers would be in a state of shortage). In the second stage, the rest of the newspapers were distributed after some time went by, in accordance with the actual demand of each retailer. Demand data were received automatically because each time a newspaper was sold, the RFID systems reported it automatically so that actual sales were recorded in the information system for each retailer. The combination of an improvement in the information system as well as a management model that is based on additional information led to a significant decrease in the quantity of printed newspapers along with an increase in sales and considerable reduction in the quantity of newspapers returned to the publishing house.

3. *Increasing quality:* The quality of the goods or services supplied by the supply chain has an effect on demand, as does the quality of the supply process itself. Therefore, quality has an effect on the income of every link in the chain as well as on the income of the supply chain as a whole. In previous chapters, we described how to measure the quality of the product and the quality of the supply process. We also saw how each link in the supply chain can define its own quality objectives as well as how to measure and obtain these objectives. In most supply chains, there is a direct connection between the levels of quality of the various links because it is impossible to assemble a high standard product from low-quality products that were supplied through the supply chain. To achieve the quality objectives, each link in the chain is required not only to define its objectives and the means to meet them, but also to define the quality standards it requires from the suppliers that provide various goods and services to that link.

4. *Increasing flexibility:* The flexibility of the supply chain can be defined and measured in different ways. For example, flexibility can be defined according to the time required to make a change in the product's structure, in the supplied quantity, or in the supply time. Flexibility in supplied quantity can therefore differ from flexibility in product design because a change in an order's quantity may take less time than a change in the ordered product's dimensions, shape, or functionality. Since flexibility impacts demand, the goal is usually to increase flexibility subject to the associated restrictions and resources. These limitations are referred to as supply chain constraints.

Other than these objectives that are common to many organizations, there may be supply chains in which the objectives of the chain as a whole or of certain links in the chain are different. For example, a supply chain that is built for the purpose of providing emergency responses to natural disasters and that entails airlifting food and medicine to a location plagued by floods or earthquakes will set a goal of saving as many lives as possible, while costs and time are merely constraints. Naturally, the chain is constructed and the supplies are prepared in advance and during normal times, financial constraints are a consideration.

For the purpose of comparison, Beamon (1998) notes several objectives for SCM:

1. *Financial objectives:* Minimum inventory levels, minimum nonused inventory levels, minimum cost, maximum profit.
2. *Objectives related to availability to customer:* Obtaining a level of service, diminishing the probability of inventory shortage, reducing supply time, reducing response time.
3. *Efficiency-related objectives:* Keeping time and cost at a minimum while still meeting a quality threshold.
4. *Flexibility-related objectives:* Maintaining maximum system capacity.
5. *Quality-related objectives:* Customer satisfaction, product quality, response time.

Supply chains operate under a variety of *constraints* that can be divided into two groups:

1. *Local constraints of a link in the chain:* One example of this type of constraint is the resources that are at the disposal of the link and that limit the potential per-period manufacturing capacity for the link. Another example is a constraint on the transport means that limits the quantities that may be transported during a given time period. Additional examples of the local constraints of a supply chain link are the budget and the information that are at its disposal.
2. *Constraints of the supply chain as a whole:* These types of constraints include, for example, time constraints that originate from the limited shelf life of raw materials. In the food industry, for example, a restaurant that serves fresh ocean fish has a limited time from the moment the fish are pulled out of the ocean and until they are served to the customer at the restaurant. If the restaurant does not cope with these constraints, the fish will not be tasty or even worse will be unhealthy to eat. This constraint is attributed to the chain as a whole, and therefore if time can be saved by certain links and utilized by others, it is important to manage the chain as a whole in order to cope with these constraints. Thus, for example, the fish can be transported by special shipment (in some cases by air) to preserve them longer as fresh ocean fish at a restaurant.

Another example involves the constraints placed by entities that are external to the supply chain, for example, constraints originating from health ministry regulations with respect to food or pharmaceuticals. All links in the chain must meet these constraints and noncompliance of one of the links creates difficulties for all the links in the chain.

The role of management is to construct the supply chain and operate it in a manner that does not deviate from the defined constraints and that optimally achieves its defined objectives.

11.8.2 Long-Term and Short-Term Decisions

We are used to think about supply chains as long-term engagements. This indeed is the case for most supply chains. However, for every rule, there are exceptions and some supply chains are only intended to operate for a short period of time. For example, supply chains for delivering supplies to a disaster area do not justify investments in construction and infrastructure and, therefore, the required means (medicine) will be stored in temporary facilities such as field hospitals.

On the other hand, long-term supply chains, such as the ones for supplying a certain brand of new cars, justify significant investments in building facilities and purchasing equipment. Investment and large changes are also justified if such measures lead to lower operating costs, shorter supply times, etc. Supply chain design deals with long-term decisions such as decisions to invest in buildings, facilities, and infrastructure. Other long-term decisions include purchasing equipment, decisions to invest in an information system, and decisions regarding manning the facilities and the information system. After carrying out these decisions, the supply chain should be operated efficiently. In this context—as discussed in previous chapters—there are two types of short-term decisions: routine repetitive decisions and one-time decisions (ad hoc).

An example of recurring decisions and their supportive information system is the basic inventory model that is based on order quantity and order point. Assuming that demand and cost will be stable for a while, the decision will be made over and over again, every time the inventory level falls below the order point. These types of decisions can be supported by an information system and sometimes can even be carried out automatically.

Similarly, the MRP system also supports recurring decisions. When the MPS, the inventory file, the bill of materials, and the capacity and routing data are released, the MRP system issues recommendations for work and purchase orders that may be operated manually or automatically.

The two above-mentioned examples describe very short-term decisions (e.g., how many raw material units need to be ordered from the supplier assuming a predicted rise in prices) compared to longer-term decisions (e.g., can the inventory model be updated following the expected price increase). Similar decisions are made with respect to transport, manufacturing, and storage.

Ad hoc decisions are decisions that provide a response for a particular case. These decisions are unrelated to the period of time. Sometimes they are related to the short term (e.g., adding a shift during a busy month) and sometimes they can affect long-term design decisions as well (e.g., an ad hoc decision can cancel a strategic decision to build an additional warehouse in a particular area). An information system that includes models that support all of the above-mentioned types of decisions, and a database that includes all the data for the models in the model base, will support the decision-making processes and will facilitate better competition for each link in the chain, and therefore for the chain as a whole.

The design of the internal processes of each link in the chain and of interconnected processes between the links is a very important decision. Designing these processes, supporting them with an appropriate information system, and building the relevant models and adapting them to the specific supply chain are among the industrial engineer's most important roles.

11.8.3 Supply Chain Performance Control and Monitoring

Any supply chain meets uncertainty and its related risks. Therefore, it is important to design a supply chain that can adapt to changing conditions (Jacobs et al., 2011). Supply chain risk management is a developing area. Mapping supply-chain processes can help enterprises understand the potential risks that exist as well as the organizations involved (Thun and Hoenig, 2011).

The model that sometimes serves as a monitoring and control system is the supply chain operations reference (SCOR) model—(Bolstorff and Rosenbaum, 2011). This is the official standard model suggested by CSCMP for analyzing supply chains. This model focuses on the following processes: planning the supply chain's activities (plan), sourcing and managing suppliers (source), manufacturing (make), distribution, and return of products (return) (Poluha, 2007).

Other than the long-term and short-term decisions listed above, proper design in this context includes two additional components: control and monitoring.

Monitoring is carried out by ongoing collection of information about actual activity and about the outcomes of the actions of the links in the chain, and of the chain as a whole. Analyzing this information, for example, by comparing between planning and actual results, enables identification of situations that require corrective action.

Control is carried out by making corrective action decisions and by implementing these decisions.

The quality control system is an example of a monitoring system. This system has three components:

1. *Definition of the relevant quality requirements and performance indicators:* This is the stage in which the decisions that were made regarding the quality requirements of products, services, and the supply process

are translated into relevant performance indicators. An objective is determined for each performance indicator, as well as boundaries that are referred to as control boundaries in which the performance indicator is expected to shift due to uncertainty. For example, a process performance indicator may be to meet the guaranteed supply time, that is, the difference between the promised delivery date and the actual delivery time. The initial boundaries for this goal would be a 1-week delay, for example. The indicator will be constantly calculated for each order by comparing between the promised supply time and the actual supply time as well as comparing between this discrepancy and the established boundaries.

2. *Quality assurance:* At this stage, the supply process will be designed in a way that will guarantee meeting the determined boundaries. Thus, sufficiently reliable means of transport should be chosen, as well as inventory levels that will protect the supply process against uncertainty.

3. *Quality control:* This is the stage at which the performance indicators are monitored, as, for example, comparing actual supply dates to promised supply dates. In any case of deviation from the predetermined dates, the reason for deviation will be examined and actions will be carried out to prevent such deviations in the future. Sophisticated systems will forecast performance indicators and corrective actions will be taken even before a deviation occurs. For example, the supply time will be distributed into manufacturing time, assembly time, and packaging and transport time, and each of these indicators will be monitored. Deviation, even in one of these times, will initiate a corrective action. For example, machine failure that causes long manufacturing times initiates effort for shortening the duration of consecutive actions: assembly and packaging. The overarching goal is that there will not be any deviation in the general performance indicator for the entire chain to meet its promised supply times.

11.9 Summary

In this chapter, we introduced the idea of a supply chain. This is a chain of organizations involved in the production of a specific end product. Their success or failure is directly related to the success or failure of the end product. Therefore, they are strategic partners and should act like a team. Each organization is a link in this chain having suppliers as the previous link and customers as the next link. SCM includes several strategic issues. The first is choosing strategic partners and suppliers, The second issue includes setting information sharing and information system policy. The third issue is developing procurement policy, and the fourth issue is developing quality management policy. The transition from strategy to tactical management entails setting goals and priorities, supervising the transition from the design of the supply chain to the supply chain operation, and deciding how to use various resources to obtain the goals of such a chain.

Our discussion on the design of supply chains, planning of their operations, monitoring, and controlling these operations integrates many of the models and concepts discussed in the previous chapters. Each link in the chain is an organization that uses resources to transform inputs into outputs in order to serve its customers- the next link in the chain. Understanding the inner workings of each link, its processes, and the transformation it performs, as well as understanding the integration of the entire supply chain, is essential to the work of the modern industrial engineer.

References

Avrahami, A., Herrer, Y., and Shtub, A. 2013. Printing house paper reel management: An RFID enabled information rich approach. *Journal of Theoretical and Applied Electronic Commerce Research* 8(2), 96–111.

Beamon, B.M. 1998. Supply chain design and analysis: Models and methods. *International Journal of Production Economics* 55(3), 281–294.

Bhattacharya, R. and Bandyopadhyay, S. 2011. A review of the causes of bullwhip effect in a supply chain. *International Journal of Advanced Manufacturing Technology* 54, 1245–1261.

Bolstorff, P. and Rosenbaum, R. 2011. *Supply Chain Excellence: A Handbook for Dramatic Improvement Using the SCOR Model.* 3rd edn. New York, NY: AMACOM.

Croson, R. and Donohue, K. 2009. Impact of POS data sharing on supply chain management: An experimental study. *Production and Operations Management* 12(1): 1–11.

Forrester, J. 1961. *Industrial Dynamics.* Waltham, MA: Pegasus Communications.

Hasan, T., Hoque, M.R., Kawsari, N., and Tomal, D.T. 2013. Reduction of bullwhip effect in auto assembly industry. *Global Journal of Researches in Engineering, Industrial Engineering* 13(2).

Helo, P. and Szekely, B. 2005. Logistics information systems: An analysis of software solutions for supply chain co ordination. *Industrial Management and Data Systems* 105(1), 5–18.

Hines, T. 2004. *Supply Chain Strategies.* Oxford: Taylor & Francis.

Jacobs, F.R., Berry, W., Whybark, D.C., and Vollmann, T. 2011. *Manufacturing Planning and Control for Supply Chain Management.* New York: McGraw-Hill Professional.

Khalifa, N. 2012. *Supply Chain Challenges in Developing Countries: A Beer Game Simulation.* Germany: LAP LAMBERT Academic Publishing.

Lambert, D.M. 2008. *Supply Chain Management: Processes, Partnerships, Performance.* 3rd edn. Sarasota, FL: Supply Chain Management Institute.

Lambert, D.M., Cooper, M.C., and Pagh, J.D. 1998. Supply chain management: Implementation issues and research opportunities. *The International Journal of Logistics and Management*, 9(2), 1–19.

Lee, H.L., Padmanabhan, V., and Seungjin, W. 1997. The bullwhip effect in supply chains. *Sloan Manage Review* 38(3), 93–102.

McClellan, M. 2002. *Collaborative Manufacturing: Using Real-Time Information to Support the Supply Chain.* Boca Raton, FL: CRC Press.

Moyaux, T., Chaib-draa, B., and D'Amours, S. 2007. Information sharing as a coordination mechanism for reducing the bullwhip effect in a supply chain. *IEEE Transactions on Systems, Man and Cybernetics, Part C: Applications and Reviews* 37(3), 396–409.

Nahmias, S. 2012. *Production and Operations Analysis.* 6th edn. New Jersy: Wohl Publishing.

Newman, R.G. 1988. Single source qualification. *Journal of Purchasing and Materials Management* Summer, 10–17.

Oliver, R.K. and Webber, M.D. 1982. Supply-chain management: Logistics catches up with strategy, In: Christopher M. (Ed.), *Logistics: The Strategic Issues.* London, UK: Chapman Hall, pp. 63–75, Outlook, Booz, Allen and Hamilton Inc. Reprinted 1992.

Olson, D. 2012. *Supply Chain Information Technology.* New York: Business Expert Press.

Poluha, R.G. 2007. *Application of the SCOR Model in Supply Chain Management.* Youngstown, NY: Cambria Press.

Stevens, G.C. 1989. Integrating the supply chain. *International Journal of Physical Distribution and Materials Management* 19(8), 3–8.

Thun, J.H. and Hoenig, D. 2011. An empirical analysis of supply chain risk management in the German automotive industry. *International Journal of Production Economics* 131(1), 242–249.

Yan, W.D. and Katok, E. 2006. Learning, communication, and the bullwhip effect. *Journal of Operations Management* 24(6), 839–850.

Zhang, X. 2004. The impact of forecasting methods on the bullwhip effect. *International Journal of Production Economics* 88(1), 15–27.

12

Introduction to Service Engineering

Educational Goals

This chapter discusses the basic components characteristic of service engineering:

1. Defining and characterizing the term *service process*
2. Becoming familiar with the characteristics and considerations involved in service system design
3. Characterizing the arrival process
4. Characterizing the service process
5. Defining a queue as the difference between the arrival process and the service process
6. Level of service
7. Service system human resources planning
8. Management of feedback and customer satisfaction surveys

12.1 Introduction

This chapter discusses service systems and service engineering. We distinguish between products and services: A product is a tangible object that can be stored before it is delivered to customers, whereas a service is non-storable and intangible. Service is defined as handling customers, or as the benefit derived by the customer while receiving service, or simply as work performed for the customer. For example, health-care services provide customers various diagnoses and treatments that are important to their health. Most outsourced projects are services performed for the client organization. This includes both projects with intangible results such as a preventive maintenance project, and projects focused on deliverables (e.g., construction projects). In this chapter, we focus more on services delivered to the end customer. It is important to note that many services are also related to the provision of products; simply supplying the products, making them accessible, and selling them is a service. Health-care services, for instance, usually provide medications as well (and these are products).

Jacobs et al. (2010) lists seven general characteristics of service processes:

1. *Everyone knows* the service characteristics (because there is no adult who has not had experience with service processes).

2. The service characteristics are *situation dependent*: what works in one situation will not necessarily work in another. For example, what is considered desirable in a fast-food chain (disposable utensils) is not desirable in a fancy restaurant.

3. *The dimensions of product quality or work quality are not equal to the dimensions of service quality.* For example, a great repair job that takes 1 week instead of 1 day is considered to be of low service quality.

4. Most of the services *offer a service pack* that includes *intangible* components as well as *tangible products and components.* The customers are interested in the entire service *pack,* and its provision requires the appropriate management approach.

5. Services are characterized by *close* and intensive *contact with the customers.* The customer experience is an integral part of service quality.

6. Service management requires an understanding of *marketing,* human resources management, and operations management.

7. Customer *encounters must be managed* in the best way possible.

Service systems integrate technology and organizational networks designed to provide services that meet the customers' needs, desires, and expectations.

A service system provides its services to a clientele comprising people or organizations. In many cases, only a small part of this clientele consumes the service at any given time, and often the level of demand for the services at any given time is uncertain.

Most of us are unaware of the wide variety of service organizations that play a role in our everyday lives. The services of these organizations are often related to the provision of products. From the moment, you open your eyes in the morning and until you arrive at work, it is likely that you have used the services of many organizations. You probably brushed your teeth with toothpaste (toothpaste is a product) that was purchased in a store or a chain that services its customers (simply displaying and selling the toothpaste is a service), rinsed your mouth with water (water is a product) provided by a company that supplies water (the transport and supply of water is a service), turned on the light and thus used the service of the electric company, warmed up coffee on a gas stove (gas is a product) supplied by the gas company (the provision and sale of the gas is a service), and put on clothes (products) that were sold in a store that services its customers (and perhaps the clothes also underwent laundry or ironing services). Some of you may have stepped outside and waited for service at the bus or train stop and some of you got into your cars and used gas (product) supplied by gas stations (selling gas is a service). On the way to work, you may have listened to music on your radio supplied to you by the radio station, or perhaps you stopped to withdraw cash from the bank that serves you. You may have used your phone or the Internet on the way, both services provided by a service provider. If you stopped on the way to get coffee and pastry (products) you used a service (the sale is a service) and if you parked your car at a parking lot, you used another service. It is probable that throughout the remainder of the day you will encounter services provided by many service organizations. Service organizations also handle special cases and emergencies, such as health-care services, fire department services, towing services, garage services, and so forth.

A service can also be provided to organizations rather than just to people. Following are some simple and popular examples of the types of services provided to organizations: insurance, banking, legal, advertising and marketing, and many others. Services provided to organizations are carried out by interacting with the organizations' representatives, and human communication is a key component in the relationship formed in the context of the service.

As previously mentioned, service provision encompasses all walks of life, from medicine to education, transportation, the wholesale and retail trade sectors, tourism and recreation, entertainment, culture, banking and insurance, and in short almost every component in our lives. Even the manufacturing organizations provide services for fixing failures and answering customer calls for various issues. Service management is comprised of the three components of every service system, as described in Figure 12.1: First and foremost are the *customers*—their preferences, expectations, and desires. The second component is the service *infrastructure* that supports the service, and the third component is the cluster of *management and operation processes* that aim at creating the customer experience (Salvendy and Karwowski, 2010).

Service system design and planning must address all three components. Providing a partial solution to any one of the components will frustrate the customers, as well as the service provider organization. The integration process that needs to operate in all the components of the organization is an important component. It is important to handle all the customer's interactions with the company, and at the same time set the actual level of service to meet customer's expectations. In other words, the customer experience must be managed.

FIGURE 12.1
Service system components.

At the same time, it is important to manage all operational processes while using the required resources and utilizing the supportive infrastructure. All this must be accomplished while meeting the required budget goals. The key to the organization's success and its comparative advantage lie in its human assets. Therefore, an additional key to the success of service processes is service-minded human-resource management. Two additional dimensions that are responsible for the success of the organization are an understanding of the regulatory environment and proper risk management. A regulatory environment change can cause a dramatic change in all the service processes. We are all aware of the importance of Federal Drug Administration (FDA) regulations for health-care providers, and Federal Aviation Authority (FAA) for the aircraft carriers. FAA regulations changes following the September 11, 2001 events were sharply felt by all air travelers in the United States. In summary, orderly and structured performance in all the aforementioned dimensions will contribute to improved performance of the service system in all service components.

The major characteristic of service processes is *the centrality of the customers*, that is, viewing the customer as the center and building *a positive customer experience* as the main objective of the company's activity. Thus, all components of customer contact become the forefront of the organizational effort.

The mix of customer experience components includes components that can usually be measured, such as punctuality, service rate (pace), and product quality. However, the mix also includes components that are difficult to evaluate, such as initiative, professionalism, and courtesy.

The speed of performing the service and the reduction of the waiting time are part of service quality and have a direct and significant impact on sales. In most cases, there is a direct and tight bond between the resources invested in a service and the service time. Nevertheless, many demand processes are characterized by fluctuations and uncertainty of how many customers will arrive, and in many cases, the customers' requirements are unknown in advance. The service time may also change from customer to customer. For example, a bank clerk's service time changes from customer to customer.

This chapter describes topics that characterize the development and design of service systems. It is clear from the above that random processes that simulate the demand process and the service process as well as additional characteristics such as malfunctions and problems constitute an integral part of the models used in service engineering. This chapter focuses on a basic description of the characteristics of these random processes.

12.2 Service Processes

12.2.1 General Characteristics of Service Processes

In the service process, service is provided to customers. The service is provided by one or more service providers to end users, for personal consumption or for organizational consumption. An organization can also be an end user, and it is represented by people who were appointed to represent it.

The purpose of the service is to provide value to the customers and to satisfy them at a cost that facilitates a reasonable profit. The customers' impressions are connected to their customer experience, which in turn is connected to the level of service. The level of service is also connected to the required effort and to the cost. Figure 12.2 describes the general relationships

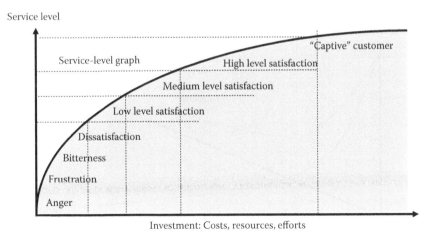

FIGURE 12.2
Graph depicting costs and efforts and level of service versus customer satisfaction.

between customer satisfaction on one hand, and level of service, required efforts, and costs on the other.

The curve depicted in Figure 12.2 indicates that the level of service increases as the invested costs and efforts increase. However, this graph has diminishing marginal utility. An addition of cost and effort will cause a greater improvement in the level of service when this level is low; however, the improvement is smaller when the level of service is high.

There are factors that impact the level of service but are not necessarily related to costs or additional payments, among them are courtesy, friendliness, attention, punctuality, and focus. However, without investment and additional costs, the service level will eventually remain low. Service engineering facilitates taking care of all service aspects, including the financial aspect.

Generally, the financial decisions in a service system are usually related to the system's capacity, that is, its ability to serve a maximal number of customers. On one hand, this capacity is critical to the customer experience and to the system's performance, and on the other hand, the capacity has a clear cost. Figure 12.3 schematically illustrates the overall cost of a service system as a function of the capacity of the service system, and of the cost of waiting in line.

Service capacity can find expression in labor, number of service stations, machines that assists the service delivery and even full automation. The left side of Figure 12.3 illustrates that in a case of low service capacity, the waiting time increases to major proportions, and therefore waiting in line can be very costly. This cost varies from one system to the next, but in most cases, it is a cost that stems from customer dissatisfaction, which harms the

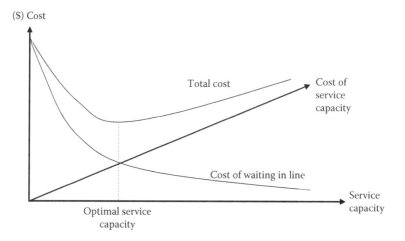

FIGURE 12.3
Scheme of the general cost structure in service systems.

company's reputation and causes customers to leave, with a consequent drop in sales. By adding capacity (labor and machines), the wait time decreases and so does the cost associated with waiting. On the other hand, the cost of capacity itself constitutes a more significant component in the total cost. The total cost is the sum of the waiting and capacity costs. When capacity is very low, the service pace is very slow, the lines are long, and the waiting costs are high. In this case, the cost of waiting during a specified time unit is greater than the cost of an employee or a machine for that time unit, and therefore the total cost may decrease when adding capacity. On the other hand, when the capacity is very high, the service is very fast paced, and as a result there are hardly any lines, and adding another worker or machine has a significantly smaller effect on the waiting time cost. Therefore, when capacity is very high, the total cost increases on addition of capacity (workers or machines) mainly according to the additional resource cost. This cost (in the case of high capacity) decreases as capacity decreases, and the waiting time cost increases. Eventually, there is a balance point at which the service capacity is optimal in the sense that the total costs mentioned above are minimal.

A basic scheme of the cost structure is shown in Figure 12.3. Yet, in most cases, it is very difficult to build this type of curve in practice, and often it is impossible. This is due to cost evaluation. For example, there is no single mathematical solution to the question what is the cost of waiting for a bed at the hospital, waiting to receive a response to an emergency call to the police, or even the cost of customers waiting in line at the bank. An additional reason is the lack of a direct connection between the waiting times and the capacity. The waiting times are caused by densely spaced random arrivals that are difficult to predict. Figure 12.4 illustrates how the random nature of service demand intersects with three different service level scenarios.

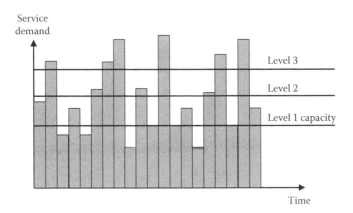

FIGURE 12.4
Example of service demand over time and three potential levels for determining capacity.

The histogram in Figure 12.4 depicts demand levels and three alternatives for determining capacity level. The higher the capacity level, the greater its expected cost. On the other hand, the percentage of demand that receives an immediate response is also greater. (In other words, there is an increase in sales, the waiting times are shorter, and the customer experience is better, and so is the reputation of the service process.) The service demand usually stems from random customer arrivals and cannot be accurately predicted. However, the randomness of the quantity of arrivals per time unit as well as the service times can be predicted by estimating the expectancy as a central indicator and the standard deviation as a dispersion indicator, and sometimes even by measuring the entire distribution. Some service systems cannot provide service to all their customers due to lack of capacity. In these cases, service is given on a first come first served (FCFS) basis, and sometimes on the basis of remaining capacity. In these cases, service capacity is the significant factor in determining the percentage of those receiving service out of those that requested service.

Service processes are differentiated from manufacturing processes by the following characteristics.

12.2.1.1 Extent of Customer Contact

Service processes are characterized by close and intense contact with the customers. This contact requires interpersonal skills and the ability to listen and understand the customer's needs. Customer contact is particularly required when the service or product is nonstandard. The service process always includes contact with customers or direct support of a process that entails customer contact. When there is no customer contact in a process, and there is no direct support of such a process, it is typically not considered to be a service process. Figure 12.5 illustrates several examples of

FIGURE 12.5
Examples of work that are not service processes—mining, agriculture, and assembly are performed away from the customers.

processes that are not service processes and clearly do not entail contact with customers.

12.2.1.2 Service Pack

As in the purchase of a product pack, customers who purchase a service pack in effect purchase a package of items. The service pack includes all the customer's points of contact with the company's products and services, and sometimes the time of the company's personnel as well. These interfaces with the customers must meet the customers' expectations. For example, purchasing a ticket for a play begins with the purchase itself and is affected by the extent of friendliness or courtesy of the salesperson and by the way in which the play is advertised. The service pack represented by the ticket includes, in addition to the play itself, the seat (and its location), the venue, and its qualities (e.g., air conditioning), the experience of entering the venue and finding the seat and even the refreshment stands and restrooms.

In Chapter 5 (Section 5.4), we listed eight characteristics or indicators of the quality of products and services (Garvin, 1987, 1988):

1. Performing the product or service's designated roles
2. Add-ons and improvements relative to the basic requirements
3. The extent of reliability
4. The extent of suitability to the technical specifications
5. The life expectancy
6. Availability of the service, the type of care, and the response speed
7. The esthetics
8. The perceived quality

Four dimensions that are unique to services can be added to this list:

1. The environment
2. The wait time (an important component to be discussed next)
3. Level of friendliness and courtesy
4. The professionalism of the service providers

The service pack includes measurable components (such as wait time, service time, and error percentage) as well as other components, such as environment (calm or noisy, orderly or chaotic, clean or dirty) and human relations (friendliness, courtesy, professionalism), which are difficult to measure. The customer's expectations are dependent on the type of service. For example, a calm environment is important in places such as libraries or malls, but is not suitable in bars, nightclubs, and other noisy venues. The service pack includes not only the service itself, but also the customer experience described next.

12.2.1.3 Customer Experience

The customer experience is the way in which the customers perceive all their interfaces with the service system in relation to their needs and expectations. Most service systems have many interfaces: various channels such as a Web page, voice mail, mobile interface, or a reception clerk. The customer may come in contact with various interfaces at the different service stages: the curiosity/interest stage, the purchase stage, the support stage, or the service, repair, and warranty stage. One of the key components in the customer experience is the wait time. Figure 12.6 illustrates the main factors affecting the customer experience in the context of the service pack.

To manage all interface in a coordinated manner, many organizations have adopted a structured customer experience system—customer experience management (CEM). Schmitt (2003) describes five stages of CEM: (1) building a platform of the desired experience and the value of the service pack for the customers; (2) analyzing the customer experience; (3) designing the brand experience; (4) building the customer interface; and (5) ongoing renewal and improvement of the customer experience. Finally, to execute the CEM policy, it is also important to understand and manage the experience of the service providers.

To take care of their customers and cultivate their relationships with them, many companies operate customer relationship management systems (CRM), which are very helpful in managing relationships with customers. These systems document every piece of information about the customers, the customer's encounters with the company, and assist in keeping in touch with the customers to offer them interesting offers. CRM systems cover parts of the interactions but usually cannot replace CEM systems. For example, the customer experience is influenced by the subjective treatment given by the service providers (see Figure 12.7): smiling, understanding, being patient, expressing concern, and being creative and resourceful. CRM systems cannot directly measure these traits. There is a tight bond between the quality

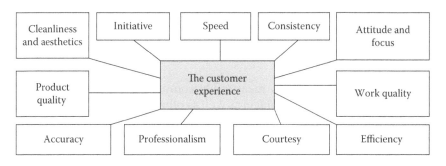

FIGURE 12.6
Customer experience components that are important and critical for service process competitiveness.

FIGURE 12.7
The customer experience is affected by the subjective attitude of service providers.

of service and the customer experience. However, the customer can separate the service from the product itself. Here is an example: "The cable car works fantastically, but the ticket purchase process is cumbersome. One of the ticket machines is almost always broken and the tickets must be scanned ten times before the scanners read them." The service is an important part, and usually the central part, of the customer experience. This chapter covers the whole interaction between the organization's personnel and the customers, that is, the service itself as well as waiting for the service.

12.2.1.4 Response Time Limitation (Availability)

One of the major differences between products and services is the ability to maintain a product inventory, and the difficulty to maintain a service inventory. Service is required at a given time, and the demand cannot usually accumulate beyond that time period. For example, a restaurant that is closed for an entire week cannot preserve the demand from that week and transfer it to the next one. Similarly, a 1-week stoppage of a theater, a bus or train line, a gas station, or a television or radio station cannot amass the demand lost during the week. Sometimes the time period in which the service can be provided can be long. However, if the service is not eventually provided during this time, the customer will use an alternative service provider or will give up on the service. In addition, a long wait for service causes customers' frustration, and therefore, the waiting time is an accepted indicator of the level of service. Availability is measured by the percentage of time that the system can immediately accept and service customers (within a reasonable time period in the eyes of the customer). Efficiency is measured by the percentage of time during which the system provides the service. Efficiency close to 100% would cause excessive waiting times and low service levels. Therefore, service systems should typically aim at efficiencies below 90% and weigh the efficiency against the related service level.

12.2.1.5 Demand Randomness

Usually the demand for service processes cannot be accurately predicted. For example, bank clerks do not know how many customers will seek their service at a given time. In addition, how much time will be required to service a customer cannot always be predicted.

12.2.1.6 Demand Dynamism

Service demand is sometimes dynamic and changes over time. The demand distribution at a particular time of day is not necessarily identical to the demand distribution at another time of day. Similarly, the demand distribution on a certain day is not necessarily identical to the demand distribution on another day. Thus, demand changes with the season of the year, the day of the week, and the hour of the day, and even more so on holidays and large events.

12.2.1.7 Level of Service

The level of service is the indicator that characterizes the quality of service. The term *level of service* usually refers to the target level of the quality of service performance. The goal is always to bring the level of service to a peak, subject to constraints such as costs. The level of service is connected to parts of the customer experience (that are usually measurable). The level of service is measured differently in different types of service. For example, at call centers the level of service can be the average time from dialing to a human answer, also called average wait time (AWT). Some other service level indicators in call centers are: (1) abandon rate (AR)—the rate of calls that are disconnected before getting an answer; (2) average handle time (AHT)—the total call time including the conversations and the transactions if applicable; and (3) first call resolution (FCR)—the percentage of first time calls about an issue with a full solution supplied by the call center. In other places, the level of service could be measured by malfunction percentage, or the number of complaints in a given time frame, relative to a fixed goal. In places that have inventory shortages that preclude providing the service, the level of service can be the percentage of service receivers, or the percentage of time during which there is a shortage. Every level of service depends on determining a reference point. In many cases, the reference point needs to not only include a 100% performance goal, but also reference to a scale and to a current or minimal performance level. For example, a large transportation organization (with consistent year-round activity) receives an average of X complaints every month and determines an optimal target for reducing the average number of complaints by half in the following months. In other words, if the current complaint level is X, the target (Y) is equal to $(0.5) \times X$. This is not enough for building a service level function and thus an additional reference point is required. For instance, the current level of complaints is 70%, and therefore, there are now two reference points:

1. Maximal level of service (100%): $Y = (0.5) \times X$
2. Current level of service (70%): $Y = X$

In other words, both reference points can be described as follows:

1. Maximal level of service (100%): $Y/X = (0.5)$
2. 70% level of service: $Y/X = 1$

We draw the axis lines so that the horizontal axis is Y/X, and the vertical axis is the level of service (Figure 12.8).

Drawing a straight line between the points (horizontal axis = Y/X; vertical axis = level of service) indicates that the vertical axis is intersected at 130% and the slope per unit (unit = 100%) is −60. In other words, the resulting service level function is:

$$130\% - (60\%)(Y/X) = \text{Service level}$$

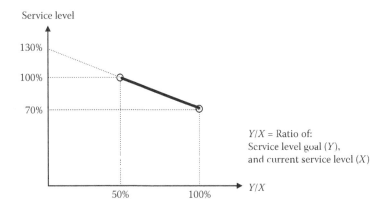

FIGURE 12.8
Graph for determining level of service in accordance with the ratio (Y/X) between the target number of complaints (Y) and the current number of complaints (X).

Thus, for $Y/X = 0.5$, the service level (100%) is computed as follows: $130\% - 60\%(0.5) = 100\%$, and for $Y/X = 1$, the service level is computed as: $130\% - 60\%(1) = 70\%$. When there is an increase in the number of complaints, the Y/X ratio grows and the service level decreases. For example, for an increase of 50% beyond the current complaint level, $Y/X = 1.5$, and therefore, the service level drops from 70% to 40%: $40\% = 130\% - (60\%)(1.5) = 40\%$.

12.3 Classification of Service Systems

In this section, we describe major processes that are common to most types of service systems. Thus, we classify the service systems in accordance with the type of organization and its customers, and in accordance with the industry to which the service system belongs.

12.3.1 Major Service Processes Common to Most Types of Service Systems

We shall use the example of air transport services in order to illustrate the major processes common to most types of service. A schematic drawing of the service process is described in Figure 12.9.

Major service processes common to most types of service systems are

1. *Arrival of customers/orders:* The arrival processes for customers and orders are characteristic of most service systems. These processes can be random (uncertain), static (do not change over time), or dynamic (change over time), and can be characterized by different paces at different times.

2. *Customer acceptance:* This is the first interaction between the arriving customer and the service system. In flight ticketing process, this is the initiation of a boarding pass. In a hotel, this is the check-in, and in a university it is registration. In many service systems, not every customer that arrives is entered into the system. Thus, in a parking lot a driver who arrives but cannot find a parking spot will continue driving and will look for another parking lot. When a long wait is expected, for a parking spot, the driver may continue looking for another place where the line is shorter. Acceptance means joining the queue to wait for the service or entering the service if there is no queue.

3. *Provision of core service:* The core service is the service on which the service organizations focus. In many service organizations, the service provided to various customers may vary significantly from

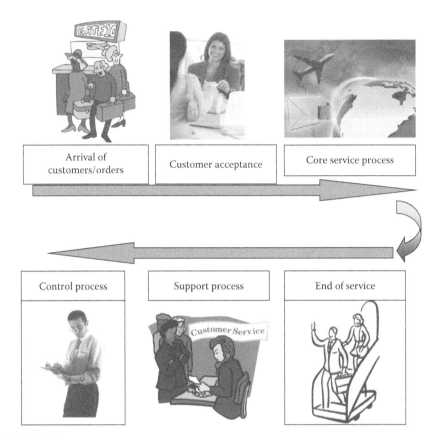

FIGURE 12.9
Schematic diagram of air transport services illustrating processes common to most types of services.

customer to customer, either in the nature of the service or in characteristics such as length of service times. Frequently, the core service is composed of a bundle of complementing services. For example, an air carrier must also provide baggage services and feeding services in long flights. A theater usually operates a cafeteria service and may have a parking lot service.

4. *Support process:* This is a process that facilitates handling malfunctions, complaints, and unusual requests. If needed, this process also includes repairing, returning, and replacing a defective product. Data related to the support process assist in measuring the quality of service. For example, municipalities collect statistical data on the number of complaints submitted to the municipal call center. A sharp increase in the number of complaints indicates a decrease in the level of service.

5. *Control process:* The control process involves monitoring service reliability and customer satisfaction. This is an active process of the service system whether it involves data collection and analysis from other processes or whether it entails distribution and analysis of customer surveys. Without a control process, the service system will not receive feedback from its customers and from its surroundings, and will lose its ability to improve the service in accordance with the customers' desires.

12.3.2 Classifying Service Processes by Type of Service-Providing Organization and Type of Customer

1. Type of service organization:
   a. *Public service:* These are publicly owned services whose purpose is to provide the service under a limited budget. Examples of public services are public reception hours at city councils and local municipalities, the court system, government hospitals, and the electric company.
   b. *Business service:* Business service aims to increase the organization's profit while maintaining a high level of service. This is the most common type of service. Banks, insurance companies, retailers, taxi companies, and gas stations are examples of business services.
2. Type of service customers:
   a. *Service for organizations and business customers:* In this case, the customers are various businesses, and the service is provided to the business's representatives. For example, the big telecommunications companies supply phone and internal communication systems to other organizations.
   b. *Service for private customers:* Service for end users. The service is provided to the customer as an individual rather than as the representative of an organization. These are the services (public or commercial) that we all receive at restaurants, at gas stations, and on public transportation.

The classification of each of the four potential types of service combinations is shown in Figure 12.10.

12.3.3 Classification of Service Systems by Industrial Sector

It is customary to classify services by their industrial sector (e.g., banking services, insurance services, health-care services, and fire department services). There are entire industries that focus on providing services. Below are some examples:

	Organizational customers	Private customers
Public service	E.g., physical infrastructure (roads, lighting, electricity, water, sewage, etc.)	E.g., utilities, health care, education
Business service	E.g., outsourcing, logistic services (3PL), maintenance services	E.g., retail, transportation, entertainment

FIGURE 12.10
Classification of service according to type of organization and type of customer.

1. Health care
2. Restaurant industry
3. Banking
4. Law
5. Insurance
6. Call centers
7. Transportation
8. Tourism
9. Entertainment
10. Cleaning

Naturally, there are many such industries and these are just a few examples. Classification by industry assists in characterizing the service system because each industry has its own unique characteristics that meet various customers' needs. Success factors in the entertainment industry, for instance, usually will not be relevant in the cleaning industry.

In addition, each industry has several different working environments in which the nature of the services is different. For example, the transportation industry includes transporting passengers in taxis as well as hauling cargo in trucks. The service skills required for each of these services differ despite their belonging to the transportation industry. Here is another example from the health-care industry: family medicine is characterized by a different service environment than emergency medical services or surgical medicine. However, clearly there are also some commonalities in how patients are treated by the service providers in these cases.

12.4 Key Characteristics and Considerations in Designing Service Systems

The planning and design of service systems must first and foremost take the customers into consideration. Therefore, it is very important to segment the market and to analyze the target audience. The purpose of segmentation is to characterize the clientele in terms of quantity, preferences, financial capabilities, and service factors that may attract customers or cause them to shy away, and of course forecast future trends. Only after the characteristics of the clientele have been sufficiently clarified is there a solid basis for effective service design.

Planning is a hierarchical process in which the vision of an optimal service system leads to strategic planning (long term), which in turn leads to operational planning (short term). Jacobs et al. (2010) identify seven characteristics of a well-designed service system.

1. The service components are integrated in and contribute to the service strategy.
2. The system is user friendly.
3. The system is sufficiently flexible and resistant to malfunctions.
4. The system allows the service to be performed consistently, easily, and at a high level.
5. In the system, service is strongly and efficiently linked to support and to the rest of the system.
6. The quality and contribution of the service are obvious to the customers.
7. The system is characterized by efficiency and by low cost relative to the level of its performance.

These authors also point out six strategic decisions that must be made in the context of designing the service:

1. Level of customer care in terms of the scope of assistance, guidance, and courtesy
2. Service speed and convenience
3. Service cost
4. Variety of services offered to customers at the place of service
5. Quality of the service system's physical components
6. Special skills required to provide a high level of service

Strategic decisions must be implemented through operative decisions and performance. A significant part of the strategy is related to the organizational culture and to employee training, and particularly to service providers. A survey conducted in 1997 in the United States (Roth et al., 1997a,b) demonstrates that service organizations emphasize 10 key considerations:

1. Service accessibility and availability
2. Thoughtfulness and openness toward service providers
3. Leadership in the field of quality of service
4. Listening to customers
5. Quality of the service system's physical components
6. Manner in which malfunctions are handled
7. Competitive service level
8. Value of service quality (including management involvement)
9. Consistency in meeting customer needs and level of customer satisfaction
10. Guidance and instruction of customers

While diligent management and training are connected with the management and operation of the service system, the system design also includes more quantitative and engineering related dimensions, as listed here:

1. *Service demand forecasting:* To provide good service, the required quantity of each type of service must be predicted for each time period. A small service quantity consumes a certain amount of resources and labor, whereas a larger service quantity usually requires a greater amount of resources and labor. Maintaining too much labor and other resources leads to losses, while not having enough of these leads to shortage and to a loss of customers and reputation. When demand is fixed, forecasting is trivial and 100% of the actual demand can be met. However, when demand is random, allocating resources for providing service determines the probability that a certain demand percentage will not be serviced. The goal is to guarantee service for a predetermined quota. Forecasting was discussed in Chapter 4 (and Chapter 5) in the context of methods to detect increasing or decreasing demand trends as well as to identify demand cycles. In addition, forecasting involves evaluation of the level of uncertainty.

2. *Level of service:* The level of service is generally connected to the ability to service a large quantity of customers in a short-time period (Daskin, 2011). This ability is measured by the number of standard deviations above the demand average that could be met. Therefore, it is very important to evaluate the demand pace for each time period, and it is important that service demand forecasting also includes standard deviation forecasting.

3. *Service location design:* The location of the service is a key issue. In the case of service where the service provider must have an encounter with the customer, the importance of the location stems from its accessibility. It is important to place the service near major roads and make sure it has nearby parking. It is also important that the service be close to the location of most of its customers. The quality of the location is sometimes dependent on how attractive it is. For example, placing a service station in an air-conditioned mall may be preferable to a location with identical accessibility, but lacking the elegant air conditioned environment of the mall. Even the decision with respect to the location of a call center or Internet provider can impact the profitability and quality of the service. For example, placing a North American company's call center in India can cut costs and facilitate the employment of many more employees and thus shorten wait times. India, however, is very large and is home to various types of populations. Therefore, it is very important to position the call center at a geographic location that is home to excellent personnel that are fluent in the required language (Figure 12.11).

4. *Determining the number of employees who provide the service (service providers):* Determining how many employees will provide the service (service providers) is closely related to demand forecasting and to the average service pace. If, for example, we forecast an average demand of 300 different customers during a 9-h shift, and servicing each customer takes 5 min on average, then an average shift comprises 1500 service minutes in a 540-min (9-h) shift. Therefore, at least three service providers are required. However, 1500 min/shift is an average load, so close to 50% of the shifts have more than 300 customers. So if the standard deviation of the customers per shift is 25, the average required time for serving 97.5% of the customers is $(300 + 2 \times 25) \times 5 = 1750$. Clearly, a fourth service provider is required to provide a maximum of $4 \times 540 = 2160$ min during a 9-h shift. If the service pace varies significantly during the shift or the differentiation of arrival quantities is high, even four service providers will result in long waiting lines. Adding the fourth service provider in such a case is not a luxury, but rather imperative in order to prevent unreasonable waiting times. Service providers are usually employed for an entire shift or at least for half a shift. The percentage of time in which the service providers are actually busy is called *capacity utilization*. In a service system that has random arrival times, the service providers' utilization should not exceed 90% where the waiting line grows massively. As the utilization approaches 100% the waiting time approaches infinity. The average capacity utilization for a single server is usually calculated by dividing arrival pace by service pace. Therefore, when systems have

Centrography map-store PCU12 customer distribution

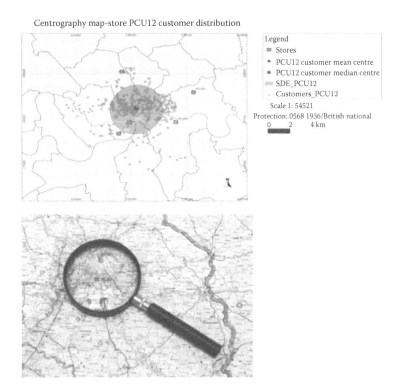

Legend
■ Stores
• PCU12 customer mean centre
• PCU12 customer median centre
▒ SDE_PCU12
. Customers_PCU12
Scale 1: 54521
Protection: 0568 1936/British national
0   2   4 km

FIGURE 12.11
Service placement and proximity to customers are critical for success. (*Source*: Adapted from Singapore Management University, https://wiki.smu.edu.sg/1112t2is415g1/IS415_2011-12_Term2_Assign4_Houston.)

random arrival times and service times, it is important to make sure that the average service pace is significantly larger than the mean arrival pace. This is beneficial for preventing long customer waits.

5. *Designing the number of waiting positions in the queue:* The number of waiting positions in the queue facilitates accepting customers who enter the queue while the service provider is occupied. It is also used for arranging a nice and comfortable waiting environment. There is usually a cost associated with preparing waiting positions in the queue, but if customers arriving at a service location do not have available waiting positions, the general assumption is that they will leave and seek service elsewhere. This is the same case when customers arrive at a queue they deem to be too long. On one hand, an additional waiting position in the queue is costly (usually a fixed amount per waiting position), but on the other hand, adding waiting positions enables accepting additional customers and increasing the revenues. The more waiting positions exist in a queue already, the

less significant would be the benefit of addition of a waiting position to the queue. For example, the waiting position that is closest to the service is occupied more than the tenth waiting position (which is not taken if there are fewer than 10 customers in line). It is beneficial to add waiting positions only when the benefit derived from adding a waiting position exceeds its cost.

6. *Ergonomic and esthetic design of the waiting system and the service system:* In most service systems, the customers are people with human needs. Therefore, the waiting positions and the service must be designed esthetically and ergonomically (i.e., considering customer convenience while they wait and while they are being serviced). This type of design is connected to the customer experience and is an important factor in the quality of service. The waiting area design is very important and has a major impact on the system's cost. Figure 12.12 demonstrates three different waiting systems. The longer the wait and the smaller the number of customers waiting, the more important it is to have a convenient waiting area and this is expressed in the location and design of the waiting areas.

7. *Customer service design:* Customers prefer to interact with a human being when they need to clarify issues and solve unusual cases and malfunctions. Therefore, a human response is a competitive requirement for almost every organization servicing a population. This is particularly important in for-profit business organizations. When deciding between organizations that provide similar services, customers prefer the company with the best customer service (the shortest wait time, dedicated care, and professional solutions). Therefore, a competitive business must carefully plan customer wait time and service time. Customer service can be crucial in public organizations as well.

FIGURE 12.12
Allocation of waiting positions impacts the cost and the customer experience. Therefore, planning the number of waiting positions (in the context of waiting for a service) requires thought and analysis.

8. *Maintenance and repair system design:* The size of the company and its volume of activity determine whether maintenance and repairs will be outsourced or performed by the company itself. Cost considerations are important, but they are not the only considerations because it is highly beneficial to be able to control the repair times (which are not reflected in cost calculations).

We distinguish between two cases:

1. In-house repairs (at the service provider facility)
2. Repairs at the customer's location or at a customer service repair lab

12.4.1 In-House Repairs

A considerable number of service-providing companies are dependent on the proper function of nonhuman resources. For example, companies that provide dry cleaning services are dependent on the proper function of their cleaning machines, taxi companies and other driving services are dependent on the proper function of their vehicles, cafes are dependent on the proper function of their coffee machines, and so forth. It is therefore evident that machine malfunctions will impede a company's ability to provide service. Hence, in the case of a malfunction, the machines must be repaired as quickly as possible. Such repairs are focused with respect to location but require great expertise. In addition, machine maintenance and malfunction repair entail an unknown malfunction component as well as a known component of preventative maintenance. These two components (known and unknown maintenance) require planning. It is usually easier to plan preventative maintenance, which is performed on predetermined dates. In preventative maintenance planning, not only are the dates predetermined, but also the various actions and the required expertise and duration are determined as well. The preventative maintenance schedule dictates the availability of professionals and the planning is certain. In contrast, malfunctions usually cannot be accurately predicted in advance. Nonetheless, it is possible to use statistical tools to describe the frequency of malfunctions or failures. The most common term in this field is mean time between failures (MTBF). If a documented history of the times between failures has been accumulated, it is possible to calculate the mean from the collected data. If data are insufficient, there are additional methods to evaluate the MTBF. It is important to understand that this is merely a mean time and that a component may malfunction immediately after it is assembled or a very long time after the MTBF. However, as a rule a higher MTBF indicates greater reliability. Following the malfunction of a machine or an expensive component, there is usually a repair period during which the machine is idle. The mean time to repair is referred to as mean time to repair (MTTR).

The customary assumption in many models is that the MTBF and the MTTR do not change over time. The sum of MTTR and MTBF is the average time between two subsequent repairs and is called the cycle time. Since during the cycle the machine works for the duration of the MTBF (and is idle for the duration of the MTTR), the time during which the machine is active is equal to MTBF/(MTBF + MTTR).

12.4.2 Repairs at the Customer's Home or at a Customer Service Lab

When a customer experiences a malfunction and there is a good chance that the company's reputation will be damaged, it is difficult to measure this damage. However, the damage would be much worse if it were not for the repair service. Planning repair service at the customer's house requires taking into account labor that is both mobile and professional, equipment that is reliable and mobile, and inventory of parts for replacement. Lab service requires evaluation of the labor needed to reduce the wait times at the lab below a reasonable threshold, as well as forecasting the quantity and duration of malfunctions. In both cases (repairs at the customer's home or lab repairs), the required labor needs to be determined.

Note: In the context of repairing a machine at the customer's home, MTBF is the average time between two consecutive calls by the customer requesting the service, and MTTR is the time period between identifying the malfunction and its repair by a mechanic at the customer's home. This type of service includes additional components, such as the time it takes the technician to arrive at the customer's home or the time it takes the customer to arrive at the lab.

12.4.3 Planning Customer Satisfaction Surveys (or Service Surveys)

Customer satisfaction surveys or service surveys aim at identifying the customer's demands and requirements, listening to and handling complaints, if any, and seeking solutions that will improve the customer experience. In such a survey, customers are requested to provide feedback on the problem-solving speed and to indicate to what extent they are satisfied with the suggested solutions. One of the most known questionnaire formats is the SERVQUAL, which has 22 questions on the expected and perceived service along five different dimensions:

1. *Reliability:* the durability of the promised service
2. *Assurance:* the knowledge and courtesy of employees and their ability to convey trust and confidence
3. *Tangibles:* the appearance of the physical facilities and equipment
4. *Empathy:* providing care and attention
5. *Responsiveness:* the willingness to help customers

12.4.4 Service Arrival Processes

Every service includes customer arrival processes and customer departure processes. Focusing on each of these processes separately enables us to analyze them mathematically and to examine the gap between them that affects the size of the queue. The processes involved in customer arrival at the place of service are known in statistics and operations research as birth processes (because of the constant arrival of new customers), whereas the departure processes are sometimes known as death processes (because the customers leave the system).

The processes of arrival at the place of service may be characterized by the following dimensions (see Figure 12.13):

1. The size of the arriving population: finite versus infinite

   a. *Finite population:* A small and finite population in which the arrival of one customer has an impact on the probability that the next customer will arrive.

   b. *Infinite population:* An infinite (or sufficiently large) population in which the arrival of one customer has no impact on the probability that other customers will arrive.

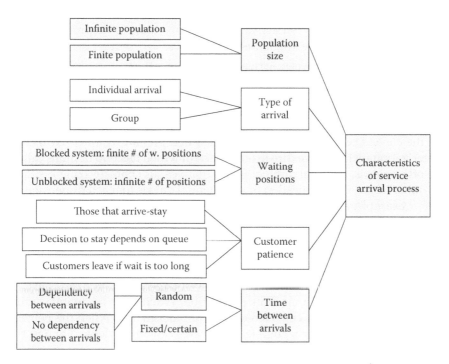

FIGURE 12.13
Characteristics of the service arrival process.

2. Type of arrival (appearance): Individuals versus groups
   a. *Individual arrival:* Each customer or item arrives separately
   b. *Arrival in groups/units:* Group arrivals (e.g., groups arriving by organized transport or in family groups) or arrivals of units/packets in which the "customers" are items or products
3. *Waiting positions:* Blocked and unblocked systems
   a. *Blocked system:* A limited number of waiting positions, and customers who arrive and see that all the waiting positions are taken, leave the system.
   b. *Unblocked system:* An unlimited number of waiting positions and all customers that arrive join the queue.
4. Staying/patience
   a. All arriving customers *stay* until receiving service.
   b. Most arriving customers stay, but some lose patience and *abandon the queue without being serviced.*
   c. Customers evaluate the status *and upon arrival decide* whether to stay.
5. Time between arrivals
   a. The time between arrivals is *fixed* or *certain.*
   b. The time between arrivals is random—there is no probabilistic dependency between the arrival times.
   c. The time between arrivals is random—there is a probabilistic dependency between the arrival times.

The characterization depicted in Figure 12.13 is very important for the statistical and mathematical analysis of the various service systems and is used in the context of queuing theory. As described previously, the arrival characterization includes five dimensions, each of which has two or more values. In other words, the total number of alternatives for characterizing a system is rather large (a multiplication of all potential values for each dimension).

For example, a standard queue at a bus stop can be characterized by an infinite population source, individual arrivals, an unblocked system, those that arrive stay, and a random amount of time between arrivals with no dependency between arrivals.

In describing arrivals, it is customary to make simplifying assumptions. For example, when the source of arrivals is a large population, it is customary to consider it to be an infinite population. When there are many waiting positions, the system is considered to be unblocked, and when the percentage of irregular cases is negligible, these cases can be overlooked in describing the general characteristics of the system and its performance over time. For example, if 99.9% of the arrivals are individual customers, the arrival process can be viewed as an individual arrival process.

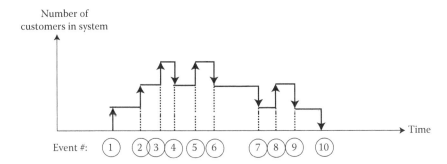

FIGURE 12.14
Random arrival and departure of customers at a particular branch (random service time).

Due to the variety of arrival processes (and the limitation of statistical knowledge), we cannot analyze all alternatives in this chapter. However, most service arrival processes are random. Therefore, our discussion will now focus on random arrival processes.

Figure 12.14 depicts random arrival times, departures following a wait, and customer service at a particular branch (post office, bank, supermarket, health-care clinic, etc.).

The following is a short description of the events depicted in Figure 12.14. The system is vacant up until Event no. 1. Event no. 1 indicates the arrival of the first customer to the place of service. The customer is serviced during the time the second customer (Event no. 2) and the third customer (Event no. 3) arrive. Event no. 4 signifies the departure of the first customer and the entrance of the second customer into the service process. Event no. 5 signifies the arrival of the fourth customer into the system (while the second customer is being served). Event no. 6 signifies the time the second customer finishes service and leaves the system, while at the same time, the third customer enters the service. Event no. 7 depicts the time the third customer finishes service and leaves the system. At the same moment, the fourth customer enters the service. The fifth customer arrives at Event no. 8 (while the fourth customer is being served). At Event no. 9, the fourth customer finishes and leaves the system, and finally the fifth customer is also finishes and leaves the system (thus, leaving the system empty).

Figure 12.14 shows that the number of customers at the branch is the difference between the number of arrivals and the number of departures. The queue develops due to the randomness of the arrival times as well as the randomness of the service times. However, if the service pace is slower than the demand pace and the system is unblocked, the number of the customers in the system will consistently grow. Such trend cannot be tolerated for long, as the waiting time become unbearable.

In the next sections, we discuss separate random arrivals of each customer. As previously mentioned, there are additional arrival processes that will not be discussed in this chapter. These include nonrandom arrival (e.g., according to a predetermined schedule), group arrivals (e.g., passengers getting off a bus or a taxi), arrival at a blocked system (limited number of waiting positions), and immediate departure, arrival and abandonment (reneging) due to impatience (abandonment [reneging] due to the initial length of the line, or after waiting for a long time).

12.4.5 Arrival Processes That Change over Time

Many service processes are characterized by arrival paces that change over time. For example, the pace at which customers arrive at restaurants is much greater during mealtimes (noon to 2 p.m., and 7 p.m. to 9 p.m.) than at other times. Customers visit car garages more frequently in the early morning hours than in the hours that follow. More customers visit car washes on Fridays than on any other day of the week. As long as the arrivals are independent of one another, it is possible to determine different time periods for analysis, according to the pace of arrival appropriate to the time period.

Characterizing the arrival pace expectancy is very important for preparation and for determining the required human resources. For example, if during a night shift, there are very few customers, then there is no need for many service providers. During the busier shifts it is preferable to increase the number of service providers. For example, the number of personnel in hospital emergency rooms is determined according to the arrival pace during the busiest time of the shift (plus a safety margin). The pace of arrival at the emergency room changes over time. For example, patients rarely arrive between midnight and 6 a.m., the pace changes throughout the day and only decreases again at night. Figure 12.15 describes the average hourly number of arrivals at an Israeli emergency room on Sundays (the first workdays in the week in Israel).

Fortunately, it is possible to divide the dynamic process into time periods in which the pace practically does not change and to focus on each time period separately. In addition, the sum of arrivals is simply the sum of all arrivals from various time periods. The expectancy of the sum of arrivals

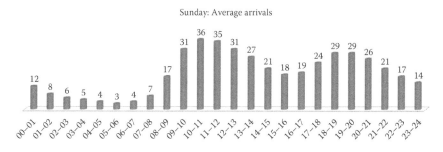

FIGURE 12.15
Sunday emergency room arrivals by time of day.

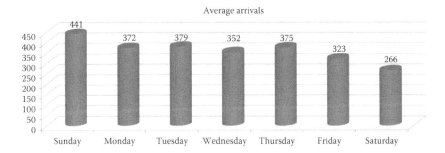

FIGURE 12.16
Expectancies of emergency room arrivals on each day of the week.

is simply the sum of the time periods multiplied by their respective pace expectancy.

It is important to identify the cyclical nature of the arrival processes. Figure 12.15 depicts a daily formation that repeats itself on Sundays.

Are Mondays or Tuesdays also characterized by this formation?

Figure 12.16 depicts the expectancies of the number of Israeli emergency room arrivals on each day of the week during 2009.

According to Figure 12.16, there are fewer arrivals on Fridays and Saturdays than on Sundays. Fridays and Saturdays are not regular work days in Israel, and examination of the hourly arrival expectancies reveals a different arrival pace formation, as indicated by Figure 12.17.

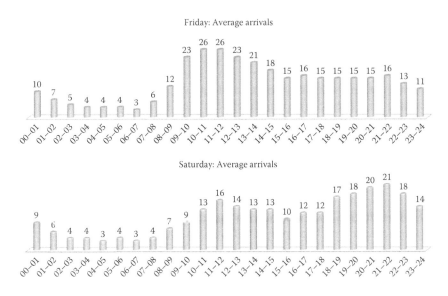

FIGURE 12.17
Emergency room average hourly arrivals on Friday and Saturday.

The expectancies of arrivals can be summed up into different time frames. For example, according to Figure 12.17, the arrival pace expectancy during a night shift that starts at midnight and continues until 8 a.m. on Friday is: $10 + 7 + 5 + 4 + 4 + 4 + 3 + 6 = 43$.

On Saturday, on the other hand, the arrival pace expectancy during a night shift is $9 + 6 + 4 + 4 + 3 + 4 + 3 + 4 = 37$.

Displaying the data according to hourly arrival expectancy and day of the week facilitates accurate analysis. Such an analysis would not be possible had we not insisted on detailing the hours (and replacing hourly arrivals with daily arrivals or shift arrivals) or the day of the week (so that the arrival expectancy is calculated for all days). Thus, we understand the importance of identifying the timeframe in analyzing the varying arrival paces. The timeframe is usually characterized by a cycle. This cycle facilitates a systematic and brief description of the service demand formation. The total demand for a given service is usually influenced by one or more of the following typical cycles:

1. Hourly cycle on each day

2. Weekly cycle (according to the day in the week)

3. Monthly cycle

4. Seasonal/periodic cycle

12.5 Introduction to Queuing Systems

Queues are typical of most service systems and constitute an important component of service quality. Temporary queues are usually the outcome of random customer arrival times or random duration of service. This randomness may be reflected by inconsistent (long or short) gaps between arrivals, as well as a variety of service durations (long or short). A random combination of several short arrival gaps with long service durations creates a queue. Therefore, any service system marked by random varying arrival gaps or varying service durations can result in the formation of a queue. Queuing theory describes the probability distribution of queue length and various parameters of a queuing system (such as mean waiting time, mean queue length, utilization, etc.) for various queuing system types as well as for a network of queues. The analysis of queues often extends for the purpose of understanding, behavioral analysis, simulation, evaluation, and decision making. The main applications of queuing theory focus on characterizing service systems and manufacturing systems. It is particularly interesting to calculate the expectancy of the wait time, the expectancy of the queue length, and the load (the percentage of time in which service personnel are busy). Following are some representative examples of information that may be provided by queuing theory.

12.5.1 Example 1: A Queue at the Bank

The bank manager wants to know how many customers wait in line during the hour following the bank's opening time, how long they wait, and how efficient the clerks are during this time period. The manager is interested in knowing how many clerks are required to ensure that 95% of the time when customers arrive at the bank they will see no more than six people ahead of them in the queue.

12.5.2 Example 2: Urban Garden Maintenance

The city of *Yew Nork* uses sprinklers and drip irrigation to water gardens and public areas. The sprinklers and drippers break down randomly, and the city general manager wants to know how many gardeners he needs to employ in fixing these systems to ensure with 99% certainty that at least 90% of the sprinklers are functioning properly at any given time.

12.5.3 Example 3: Car Interior Cleaning Stations

The cars at a car wash wait in one queue for the car wash tunnel and then in a second queue for the interior cleaning stations. The car wash owner wants to know how many interior cleaning stations must be available and manned on Fridays so that 99% of the time, the queue for the interior cleaning stations will not be more than three cars long.

12.5.4 Example 4: Self-Service Airline Check-in Kiosks

The ticketing manager of the airline *Wings* wants to relieve the load at the check-in desks and allows some of the passengers to use self-service check-in kiosks to issue boarding passes. How many kiosks must he place at the check-in area in order to reduce the load at the check-in desks by 30%?

A queuing system includes (1) customers who are characterized by the distribution of their arrival pace; (2) service personnel characterized by the number of servers, type of service, and distribution of service duration; (3) the queue regime and the preference rules (e.g., FCFS or service in random order [SIRO]); (4) the queue capacity; and (5) the system structure: the order in which the customers move between the service personnel.

A simple queuing system with one server is described in Figure 12.18. The illustration demonstrates that the queuing system consists of the service station and the queue. It depicts the person providing service, the people waiting in the queue, and the customer receiving service. The distinction between the service system and the queue also exists in more complex

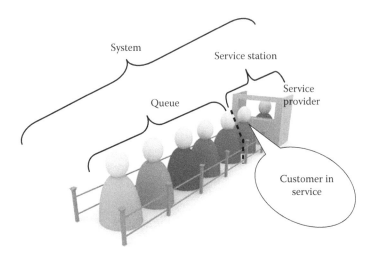

FIGURE 12.18
Simple queuing system with one service provider.

systems, for example, systems that have more than one service provider, systems in which the customers in the queue do not wait according to the order of their arrival, or a service provider that alternately services two queues.

12.5.5 Kendall's Standard Queue Notation

Kendall's queue notation is the standard system used to describe and classify a single queuing system. Single queuing system refers to the case where there is one sequential waiting line from which the customers are chosen for service. The notation is called by the name of its inventor D.G. Kendall (1953). Initially the notation described queuing models using three factors: arrivals, service, number of servers—written as $A/S/c$:

   A—describes the arrival process
   S—describes the service process
   c—the number of servers

   While c is a positive integer number, A and S values denote inter-arrival and service time distributions, correspondingly. Some popular values for these distributions are as follows:

   M—memoryless Markovian (any independent process)
   D—deterministic times
   E_k—Erlang distribution with k stages
   G—general distribution (could be anything!!!)

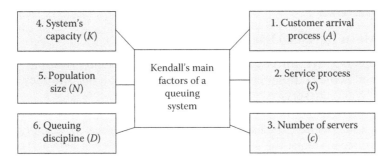

FIGURE 12.19
The six factors of Kendall's notation.

For example, according to Kendall's notation, $M/M/1$ is a single server queue with independent arrivals and independent departures (both arrival and service are Markovian). An automatic vending machine with fixed service time and independent arrivals would be noted as: $M/D/1$. A single server queuing system with independent arrivals and any possible service distribution is denoted as: $M/G/1$. A line that is waiting for the first available counter of three counters each with E_5 is denoted: $M/E_5/3$. Figure 12.19 summarizes Kendall's notation.

Later on, three other notations were added (Taha, 1982)—$K/N/D$:

K—number of waiting positions in the queue (queue capacity)

N—population size (the number of entities in the system)

D—queuing discipline (the order in which the customers are selected for service)

While K and N are positive integers or infinity, D has several popular disciplines that are abbreviated in Figure 12.20.

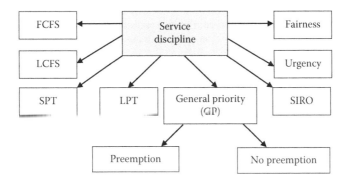

FIGURE 12.20
Various service disciplines.

12.6 Main Service Disciplines

In Chapter 7, we introduced priorities for scheduling waiting jobs to machines. Some of these could be used for deciding on the priorities for customers to get into service.

1. *FCFS or first come first served:* This is the most prevalent discipline in service systems: the service is given in the order of arrival. People usually view this discipline as justice for all. It is a fair and equal opportunity given to anybody who wishes to come early.

2. LCFS-Last in first out or last come first served is giving the priority to the last arrival. This usually refers to a stack type of queue, where items are piled on top of each other, and the first arrival is served last.

3. *SPT or Shortest processing time:* The customer with the shortest service duration is served first (at the first time that the server becomes available).

Some other service disciplines are:

4. *LPT or Longest processing time:* The customer with the longest service duration is served first (at the first time that the server becomes available).

5. *SIRO or service in random order:* A service order that is randomly chosen as a lottery among the waiting entities.

6. *Urgency:* The most urgent customer gets the next service (this is suitable for emergency rooms and places where urgency is a major factor).

7. *General discipline:* This is a notation for a general queue without any defined discipline (as the discipline can be any discipline).

There are two cases for giving priority:

1. *No preemption:* On arrival of a preferred customer, he or she joins the queue in the first place waiting for the service to finish, before entering service.

2. *Preemption:* On arrival of a preferred customer, the current customer in service is ejected and replaced by the preferred customer, which immediately starts service.

We use the following example for illustrating Kendall's notation.

A single server queue with a general arrival process and deterministic service times, with no limit on the number of its waiting positions, and infinite population, has a discipline of FCFS. Find the Kendal notation for this system.

The answer is: *G/D/1/∞/∞/FCFS*

However, the last three values ($K = \infty$, $N = \infty$, $D = $ FCFS) are default values so:

The answer is also: *G/D/1*

Second example for a full Kendall's notation:

Queue for fixing failures of a fleet of 100 trucks has three fixing positions and is characterized by a parking bay for six trucks, and priority based on arrival times (FCFS). The failure times are independent of each other, and so are the fixing times. The queue notation is:

$$M/M/3/6/100/\text{FCFS}$$

12.7 Service System Simulation

The mathematical analysis of a single queue with one service provider is relatively simple, while analyzing many service providers and many queues could be very complex. Real-life systems have many characteristics that complicate the analysis, such as group arrival, limited number of waiting positions, batch processing, etc. Therefore, it is customary to use simulations in order to analyze complex service systems.

For a single queue, there are many analytical results for many models. The analytical results consist of formulas that typically reflect the distribution of the queue length. This subsequently makes it possible to calculate other queue parameters, such as queue life expectancy, wait time expectancy, and queue efficiency. Yet when analyzing queuing networks, the situation is more complex (e.g., Figure 12.21). These networks are marked by complex

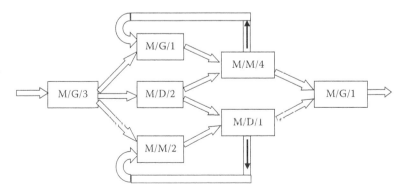

FIGURE 12.21
A queuing network system for simulation (each node is a queue).

constraints and rules as well as dynamic behavior. Systems such as those depicted in Figure 12.19 can be analyzed by running simulation models.

12.7.1 What Is Simulation?

Simulation is a tool that can assist in building a model that mimics the operation of an actual system. A model is a partial description of reality built for a specific purpose, while disregarding aspects that are irrelevant or complicated with respect to the description of reality. The advantage of this type of model usually stems from its cost and from its ability to provide insights— even if only partial—about the behavior of a system in various situations that an actual system cannot provide in a reasonable time or at a reasonable cost. Simulation mimics the behavior of a system and therefore often makes it possible to learn about the system and its traits, behavior, and responses to various scenarios. It also facilitates examining and comparing the effects of changes in the system structure or of selected factors and making decisions regarding those factors. Figure 12.22 schematically describes the method of operation of a computerized simulation.

Model validation is designed to ensure that the model is producing correct predictions and that the factors that were disregarded or filtered while building the model do not cause the model to deviate significantly from reality.

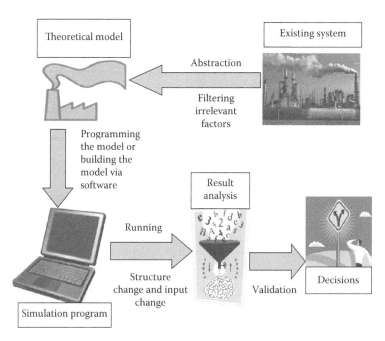

FIGURE 12.22
Schematic description of a computerized simulation.

Heizer and Render (2014) list the advantages and disadvantages of simulation.

The advantages:

1. Simulation can be used to analyze complex and large systems with uncertainty that cannot be analyzed otherwise.
2. Simulation can include nearly any constraint or complication with which other performance analysis models cannot cope.
3. Time can be "shrunk" by simulation models, thus providing examples of behavior that would have taken many years if using other methods.
4. It is possible to perform scenario analysis and check the system's response (even in risky situations) in order to make a decision about the optimal mode of operation.
5. Simulation is not integrated into the actual system and does not disturb its normal activity.

The disadvantages:

1. A relatively long period of time is needed to develop a good and valid simulation model on which decisions can be based.
2. Simulation is not optimization and does not necessarily provide optimal or even close to optimal solutions.
3. The quality of the model's results is dependent on how well its developers built it: To what extent does it represent the process being modeled? Does it make reasonable and correct assumptions? Were all the constraints properly entered into the model? Does the model include all the characteristics? In addition, the quality of the model's input determines the quality of its output.
4. Each model is built to simulate a specific system. Therefore, usually it is impossible to reuse one simulation's solution and implement it for another system.

12.7.2 Major Types of Simulation

Simulation simulates reality. Simulating reality can be carried out using physical models, visual models, and abstract models.

Simulation using physical simulators: The public is very familiar with physical simulation systems. For example, gyms have treadmills that simulate walking and running at various speeds and inclines, stationary bicycles that simulate riding a real bicycle, and also machines that simulate rowing, skiing, and climbing stairs. Some physical simulators—flight simulators and astronaut simulators, for example—can be very expensive.

Visual simulation—animation: Visual models produce illustrations, photographs, or virtual reality films. For example, many computer games simulate a reality in which the player must operate in a dangerous, surprising, or interesting environment. The Wii system (a game console that is a computer designated for video games) simulates sport activities such as tennis, golf, and boxing. There are flight simulation programs that simulate flights on the screen. Furthermore, some industrial control systems (mainly chemicals) use animated process control to monitor processes.

Abstract simulation (usually numerical simulation): Abstract simulation involves building a model that combines logic and mathematics to simulate the functioning of systems. For example, a model that simulates how the stock exchange operates can eventually depict increases or decreases in the stock exchange's major indices. Sometimes it is possible to connect the abstract simulation to an animation model that uses the abstract simulation's results for the purpose of visual simulation. Industrial engineers usually use simulation as a supportive tool for making decisions. Since abstract simulation is the most useful tool in such cases, we will focus on it. Abstract simulation is typically divided into two types: static simulation and dynamic simulation.

Static simulation: Static simulation simulates a complex situation. It represents a point in time or summary of a period but does not simulate situations that develop over time.

Dynamic simulation: Dynamic simulation simulates situations that develop over time. At any point during the simulation, the next situation is dependent on the history of the simulation.

Dynamic simulation is typically divided into two types: continuous simulation and discrete simulation.

Continuous simulation describes system states using a continuous time function, as in the level of gas in the tank, the temperature at a given location, or the pressure in a certain pipe. Additional examples are the location of vehicles while driving on the road or the location of aircraft in flight.

Discrete simulation (or discrete event simulation) describes system states through discrete events, as in cars entering and leaving a parking lot or customers arriving at and leaving a place of service. Assembly, maintenance, and service processes are usually processes that include events at distinct points in time and, therefore, discrete simulation can be used to simulate their behavior.

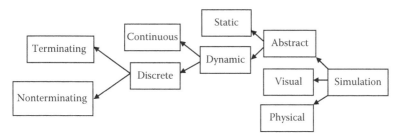

FIGURE 12.23
The major types of simulation (Pollatschek, 2007, The Open University).

Discrete simulation is typically divided into two types: finite simulation and ongoing simulation.

Terminating system simulation is system simulation that has a natural beginning and end. For example, a store that opens at 8:00 a.m. and closes at 6:00 p.m. every day is a finite system. Another example is simulation of a call center that operates two 8-h shifts each day. The call center opens every day at a certain time and closes 16 h later.

Nonterminating simulation is system simulation that does not have a natural and clear beginning and end. For example, simulation of an emergency room is ongoing, and the number of daily patients is affected by the events of the previous day. A car park that operates continuously is an ongoing system, as is a manufacturing factory with inventory that moves from day to day and from shift to shift, thus creating constant dependency on the previous day or shift. A nonterminating simulation usually requires a "trial" or a "warm-up" period enabling it to reach a state of normal functioning from a nonnatural starting point. In addition, an artificial decision needs to be made regarding the time to end this type of simulation.

Figure 12.23 describes the major types of simulation.

Industrial engineers usually use simulation for making decisions in times of uncertainty (by describing and analyzing the behavior of systems). The main technique used in uncertainty-oriented simulations is the Monte Carlo simulation, described in further detail next.

12.7.3 Monte Carlo Simulation

The Monte Carlo method developed in the twentieth century is based on generating random numbers that simulate lotteries from distributions selected by the simulation operator. These random numbers make it possible to include the uncertainty component in the simulation.

Scientists at the Los Alamos Nuclear Laboratory first invented the Monte Carlo simulation in 1946. They sought to analyze alternatives for defense against neutron radiation. However, the movement and direction of the neutrons were random, and the scientists did not possess a tool that could calculate their behavior. Scientists Stanislaw Ulam and John von Neumann decided to emulate random lotteries in order to model the process. Since this was a secret project, it was given the code name Monte Carlo, named after Monaco's Monte Carlo Casino, one of the world's most well-known gambling venues. The Monte Carlo method makes it possible to simulate randomness by "random number" lotteries, thus allowing introduction of the randomness factor into the simulations. The use of number lotteries from various distributions makes it possible to simulate many processes that include different levels of randomness.

Service system simulations are typically quite random. This randomness may be reflected in the number of customers, their arrival time, the customers' type of demand, the service times, as well as in malfunctions and unforeseen events. The Monte Carlo simulation makes it possible to simulate the uncertainty and randomness of these systems.

The Monte Carlo simulation has various definitions, all of which involve number lotteries. Here we will adopt the definition of Law and Kelton (2000). According to Law and Kelton, the Monte Carlo simulation is a problem-solving method that selects numbers that are randomly scattered in a range between 0 and 1 so that it is equally probable that any value between 0 and 1 will be selected.

The distribution described in Law and Kelton's definition is a *uniform distribution* between 0 and 1 and is indicated by Uniform (0,1) or simply U(0,1). This distribution plays a key role in the Monte Carlo simulation and in simulations of discrete events. The Monte Carlo method generates random numbers that are randomly scattered in a range between 0 and 1 and then converts the values for each desired distribution function and integrates them into calculations that simulate the system behavior.

The three major stages of the Monte Carlo simulation are as follows:

1. Generating random numbers from a uniform distribution between 0 and 1 (U(0,1)).
2. Converting these numbers to a desired distribution.
3. Integrating the random numbers into the simulation.

12.7.4 Discrete Event Simulation Programs

There are complex and random systems whose behavior cannot be modeled without simulation. Most service system simulation programs are discrete event simulation programs. In other words, the simulation program progresses from one event to the next in a sequential and calculated manner.

The first discrete-event simulation language was developed in 1961 by Geoffrey Gordon and was named General Purpose Simulation System (GPSS). Immediately thereafter, GASP and SIMSCRIPT also emerged. Discrete simulation programs usually describe the movement of items, customers, or entities, in a network of automated decision points, probabilistic branching junctions, and queues that include (1) Use of Monte Carlo method for choosing numbers from random distributions for various purposes. These purposes typically include the arrival times of customers to the system, their service times, and their choice in probabilistic branching point. (2) Branching junctions, or junctions that activate a certain action, including calculations each time the entity or customer or item passes through them. (3) Servers or machines or various resources. Such a server or resource can only serve one customer/entity at a time, and thus is a cause of a queue. (4) Logic for modeling various phenomena such as machine failures, signaling, prioritizing customers in a queue, or batching: gathering or packing several entities into one group and treating this group as a batch (e.g., gathering people into a taxi cab).

Many discrete simulation programs have since been developed that allow users to define the systems to be simulated with relative ease. Following is a short list of discrete simulation programs:

1. *ARENA:* A discrete simulation program based on the SIMAN language.
2. *ExtendSim:* A discrete and sequential simulation program.
3. *Promodel:* A collection of discrete programs that are geared to various sectors individually: service, industry, medicine, business, and economics.
4. *SAS simulation studio:* A SAS procedure library for discrete simulations.
5. *SimEvent:* A discrete simulation plug-in for MATLAB.
6. *SIMUL8:* An object-oriented simulation program.

These simulation programs are very flexible and enable building a model for nearly all types of queues and queuing networks. These programs can be used for building models and run trial runs for nearly any possible queuing network and subsequently for receiving various estimates (including mean efficiency, queue length, and wait time).

The simulation programs facilitate the analysis of complex models that include dynamic decision making according to system status, fluctuating service, and arrival paces, unification and division of items and complex logic. These programs are equipped with animation capabilities as well as control over the speed of the simulation, making it possible to follow and validate the simulation model.

The flow and logic of a simple queue simulation model in ARENA are explained next (Figure 12.24).

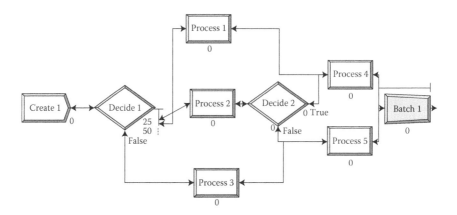

FIGURE 12.24
Example of an ARENA queue simulation model.

Rectangular or diamond-shaped blocks that are connected by lines are depicted in Figure 12.24. The simulation entities (customers, items, etc.) are created in the block on the far left and advance along the lines between the model's blocks. Each block in the ARENA model symbolizes a process or a complex activity performed on the entities passing through the block. The left block simulates random arrival of items into the system and directs the entities toward the next stages of the simulation. In addition to the left block, there are diamond-shaped blocks that serve as decision junctions for each entity that arrives, as well as rectangular process blocks that can include a complete queuing system. The block on the far right unites an entity that underwent Process 4 with an entity that underwent Process 5.

The various simulation programs usually enable viewing the advancement of the entities through the system during the simulation run. They also enable monitoring the accumulation of entities in the queues. A basic example of this is described in Figure 12.25.

Figure 12.25 demonstrates the situation frozen at a given moment. At this particular moment, 397 entities arrived in the system via the left block, of which 285 entities left the system via the right block. Of the entities that remained in the system, 23 are in the left (first) queue and 89 are in the second queue.

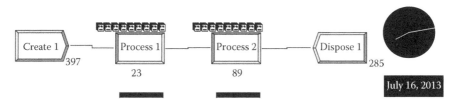

FIGURE 12.25
A basic example of monitoring entities during ARENA program simulation.

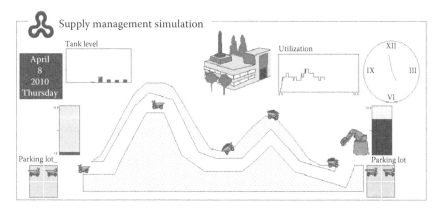

FIGURE 12.26
Example of an animation screen illustrating the simulation process.

To better illustrate the simulation process to those who are not knowledge-able in simulation logics, the simulation programs allow developing anima-tions of the processes simulated by the simulation models.

Figure 12.26 depicts an animation screen of the process of shipping and transport by trucks.

The duration of the simulation run and the number of times it is repeated are important decisions in the operation of simulation programs. It is impor-tant to separate between a system that runs continuously (nonterminating system) and a system that shuts down at the end of each day (or time period) and starts over again with no customers (terminating system). For example, an emergency room is a system that runs continuously, while a store opens each day with zero customers and shuts off at the end of the day. A system that shuts down at the end of each time period is called a *terminating sys-tem*. In such a system, the number of periods in a single simulation run is an important decision parameter. On the other hand, a system that operates continually is called a *nonterminating system*. In such a system, the initial trial run period (or warm-up period) and the total run time are important decision parameters. To achieve results with a strong statistical validity, it is common in both cases to run the simulations many times and to examine the average and standard deviation of the results. The combination of mean and standard deviation enables the estimation of statistical likelihood of real-life results.

12.7.5 Important Characteristics of Discrete Simulation Programs

Jacobs et al. (2010) note the important characteristics of simulation programs:

1. Interactive capabilities (user intervention during the simulation) and noninteractive operation capabilities (level of interaction is con-trolled by the user).
2. User friendliness.

3. The ability to separately build and operate the components of the model and the potential to assemble these components.

4. The ability to incorporate code written by the user.

5. The ability to use procedure "blocks" controlled by short orders and minimal input.

6. The ability to build repetitive "macro" procedures.

7. The ability to model the movement of items and the transport of materials.

8. Producing standard statistical output that includes elements such as utilization, queue length expectancies, waiting times, and costs.

9. The ability to facilitate various alternatives and to issue reports for analysis of input and output data.

10. Animation capabilities for the purpose of illustrating dynamic processes.

11. The ability to identify and correct errors (debugging), and the ability to monitor the model variables in cases of errors.

Several important characteristics can be added to this list:

12. The ability to change the values and attributes of customers, items, and entities during the simulation and to make decisions based on the system's values and the attributes of customers, items, and entities at a given time.

13. Facilitation of building, handling, and computing mathematical and statistical functions (or expressions).

14. The ability to monitor a customer/item/entity in the system.

15. The ability to externally enter values for variables and qualities into the simulation or to enter customers/items/entities.

16. The ability to link to other programs for data input, adding code, exporting data, data analysis, and presentation of outputs.

17. The ability to present graphs and dynamic statistics during the running of the simulation.

12.8 Customer Patience: Level of Service

12.8.1 Level of Service

The purpose of service is to benefit customers. Therefore, the level of service is related (sometimes indirectly) to the extent of the customer's satisfaction with the guaranteed service. A typical customer is dissatisfied if the service that was promised is unavailable or requires a long waiting time. A typical customer is also dissatisfied if the level of service is inferior (in comparison to the promised quality). The level of service normally does not assess the level and sophistication of the service, but rather the level of performance and availability of the promised service.

Service level in call centers is probably the most famous closely monitored parameter. However, even in call centers, which only focus on answering calls, service level is usually composed of a combination of measured parameters, such as average holding time (AHT) overall call time, beginning to end, average waiting time (AWT), beginning to human answer, ASA—average speed of answer (time from final menu selection to human answer), first call resolution (FCR) the percentage of issues resolved in first call, threshold service frequency (TSF)—percentage of customers waiting less than a threshold value, and abandon rate (AR).

The usual indicator of level of service is the expected/average waiting time, although decreasing the variance of wait time is also an indicator. High variance indicates the possibility that customers will wait for a long while. Service threshold level or service target level can also be measured in percentages. For example, it is possible to establish a threshold percentage for phone calls answered/handled at a call center. A threshold percentage can also be determined in the context of calls for which customers waited less than a predetermined wait time. The average queue length is also an accepted measure of the level of service at a call center. In a repair service system, the level of service is measured by the time that passed from the time a malfunction was reported until it is fixed. It is also customary to periodically measure customer satisfaction through questionnaires. This, however, can only give feedback for short periods of time.

Contrary to the waiting time indicators at call centers, other fields have different indicators for level of service. For example, when inventory is managed under random demand, sometimes there are shortages, and the level of service is connected to the level of certainty that a customer will not encounter a shortage. There are two main versions for measuring the level of service in inventory management:

Type 1 service level (inventory): In this case, we indicate the probability that there will be no shortage during the supply period by the letter α, which also indicates the percentage of cycles (from the arrival of ordered inventory until the next arrival) during which there is no shortage.

Note that when we say that we will strive to supply 95% service, our intention is to supply 100% of the demand in 95% of all the order cycles. In addition, since different items have different cycle lengths, this measure will not be consistent over different items. For example, there may be a shortage of one item for 1 week during a 1-year span, whereas there may be a shortage of another item for 1 week during a 10-year span.

Type 2 service level (inventory): Type-two service level measures the percentage of demand that is directly supplied from the inventory. We use the letter

β to note this percentage. Naturally, the aim is for β to be as close as possible to 100%. In the case of a shortage when customers are unwilling to wait until the arrival of the next order and turn to the competitors, service level type 2 can be described in terms of loss because $1 - \beta$ of the potential sales was not fulfilled due to unavailability.

12.8.2 Customer Patience

The success of fast food chains is an indicator that people do not enjoy waiting. Customer patience is a critical factor in the success or failure of a service. However, customer patience cannot be measured directly. It is possible to measure how much time customers spend waiting in the system. The longer this wait is, the higher the chances are that customers will get frustrated and that customers who are already frustrated will become even more so. The shorter the wait, the more satisfied the customer will be. This is reflected not only by customer loyalty, but also by positive word of mouth reputation that increases sales. Therefore, in many cases, the customers' waiting time is quantified in terms of money in order to make practical decisions regarding investment in service and in service personnel.

Impatient customers are the ones that disrupt the system and cause problems and losses. This impatience can be expressed in several ways:

1. Giving up in advance before entering a busy queuing system.
2. Reneging–abandoning the queue after a wait.
3. Complaints and incitement while waiting in the queue.
4. Avoiding revisiting the same place where the waiting took place.
5. Damaging reputation by giving the service a bad name (publicly spreading experiences of waiting in the queue).

In summary, the customer's waiting time is expensive. Therefore, it is important to try to prevent situations of a long wait by increasing the number

FIGURE 12.27
Waiting wears down customers' patience.

of service personnel, increasing their service pace, or limiting the number of waiting spots in the queue (Figure 12.27).

12.9 General Approach to Planning Number of Service Personnel

The following quote portrays a typical major error in calculating the number of service personnel required at a call center:

> My calculation is simple: we have an average of 5400 incoming calls every day, during the course of 9 office hours. In other words–600 calls per hour. On average, each call is 2 minutes long, and therefore every clerk can handle 30 calls an hour. To provide a response to 600 calls an hour, we need 20 positions (so that 30 calls are handled at each position). Therefore, we must recruit 20 service employees for handling the calls.

The error naturally stems from ignoring the fact that a significant percentage of cases are above the average and that random distribution is not necessarily uniform. There are days with much more than 5400 calls and rush hours with much more than 1/9 of all calls. Every day has its rush hours, and in every rush hour there are busier time periods or less busy time periods. Furthermore, not all calls are exactly 2 minutes long. Therefore, the work plan suggested in the above quote is likely to create excessively long queues and waits.

This section explains the principles behind establishing the number of service personnel:

- Rush hours and their importance.
- The term *utilization* and its connection to wait time expectancy.
- Variations in arrival and in duration of service and their impact on determining the number of service personnel.
- Cost considerations in determining the number of service personnel.

12.9.1 Rush Hours and Their Importance

Rush hours are the hours in which the most customers arrive, and it is usually very advisable to plan human resources and system service capacity for rush hours. Planning that does not include rush hours neglects significant percentages of the customer public and causes a loss of reputation and customer defections. Planning must take rush hours into consideration, and the decision on the level of service for customers during rush hour needs to be an educated one. This principle dictates that there will be hours that are not as busy, during which service personnel will not be optimally exploited. Naturally, when work is based on shifts, the planning is carried out according to the busiest hour in the shift. In organizations where it is not customary or impossible to split shifts into time segments, planning according to the busiest hour seems like a waste of money and resources. This is especially apparent when there is a large variation between rush hours and the "dead" hours, in which many service providers are not doing any work. In such cases, some organizations use permanent personnel for temporary rush-hour reinforcement. In places where shifts can be split into time segments (e.g., 4-h segments), the capacity and number of workers are determined according to the rush hours in each time segment.

12.9.2 The Term *Utilization* and Its Connection to Wait Time Expectancy

Utilization is a measure of the percentage of time in which the service providers are busy. The percentage of time in which the service providers are busy is the ratio between the arrival rate and the service rate. It is customary to calculate this ratio according to the expectancies of these paces, under the assumption that the service pace is based on all service providers being busy. The purpose of the graph in Figure 12.28 is to illustrate the impact of utilization on customer waiting time. It describes the relationship between waiting time and utilization for a single server queue with statistically independent arrival times and service times.

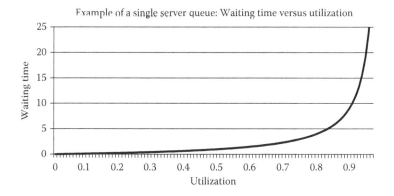

FIGURE 12.28
Example of waiting time as a function of utilization (ρ) in a queue with only one service provider and independent random arrivals and service times.

Figure 12.28 illustrates that the average waiting times start an exponential climb after utilization of 70% and reach infinity as the utilization approaches 100%. When utilization is lower than 50%, the waiting time is negligible. Beyond 50%, the waiting time begins to increase and reaches values that normally would be considered reasonable until utilization reaches 70%. Starting from 70% utilization, the waiting time begins to increase exponentially. It is typical to plan for customers' waiting time expectancy values during rush hours to be in the 70%–90% utilization range. This range reflects the trade-off between utilization of the service providers and the customers' waiting times. When utilization surpasses 90%, the queue length begins to reach its reasonable limits, and when utilization approaches 100%, the queue "explodes" and the waiting times become unreasonable (infinite).

12.9.3 Additional Methods of Reducing Service Time

In addition to the number of workers and machines, there are additional methods to improve service pace and quality:

1. *Training* to increase the service pace, quality, and decrease the variance in service times.
2. *Time slotting*—booking appointments in advance (for a specific time slot)—this method reduces the variance in the arrival process.
3. *Call routing system* automatically directs incoming calls to various queues.
4. *Dynamic labor allocation* from other tasks to service stations during rush hours.
5. *Demand regulation*—encouraging customers to arrive during the less busy hours.
6. *Offline service*—recording customers' information and calling them back after rush hours.
7. *Complete offline service*—the ability to leave a voice message that will be attended to after rush hours.
8. *Use of the Internet and self-service*—filling out an Internet form if it can replace or assist with receiving or ordering the service.

In practice, most methods of improving service cope with the variance of customers' arrival into the system and the variance of service times. Variance is what causes queues and temporary system overloads. However, there are also improvement activities that may shorten the expectancy of service duration, among them training that improves the skills of the service employees and the addition of technical aids and computers that spare human effort.

12.10 Feedback: Customer Satisfaction Surveys

Without customer feedback, the quality of service cannot be measured and improved. Feedback is a critical stage of service and closes the feed back loop between the service providers and the customers. Without feedback, it would be impossible to know that customer dissatisfaction and defection could have been prevented by basic preventative measures. Without feedback, it would also be difficult to know how to effectively improve service (Johnston and Clark, 2008).

The accepted feedback method in service systems is to conduct customer satisfaction surveys, also known as customer service surveys. The reason surveys are used as a feedback method is that most customers will not complain even if the service is defective. Very few customers actually complain, so management only finds out about problems in service after the damage has been already done and at that point it is difficult (and sometimes impossible) to make things right. Therefore, waiting for the complaints to arrive is not a good idea. An additional important point to take into consideration is that customer satisfaction may change over time, and therefore, it is important to conduct periodic customer satisfaction surveys. Conducting a periodic survey is important for finding service malfunctions and facilitates making repairs and preventing the defection of silent customers who do not complain.

A customer satisfaction survey—as its name indicates—measures the extent of customer satisfaction. Customer satisfaction stems from the gap between expectations (which are subjective and vary from person to person) and the perception of actual performance (which sometimes is also subjective). Customer expectations are related to the product, the service, or the customers' relationship with the organization. When the perception of the product, the service, and the relationship exceeds expectations, customer satisfaction is high and the organization is competitive. When that perception only fits customers' expectations, customers' satisfaction is met, but their loyalty is not as strong and competition could be on the rise. When the perception is lower than expectations, customers are dissatisfied, and often look for alternative suppliers.

The purpose of the survey is to focus the customers on details that assist in understanding their expectations of the organization, the service provider, or the products they received. We classify the satisfaction surveys according to: (1) surveys about service processes and (2) surveys about products.

Examples of the main topics in a survey about the service process are as follows:

1. *Professionalism:* What does the customer know about the requested service?
2. *Courtesy and politeness:* The extent of attentiveness, empathy, calmness, and personal care (positive communication) experienced by the customer in the context of service.

3. *Availability:* Was reaching the company quick, convenient, and easy, and is there a variety of ways to reach the company?

4. *Understanding the customer:* To what extent were the customer's demands understood?

5. *Willingness:* Is the service provider willing to perform the service and to solve problems?

6. *Reliability:* Is the service efficient and can the company be trusted?

In surveys that are about products, it is customary to ask, among other things, about Garvin's eight dimensions of quality:

1. *Product performance:* How does the product live up to its role (in the case of a car, for instance, performance measures will include power, convenience, safety, and petrol consumption)?

2. *Features:* What enhancements reflect the customer's expectations?

3. *Reliability:* Can the product accomplish its functions within a specific time frame and without malfunctions? Are there any product failures?

4. *Conformance to design:* Is there consistency in conformance with specifications over time and is the variance in the measured sizes of product small and within the limits of the specifications?

5. *Durability:* The length of a product's life.

6. *Serviceability:* The extent of service availability, the quality of service, and the responsiveness of service (usually with respect to repairing the product).

7. *Aesthetics:* Does the product's aesthetics meet or exceed the customer's expectations? Aesthetics refers to how the product looks, tastes, and smells—aspects that are difficult to measure.

8. *Perceived quality:* How does the customer perceive the quality of the product? Perceived quality is not necessarily connected to the other measures, but rather stems from the reputation of the manufacturer and the reputation of its products. This reputation can also be strengthened by advertising.

It is customary to rank customers' responses on a five-point Likert scale. For example:

Rank the sentences below according to the following scale. 1 – Poor, 2 = Fair, 3 = Average, 4 = Good, 5 = Excellent.

1-2-3-4-5 Professional level of the employee who handled your complaint
1-2-3-4-5 Employee availability
1-2-3-4-5 Overall score of the service
1-2-3-4-5 Courtesy of the employees who handled your complaint
1-2-3-4-5 Diligence of the employees who handled your complaint

It is also possible to leave room for customers to write their opinion on matters that were not mentioned in the Likert scale, for example, comments/ recommendations for improvement:_____.

It is also important to validate the responses by repeating each question using different wording and checking that the ranking provided for the same questions is identical or at least similar. Conducting surveys is a field of study in and of itself.

The first major role of the customer satisfaction survey is to measure the relative importance of each topic in the eyes of the customers. Once the various topics are evaluated and their relative importance is understood, the second major role is examining customers' satisfaction regarding each topic. Thus, it is possible to propose an improvement plan that focuses on matters that are both important to customers and need improvement (when customers are dissatisfied). In this manner, it is possible to propose an improvement plan that is suitable for every type of service.

Customer feedback can indicate the importance of how various parts of the organization function, and thus help identify the need for improvement at various locations in the organization. The survey also helps identify the types of customers who have higher expectations about the level of service than other customers. These customers require more attention or are more aware of the importance of service. It is also possible to use a customer satisfaction survey to identify and reward exceptional employees that are very service oriented. This is done to encourage courtesy and high service level. In such cases, the reward is typically publicized within the organization to increase employees' awareness of the importance of service to the organization.

A survey can be conducted by several means:

- A service survey of a focus group (of volunteer customers).
- A service survey conducted with customers at the place of service.
- A phone service survey.
- A service survey sent to the company's customers by email.
- An online service survey to which the customers are directed.
- A service survey sent to the customers by snail mail including a return stamped envelope for customers to return the survey.

Quality standards require organizations to actively monitor their customers' satisfaction. For example, Section 8.2.1 of the ISO-90001:2008 Standard requires the following.

Customer Satisfaction (ISO-9001:2008-8.2.1)

> One of the measurements of quality control performance conducted by an organization will be to monitor information related to customers'

perceptions with respect to whether the organization is meeting their expectations. The methods of obtaining and using this information will be determined.

Note: Monitoring customer perception may include obtaining inputs from sources such as customer satisfaction surveys, customer data about the quality of a supplied product, public (users) opinion surveys, lost business analysis, words of praise, liability claims, and dealer reports.

12.11 Summary

Service processes are characterized by a great deal of uncertainty with respect to the exact timing of demand and the duration of service. Many service processes are performed in front of customers whose time is valuable. Therefore, the quality and efficiency of the processes and the manner in which they are performed are all very important. Despite uncertainty, it is possible to characterize and analyze the pace and the duration of most arrival and service processes. We began by presenting the general characteristics of service processes, such as the response time limitation, the randomness of demand, the customer experience, and the level of service. We described the major types of service processes, such as arrival of customers, receiving customers, providing a main service, support, repair and replacement, and reimbursement. In addition, we described major characteristics and considerations in the planning of service systems, such as forecasting the service demand, the location and distribution of service, the number of service providers, the number of waiting positions in a queue, ergonomic and aesthetic considerations, customer service, maintenance, and repairs.

Special emphasis was placed on characterizing the arrival to service processes. Since in most services customer arrival is uncoordinated and the arrivals are not dependent on one another, the emphasis was placed on independent arrival processes. In such processes the interval between every two consecutive arrivals is exponentially distributed and characterized by a single parameter of the mean inter-arrival time. This parameter is the inverse of the process's mean arrival pace.

A queuing system is comprised of: (1) arrival at the place of service, (2) waiting for the service, (3) the service itself, and (4) departure. Queuing was introduced and discussed shortly, while emphasizing the characteristics and classification of queuing systems. Single service provider models are usually simpler to analyze than models with multiple service providers. Discrete event simulation is typically used for analyzing systems with multiple service providers as well as queuing networks. Discrete event simulation can model and help analyze sophisticated and complex systems with a variety of uncertainties and risks. The uncertainties are modeled by the simulation using lotteries according to the user specified probability distributions. The process of generating these lotteries is based on Monte Carlo simulation discussed in this chapter. The simulation is iterated many times: each time with different values from the same probability distributions. Thus, the probability distribution of the results is estimated.

We concluded the chapter with a short discussion of a number of topics accompanied by examples:

- Level of service and customer patience (including mathematical examples)
- Planning human resources or number of service providers (including a general approach to planning the number of service providers)
- Surveys of customer/service satisfaction in service and product sales environments

References

Daskin, M.S. 2011. *Service Science: Service Operations for Managers and Engineers.* Hoboken, NJ: Wiley.

Fitzsimmons, J.A. 2005. *Service Management Irwin/Mcgraw-Hill Series in Operations and Decision Sciences.* 5th edn. McGraw-Hill.

Garvin, D. 1987. Competing on the eight dimensions of quality. *Harvard Business Review (HBR).* November.

Garvin, D. 1988. *Managing Quality: The Strategic and Competitive Edge.* New York: Free Press.

Heizer, J.H. and Render, B. 2014. *Principles of Operations Management: Sustainability and Supply Chain Management.* Prentice Hall.

Jacobs, R.F., Chase, R.B., and Aquilano, N.J. 2010. *Operations and Supply Chain Management.* 14th edn. New York, US: McGraw-Hill.

Johnston, R. and Clark, G. 2005. *Service Operations Management: Improving Service Delivery.* Pearson Education.

Kendall, D.G. 1953. Stochastic processes occurring in the theory of queues and their analysis by the method of the imbedded Markov chain. *The Annals of Mathematical Statistics* 24 (3): 338.

Law, A.M. and Kelton, W.D. 2000. *Simulation Modeling and Analysis.* 3rd edn. Boston: McGraw-Hill.

Roth, A.V., Chase, R.B., and Voss, C. 1997a. *Service in the U.S.: Progress Towards Global Service Leadership.* Birmingham, UK: SevernTrent Plc.

Roth, A.V., Chase, R.B., and Voss, C. 1997b. *Service in the U.S.: A Study of Service Practice and Performance in the United States.* Birmingham, UK: Severn-Trent Plc.

Salvendy, G. and Karwowski, W. 2010. *Introduction to Service Engineering*. Hoboken, NJ: Wiley.

Schmitt, B. 2003. *Customer Experience Management*. Hoboken, NJ: Wiley.

Taha, H.A. 1982. *Operations Research: An Introduction (For VTU)*. Pearson Education India.

United Nations. 2012. *UNCTAD Handbook of Statistics*, New York.

Index